Autonomous Driving Changes the Future

Zhanxiang Chai · Tianxin Nie ·
Jan Becker

Autonomous Driving
Changes the Future

Zhanxiang Chai
Automotive Committee
China Council for the Promotion
of International Trade
Beijing, China

Tianxin Nie
AIKAR Technology
Los Angeles, CA, USA

Jan Becker
Apex.AI, Inc.
Palo Alto, CA, USA

ISBN 978-981-15-6730-8   ISBN 978-981-15-6728-5   (eBook)
https://doi.org/10.1007/978-981-15-6728-5

Jointly published with China Machine Press, Beijing, China
The print edition is not for sale in China Mainland. Customers from China Mainland please order the print book from: China Machine Press, Beijing, China.
ISBN of the Co-Publisher's edition: 9787111581871

This Springer imprint is published by the registered company Springer Nature Singapore Pte Ltd.
The registered company address is: 152 Beach Road, #21-01/04 Gateway East, Singapore 189721, Singapore

Foreword by Zheng Xiancong

Self-driving and Jolly Lifestyle—The Future You Deserve!

AI-powered driverless is the foreseeable future, the prelude to changing the way we live.

Every advance in technology will bring about changes in productivity and in a most critical aspect of the human life; the emergence of artificial intelligence represented by "driverless" will fundamentally change our lifestyles.

We spend more of our valuable time on the road while busy in driving will also be a continuation of personal disposable time. All will be returned to us via technology, and this high-quality time can make us enjoy life and happiness more.

This is the greatest significance of science and technology!

This book will talk about autonomous driving in terms of technology development, evolution, infrastructure, social collision, industry search, opportunities in China, future prospects and other aspects including driverless past and present, people's desire and anxiety, and how we describe the twenty-first century a new way of life. NextEV is very pleased to be able to provide intellectual support for this book and to lay the groundwork for an intelligent future for the automotive industry. For the whole industry to grasp the trend of thinking, doing a bit of work in a straight book, for the whole society to embrace the advent of self-driving cars, does a little consumer science education work. We will harvest your thinking and appeal to our reason; we will harvest your emotion and appeal to our sensibility.

Auto industry has more than 20 years experience in Ford, Fiat-Chrysler, and other brands. As a practitioner and observer, I have witnessed the development of the automobile industry in the twentieth century and the process of creating a new technological revolution henceforth. Today, I am also a motorist. Looking back at the development of human industry represented by automobile industry, I increasingly believe that the eternal theme of life is "to enjoy life." It is also to spend our own precious time to create good and enjoy the good. After many years of struggle for the auto industry, we finally saw the moment when cars began to fulfill their promises to people. Because of the emergence of "driverless," we will finally return to life itself, to create and enjoy the beauty of every single moment!

"Mist lock nine cities, thousands of years of jade Pan Qingyi. Weiran Tiangong re-appear world, more Xiangyi mountain green water."

This is our life full of joy, but also our inexorable future.

Through reading this book, I hope that our readers can also, much like me, look forward to the future at hand!

Shanghai, China   Zheng Xiancong
                                                  Co-founder of NextEV

Foreword by Dr. Daniel Kirchert

Embracing the New Era of Autonomous Vehicles

It is a great privilege for me to be invited to preface the English version of *Autonomous Driving Changes the Future* authored by James Chai (Chai Zhanxiang). In this book, James Chai elaborates what autonomous driving is from multiple perspectives like history, technology, and market, and how it will influence the development of the auto industry. Although the author modestly defines it as a "popular science reading," this book, in my opinion, can be used as a very useful reference book in the autonomous driving field, given its authoritative and comprehensive content.

Reading through this book, I feel that autonomous driving will not only change people's travel habits and needs, but fundamentally transform the competition pattern of the automotive world, thus leading to unprecedented changes to the global automotive industry.

Since my compatriot Karl Friedrich Benz was granted an invention patent on automobile in 1886, the automotive industry has evolved, as a pearl on the crown of the modern industry, into one of the most globalized industries, driven by generations of industry leaders, such as Henry Ford, Alfred Pritchard Sloan, Jr., Kiichiro Toyoda, and Soichiro Honda.

Over the century, however, despite enormous technological advances, for example, engineers developed automatic transmission and steering assist in an endeavor to improve driving and riding experience, the basic attribute of automobile as a machine driven by people remains unchanged. As a result, it is still impossible for people to choose to work, take a rest, or enjoy entertainment when they are driving their car as they do in a plane or high-speed train.

Good news is that the development and maturing of autonomous driving technologies will bring fundamental changes: With high-level autonomous driving, a car can take full control and drives itself, freeing driver's hands, feet, and eyes. When a car is no longer just a vehicle, the value it brings to driver and passengers will totally different from what it used to offer over the past 100 years. Perhaps in a few decades, a car will be considered as the extension of people's mind rather than their feet. By then, the definition of a good car will be completely different.

Based on my understanding, I believe the development and maturing of autonomous driving technologies will lead to unprecedented changes to the global automotive industry, thoroughly disrupting the global automotive competition landscape formed over the past century.

In light of this tendency, I chose to leave traditional automakers in 2016 and founded the premium smart EV brand "BYTON" together with many like-minded partners. While I am writing this preface, BYTON is proactively boosting the mass production of its first model M-Byte. We expect it to break traditional auto designs and architectures, and become an emotional smart device as an extension of the minds of the driver and passengers to empower mobility. M-Byte is expected to hit European and the US markets in 2021–2022, when readers will have the chance to experience our insights into autonomous driving and smart driving.

BYTON is not the only company vigorously preparing for the autonomous driving era, and China is not the only country paying close attention to the autonomous driving technology and its potentials. Readers will learn from the book how major car-making countries and companies are painstakingly developing autonomous driving technologies. The strategies they take vary greatly, but there is one thing in common: They all consider autonomous driving technologies the heart of the next-stage competition.

Finally, I'd like to express my thanks to James Chai again for presenting such an authoritative and readable book that introduces the autonomous driving technology and autonomous driving era with detailed, accurate data in a concise and easy-to-understand language.

Let us look forward to the advent of this wholly new era!

Nanjing, China   Dr. Daniel Kirchert
September 2019   CEO of BYTON

Preface

Autonomous Driving Technology Empowers Future Travel: Safe, Efficient, and Intelligent Mobility are Coming Soon

Boldly imagine an innovative way of traveling. For every journey, a passenger only need to step into the vehicle and relax—their route already plotted by using big data and downloaded to the car through the cloud, and the vehicle would automatic driving alone this route. Instead of wasting time behind the wheel, passengers can use their time on the road to plan their day's work, or just as easily shop, or chat with family and friends. The trip is easy—even pleasant—and the passenger disembarks, well-rested to face the rest of their day. After completing its journey, the vehicle finds itself a parking space to rest until another journey is called. Children no longer need to walk to school on their own; instead, an autonomous school bus can escort them. These smart vehicles can remotely access big data on virtually every road condition, gaining experience in increasingly complex road conditions to ensure safe driving. This is science fiction which comes to life—an imminent reality that is the basis for self-driving technology.

Our generation exists at a fortuitous time in history; we have reaped the benefits of large-scale industrial production and lucky to witness the arrival of such innovative technology. When I was working in the automotive industry twenty years ago, most vehicle technology breakthroughs focused on improving the mechanical performance of the vehicle itself. Enter increasingly complex suspension mechanism in strengthened vehicle bodies, matching more shifts transmission. As global attention turned to the energy crisis, climate change, and environmental emission regulations, research and development shifted toward turbochargers and sophisticated precision electronic fuel injection systems, squeezing more power from engines and reducing air resistance for improve fuel consumption. Recent breakthroughs in battery technology have increased the energy density of a single-battery cell enabling fully electric vehicles to roam the streets. We are so fortunate to participate in and witness the exciting development and popularization of new energy vehicles. These new energy vehicles promise not only to be highly electrified, but also interconnected, and driven by data-based research. The deep integration of the products and service from Internet and electronic tech companies into these vehicles will also enable the development of autonomous driving. The safety, efficiency, and intelligence of autonomous driving will completely change our way of life.

Thoughts of this revolutionary autonomous technology are sure to excite even the most stoic of automotive industry professionals. Some may even find inspiration, as Mr. Chai Zhanxiang, Dr. Jan Becker, and I have. Thus was the genesis of this book: *Autonomous Driving Changes the Future*. We three are professionals of the automotive industry, my partners, experts, and scholars of their respective fields. This book aims to provide readers of all ages with a clear and digestible state-of-the-field introduction to autonomous driving. This innovative technology will be discussed in terms of technological R&D reform, development and evolution roadmap, infrastructure improvement, multi-dimensional challenges to new technology, corporate developments, and anticipated prospects. Chapters of the book further introduce the strategic planning of international automotive OEMs in the field of autonomous driving, and the participation of intelligent manufacturing in China. In addition to these technological introductions, this book will also discuss hot topics in the popularization of autonomous driving technology, such as social ethics, privacy, hacking, sharing economy, business models, and automobile insurance.

Mr. Chai Zhanxiang is a graduate of the Department of Automotive Engineering of Tsinghua University and the Business School of Colorado State University. He has worked in the automotive industry technology, market research, and management for more than 20 years, and has rich and comprehensive industry experience. He has penned most of this book. Dr. Jan Becker is an internationally renowned expert on autonomous driving and a lecturer at Stanford University. Dr. Jan Becker authored most of the second and third chapters of this book. As for my contributions, I undertook the research and writing of chapters in the book, completing the book's editing, compilation, and English publication with the team. My gratitude goes to Mr. Huang, Chendong, CTO of XPT Technology Co., Ltd. belongings to NIO Inc., who personally approved Chap. 4. I am also appreciative of Dr. Sinisa Durekovic, Faraday Future, who provided content for the High-definition Digital map in Chap. 2. My sincerest thanks to the experts and executives who enabled the creation of this book through their unwavering support.

Where possible, this book uses easy-to-understand text to describe advanced, complex, and sometimes obscure industry technologies, improving accessibility for readers of all backgrounds. This format aims to provide readers keen on new technologies with a state of the field from the perspective of longtime automotive industry insiders. It is my hope that readers of this book will be able to identify and feel the unprecedented change quietly but unmistakably happening within society. The automotive industry is moving rapidly. In the era of electrification and intelligence, autonomous driving will soon become a reality. We firmly believe that in the future, automated driving technology will change human ways of life through widespread use of 5G communications, Industry 4.0 manufacturing, and biotechnology.

After participating in writing this book, I feel deeply that autonomous driving technology is not a mere theory, though the entire industrial chain cannot be established overnight. Cooperation between OEMs companies and suppliers within the supply chain is required to popularize and realize the wide use of autonomous driving, whether level 2 or level 5. This must be further supplemented by information-based road network infrastructure, High-definition Digital map, and other resources. This future-oriented driving technology is on the cusp of industrialization. Autonomous vehicles need driving system with full controlled by wire capabilities to enable faster responses and higher efficiency. Multiple control redundancy is also necessary to ensure the safety of the system. At present, products on the global market are not yet able to meet all these requirements. To that end, I founded AIKAR Technology in early 2018 in Los Angeles, bringing together professionals from the North American Bay Area and the world widely to work on an innovative, integrated, drive-by-wire system using a mature supply chain. This system features a highly integrated drive unit, brake unit, and control system that responds in real time to adjust the distribution of driving and braking forces by software. This in turn enables efficient energy recovery, reducing energy consumption, and significantly improving the mileage of electric vehicles. In addition, the integrated system uses existing vehicle-grade sub-components and modular design concept to reduce product development risks and shorten development timing. Our global team is accelerating the integration of supplier resources and launching products. We are committed to enable our product for service autonomous driving technology. In doing so, we hope to usher in an era of safe, efficient, and intelligent mobility.

Los Angeles, USA

Tianxin Nie
Former Faraday Future
Senior Vice President and Co-Founder
AIKAR Technology Founder and CEO

Contents

Chapter 1
The Centennial Automotive Industry and the Looming Transformation

Inventor, entrepreneur and visionary Ray Kurzweil is a significant character in our digital age. As a scientist at Google, he believes that the more developed the society, the stronger its ability to develop and iterate faster. As human society of the nineteenth century collectively master more knowledge than people in the fifteenth century, so the speed of human development in the nineteenth century is naturally faster than in the fifteenth century. He calls this phenomenon of accelerated human development the Law of Accelerating Returns.

By this logic, Kurzweil also concludes that the progress of the entire past century could be achieved in only 20 years time going by the speed of development of the year 2000. (This would mean that the development speed of 2000 is five times the last century.) Based on this yearly acceleration, it would take 14 years from 2000 to achieve the progress of the whole twentieth century. Then in 2014, it would take only seven more years (2021) to achieve another century of progress. Within only a few decades, we would achieve several times the progress of the twentieth century, until each century takes only one month. In accordance with the Law of Accelerating Returns, Kurzweil believes that human progress in the twenty-first century will reach 1000 times the speed of that in the twentieth century. If Kurzweil is correct, then the world of 2030 may become shocking to those in 2000, and the world of 2050 will be beyond recognition.

These changes that have shocked us will undoubtedly come from different factors. The most shocking of these changes will likely be from autonomous driving vehicles (including driverless cars) that will fully drive into our lives, even faster than we expected.

For the Chinese people, the car should hold special significance. The wheel was first invented by the ancestors of the Chinese nation. The first cart in human history was also made by the dexterous hands of our ancestors. In ancient Chinese mythology, the Yellow Emperor made a cart, resulting in his moniker *Xuanyuan*. *Xuan* refers to an ancient carriage with a shed, and *Yuan* is the basic building block of a vehicle. Even from the title, we can conclude that the Yellow Emperor is the inventor of the

© China Machine Press, Beijing and Springer Nature Singapore Pte Ltd. 2021
Z. Chai et al., *Autonomous Driving Changes the Future*,
https://doi.org/10.1007/978-981-15-6728-5_1

carriage. (Of course, the carriage mentioned here refers to a kind of hand-held trolley, and only later developed into the horse-driven carriage.)

The automotive industry is facing unprecedented and profound changes. Those changes are mainly reflected in the following four aspects. The first is interconnection. Many IT companies enter the automotive industry because they see automobiles as smart mobile terminals. The second is autonomous driving. Even people who don't drive can also enjoy the convenience of car travel in the future. The third is "Transportation as a Service" (TaaS). The future car not only serves as a product, but also gradually becomes an indispensable transportation service in the intelligent age and for future society. The fourth is the sharing economy. The twilight of car sharing will shine on the reality of car use, which will inevitably lead to changes in the ways cares are sold and used.

In a word, the future of the automotive industry is greatly reliant on the evolution of connectivity, intelligence and car sharing, which serve as the ultimate core of autonomous driving. Self-driving cars will change every aspect of our lives, like never before. Although current autonomous driving technology has not yet matured, opinions differ on the prospect of self-driving cars entering thousands of households. However, car makers and emerging technology companies have shown overwhelming enthusiasm for autonomous driving, making the new trend of this technology development crystal clear. The century-old automotive industry plans to use autonomous driving to achieve new feats of rapid transformation. We should prepare ourselves from this revolution accordingly.

1   From Feature Cars to Smart Cars

1.1   From the Birth of a Car to the Boom of Automotive Electronics

Karl Benz invented the car and brought mankind into the car era. Henry Ford made mass production a reality through his assembly line, which enabled the car to enter thousands of households at an affordable price. Ford's production line is a revolutionary and epoch-making innovation of large industrial production methods. The assembly line production is based on standardization, which facilitates mass production and becomes a form of production organization with high labor productivity. It has greatly changed the form of industrial production, promoted the development of the automotive industry, and brought the industry into a promising and practical field, changing society.

The Toyota Production System (TPS) can be said to be another great milestone in the history of world manufacturing. The leading position of the Japanese manufacturing industry in terms of "mode of production", "organizational capabilities" and "management methods" represented by Toyota Production System has changed the form of global manufacturing in the twenty-first century, promoting the lean production of the global automotive industry. An even leaner era comes.

For more than a hundred years, the car has not changed its basic structure as a box with four wheels, one steering wheel and several rows of seats. Even the basic structures of chassis and engine remain similar to how it was first conceived. However, the application of automotive electronics has been constantly improving.

The development of automotive electronics technology can be divided into three stages.

The first stage is one of separate control. The 1950s and 1960s marked initial stage of the development of automotive electronics technology. The mechanical designs of electronic devices such as the electronic engine ignition module were divided into independent components to improve performance and cut costs. From the late 1960s to the 1970s, electronic systems combined various discrete electronic devices formed, improving energy consumption and safety.

The second stage is characterized by centralized control. In the 1970s and 1980s, the adoption of digital circuits and large-scale integrated circuits and the increase in CPU computing speed and storage capacity expanded the control functions of automotive electronic control units (ECUs). In addition, the sensors used in the various controllers were universal, so multiple controls were concentrated on one ECU. This type of control is called a centralized control system, also known as a car microcomputer control system.

The third stage consists of the network control stage. Before 1990, most of the car's electronic control systems were run independently, resulting in bulky wire harnesses. In a high-end car designed the traditional way, the wire length can reach over 2000 m, and with more than 1500 electrical nodes. Only after the gradually adoption of network control and use of data sharing by the ECU was the overall performance of the system optimized and electrical wiring simplified. The network technologies that were widely used in automobiles at this stage were CAN bus and LIN bus.

Extended Reading

CAN Bus and LIN Bus Technologies

Bosch originally developed the Controller Area Network (CAN) in 1985 for in-vehicle networks, which reduced wiring cost, complexity, and weight. CAN, a high-integrity serial bus system for networking intelligent devices, emerged as the standard in-vehicle network. The automotive industry quickly adopted CAN and, in 1993, it became the international standard known as ISO 11,898. Since 1994, several higher-level protocols have been standardized on CAN, such as CANopen and DeviceNet. Other markets have widely adopted these additional protocols, which are now standards for industrial communications. Its emergence provides powerful technical support for distributed control systems to achieve real-time and reliable data communication between nodes. However, the CAN bus still has some drawbacks. The CAN bus communication adopts the carrier monitor lossless arbitration technology. When the network load is small, the CAN bus real-time performance can meet various requirements, but as the network load increases, the probability of information collision on the bus also increases. If you continue to use the basic CAN protocol, the

real-time performance of lower-priority information transmission will be affected.
After the network load reaches a certain level, it will even exit the bus competition.
The CAN bus protocol adopts a static fixed priority allocation method, making it
difficult to share the bus usage rights fairly with information of different priority.
These defects become factors that restrict its further development.

Initially developed in Germany in 1998, Local Interconnect Network (LIN) auto-
motive communication protocol offers a cost-effective way to deliver low-speed
communication. It is used for all comfort and convenience applications in modern
cars. LIN has the capability of addressing all applications which cannot be directly
connected to the controller area network (CAN) because of the high costs involved.
In addition to defining the basic protocol and physical layer, the LIN specification
defines development tools and application software interfaces. LIN communication is
based on the SCI (UART) data format, using a single master controller/multiple slave
device mode, and a single 12 V signal bus and a node-synchronous clock line with no
fixed time benchmark. This low-cost serial communication model and corresponding
development environment has been standardardized by the LIN Association.

The application of modern automotive electronics technology can be reflected in
the following aspects.

First, in dynamic performance and environmental protection, including CPU-
controlled ignition, idle speed control, engine deflagration control, automatic gear
shifting, engine electronic fuel injection, air–fuel ratio feedback and exhaust gas
recirculation, etc.

Second, in safety, including acceleration anti-slip control, automatic power distribution, anti-lock braking systems, traction control and automatic brake differential,
etc.

Third, in trafficability, such as differential lock control, adjustable shock absorber,
etc.

Fourth, in vehicle handling and stability, including cruise control, stability control,
body active control, dynamic stability control, etc.

Fifth, in the vehicle body electronics, including seat automatic adjustment system,
intelligent headlightsystem, night vision system, electronic door lock and anti-theft
system, etc.

Sixth, in communication and infotainment systems, including intelligent car navigation system, automatic speech recognition system, car maintenance data transmission system, car audio systems, real-time traffic information consultation system,
dynamic vehicle tracking and management system and television entertainment
system.

As the vehicle performance becomes more dependent on electronic technology,
the future development of the automotive electronics industry must be more cognizant
of environmental protection, safety and communication connectivity.

The first issue is environmental protection. The current trend in the automotive
industry is to develop engines with high efficiency and low carbon emissions. The
current options are as follows: First, an advanced diesel engines with electronic
control systems, can improve fuel economy on the highway by 30–40%; second, pure
electric vehicles or hybrid vehicles can improve the fuel economy by 30–40% and

reduce carbon emissions by 60%; Third, cylinder pressure sensing and homogeneous charge compression ignition (HCCI) systems can be used to improve fuel economy and reduce emissions. These innovative powertrain technologies will add substantial electronic contents to cars around the world in the next five to fifteen years.

This is followed by safety. Currently, passive safety technology provides protection technologies and products for drivers and passengers in the event of a collision, such as collision sensors, airbags and safety belts. The latest development direction is to pursue active safety. Sensor technologies such as LiDAR, radar, and camera allow Advanced Driver Assistance System (ADAS) to actively avoid collisions.

Finally, communication connectivity. Today, consumers enjoy the benefits of digital electronics and internet infrastructure such as mobile video, digital TV, Wi-Fi and GPS. Consumers also want to enjoy the same technology and communication benefits in cars to make the driving process more efficient, convenient and entertaining.

1.2   The Era of Smart Cars is at Your Fingertips

Following these automotive electronics technology trends, cross-border cooperation between automakers and IT companies has also begun. The huge amount of information that cars can collect is very attractive to tech companies like Google, who mine personal data. Such a philosophy existsi nside Toyota: "If you can father data from all the windshield wipers, you can understand the detailed weather information of each place." If you can collect various kinds of information from tens of millions of vehicles on the road, that will generate unlimited business opportunities. Be it Toyota, other global leading automakers or parts companies, or even IT giants, everyone wants a slice of the pie. Through their exploration, automobiles are led to pursue intelligent technology. At this point, the hundred-year-old automotive industry is experiencing an exciting transformation from the era of feature cars to the era of smart cars.

What is intelligentization?

What is intelligentization? Intelligentization refers to the application of a certain aspect by a combination of modern communication and information technology, computer network technology, industry technology and intelligent control technology. The process of transforming feeling to memory, then to thinking is called "wisdom." The result of wisdom produces behavior and language, and the process of expression of behavior and language is called "ability". The two are collectively called "intelligence." Intelligence generally has the following characteristics:

Firstly, it is able perceive the outside world and through external information. This is a prerequisite and a necessary condition for generating intelligent activities.

Secondly, it has the ability to remember and think, to store the perceived external information and the knowledge generated by the thinking, and to use the existing knowledge to analyze, calculate, compare, judge, associate and make decisions.

Thirdly, it has the ability to learn and adapt through the interactions with the environment, continuous learning and accumulation of knowledge, so as to adapt to environmental changes.

Fourthly, it has the ability to make decisions, to respond to external stimuli, to form decisions, and to convey corresponding information.

The system with the above characteristics is an intelligent system or an intelligentized system.

With the overall development of the automotive industry, intelligent technology will be applied more frequently. In recent years, we have witnessed the trend in which a contemporary "Feature Car" gradually evolves into a "Smart Car." Automobile intelligentization has become a central issue, and smart cars have become one of the future development trends of the automotive industry. Intelligentization will bring more and more changes to the manufacturing technology and business models of the automotive industry, making the operation of a car simpler, the engine more and more powerful and efficient, and the driving safety more and more reliable.

There are many intelligent applications for automobiles, such as vehicle power control systems. The system detects the speed, angular velocity, steering wheel angle and other motion states of the vehicle through various sensors, and changes the motion state of the vehicle by actively braking one side wheel as needed. The car is optimally driven and maneuverable, increasing its wheel adhesion, the handling and stability. Other intelligent applications include automatic emergency braking system(AEB), adaptive cruise control (ACC), etc., which will be discussed in more detail later in this book.

The continuous application of new materials and new technologies is also improving the level of intelligence for automotive parts and materials, such as smart glass, which is often overlooked. There are many types of smart glass, including light-proof and rain-proof glass, electric snowmelt glass, image display glass, shatter-proof glass, dimming glass, and photoelectric sunshade glass. Light-proof and rain-proof glass is made of new materials and new surface treatment methods. When the rain falls on the glass, it will flow away without leaving water droplets, meaning no need for windshield wipers. The inner surface of the glass is low in reflectivity, leaving no inverted images of the instrument panel and car interior, so that the driver's sight is undisturbed. The image display glass is coated with a transparent reflective film on a part of the windshield, so that the film displays the instrument panel via a projector. If combined with an infrared image display system, it would allow the driver to see the objects about 2 km away in a foggy day. The photoelectric sunshade glass can absorb, accumulate and use solar energy to operate the fan inside the car when driving or parking, and to recharge the car battery. The development and application of smart glass has been valued by many car companies. Tesla has a glass technology research and development team dedicated to the development of special glass manufacturing technology, which could allow it to absorb heat and melt snow on its surface through the heat of the sun.

Automotive intelligent applications will eventually lead to intelligent driving. The scientific question involved in intelligent driving include how to simulate human driving behavior, including how to formulate the driving behavior of different people, how to use sensors to characterize the visual and auditory sense of the driver, and how to use machine intelligence to describe the driver's driving process, how to achieve automatic switching between autonomous driving and manual driving, etc. In the future, intelligence will redefinition of automobiles. Electronic technology and software will overwhelm traditional mechanical and electrical systems. A car will become a moving intelligent space and serve as a new carrier for mobile work and entertainment.

2 Globalization has Made the Automotive Industry Flourish

2.1 The Automotive Industry is the Most Typical Global Industry

The predicament of industrial civilization is a prerequisite for the arrival of globalization. Globalization is a new phenomenon that has become increasingly prominent in the world since the 1980s, and is a basic feature of the modern era. The automotive industry has since become one of the most globalized industries.

The focus of globalization lies in economic globalization, and the core of economic globalization is industrial globalization. As one of the most important industries in the economic field, the automotive industry, witnessing a rapid development in the past 30 years, has undoubtedly benefited from globalization.

2.2 Brilliant Manipulations in the Auto Sector

In the context of industrial globalization, the world automobile industry has undergone two very significant and interrelated changes: First, auto giants continue to do cross-border mergers and acquisitions, forming a huge automobile enterprise group; second, the industrial chain of the automotive industry extends worldwide. These two major changes have fundamentally reshaped the traditional resource allocation of the automobile industry, the competition between and organizational structure of enterprises, intensifying global competition in the auto industry. The auto industry is also an industry with high correlation between upstream and downstream sectors. Its development can lead to the prosperity of a series of related industrial segments such as from steel, chemicals, rubber, glass and electronics down to petrochemical, road construction, automobile repair and maintenance, tourism and auto finance. Therefore, the competitiveness of the automotive industry is an important part of a country's comprehensive national strength.

Horizontally, the integration and competition among various multinational automobile groups will inevitably lead to industrial transfer. The rapid sales growth and the quick rise of local vehicle manufacturers in emerging markets have led to the shift of vehicle manufacturing to emerging countries; the thrift of emerging markets and the demand for cost reduction in automotive R&D have led to the migration of global automotive R&D resources to emerging markets as well.

The automotive industry has experienced many structural changes, changing from hundreds of brands to several well-known multinational companies. With their technical and capital advantages, each group has launched all-round fierce competition in product design, manufacturing, information technology, e-commerce, sales and after-sales services and capital operations. On the one hand, it exports surplus capital and technological know-how to emerging countries. On the other hand, it absorbs global resources and expands global market share through mergers and acquisitions, further promoting the globalization of the industry.

In 1998, Daimler-Benz and Chrysler merged to form the Daimler-Chrysler Group. In 1999, Ford acquired the Volvo Car Division, and Renault invested in Nissan. Since then the pattern of "6 + 3" has appeared, namely, GM, Ford, Daimler-Chrysler, Toyota, Volkswagen and Renault as the Big Six, and Honda, BMW and PSA (Peugeot Citroen Group) as the Small Three. The financial crisis has accelerated the pace of global auto map adjustment, mainly reflected in the changes of the Big Three in the North America: Chrysler could not survive independently. After two years of separation, re-integrated by Fiat, GM and Ford continued to spin off their own subbrands. Rolls-Royce was resold by the British to BMW, Land Rover and Jaguar have thrown themselves on the Tata Group of India, and Volvo was sold again by Ford to be owned by Geely of China. A series of changes have led to a new "6 + 3 + X" pattern in the global automotive industry. The new six groups include Toyota, Volkswagen, new GM, Ford, Renault-Nissan Alliance and the new Fiat-Chrysler Alliance. The three small groups include Hyundai—Kia, Honda and PSA. In addition, Daimler-Benz, BMW, a number of Japanese car companies including Suzuki, and growing car companies in emerging markets such as China and India are also indispensable forces in the global auto map.

Vertically, the world's major auto makers have implemented global production and global sourcing, from sourcing from multiple auto parts manufacturers to sourcing from a few system suppliers; from single auto part procurement to module procurement; from domestic procurement to global procurement. The globalization of auto industrial chain is increasingly evident, and the re-division of the value chain is increasingly prominent. The globalization of the industrial chain has resulted in a refinement of the division of labor in the automotive industry, aggravating the dissociation between vehicle assembly and parts production.

With the degree of specialization, it has become possible for a parts and components company to present OEMs with multi-series and large-scale production products. The auto parts are developed from a single component to a modular and integrated product. The regionalization of the component industry is shifting to internationalization, and more and more multinational parts manufacturing companies have emerged. European auto parts manufacturers have invested overseas to carry out international production; North American auto parts manufacturers have invested in Europe to expand the Eastern European market; Japanese auto companies are also constantly establishing their own branch production facilities on a global scale. The Asian market is the focus of competition among major auto parts multinationals, where China is a major player.

The reform of the procurement system of OEMs requires auto parts manufacturers to constantly keep up with changes. It not only requires auto parts manufacturers to expand their own strengths, improve product development capabilities and design in the form of sub-assembly, but also requires them to shorten development cycle and provides quality and cheap products. Under the circumstance of this internationalization, component suppliers are both competitors and partners. In order to meet the ever-increasing demands of customers, sometimes they have no choice but to work together to provide innovative solutions. The global auto parts industry is increasingly engaged in strong alliances, acquisitions, cooperation, transformation, and the divestiture of traditional businesses to betterserve customers.

In May 2015, ZF completed the acquisition of TRW. TRW was incorporated as a new business unit–the Active and Passive Safety Technology Division, fully operated by ZF. The deal has made ZF a larger and more diversified global industry giant, with a better focus on future powertrain, transportation and safety solutions for the automotive industry. Two years later, in early May 2017, two "friends" of ZF and Faurecia announced to cooperate in a strategic partnership for the development of disruptive and differentiating interior and safety technologies for autonomous driving. Within this special advanced engineering partnership the two companies will identify and develop innovative safety and interior solutions linked to different potential occupant positions. One of the highlights of this cooperation is that it will be based on the sharing of expertise between the two parties, and does not involve capital transactions.

On May 3, 2017, Delphi announced in the UK that it would fully spin off its powertrain division. The original company would focus on the electrical and electronic business, especially in the areas of autonomous driving, smart technology and safety technology. It employed 15,000 engineers and 145,000 employees worldwide. After Delphi announced the news, its share price soared 12%. The global auto parts giants have realized that the electric wave is hard to contain, so they decided to seek change actively and embrace industrial transformation. Almost simultaneously, Bosch also announced the sale of its starter and generator business, together with 7000 employees in the business unit, to a Chinese consortium.

This evolution of globalization has directly produced two outcomes. On the one hand, the production of auto parts has become increasingly specialized, and the resulting economies of scale have significantly reduced production costs and promoted the allocation of global resources. On the other hand, the links between auto parts manufacturers and automakers are closer, especially in terms of technical cooperation. Auto parts manufacturers have begun to assume more R&D and manufacturing responsibilities, and are able to intervene more deeply in the development of new models of vehicles, so that the vertical depth of production is continuously improved to meet the needs of OEMs. For example, Audi's development depth in 1996 was 80%. By 2000, its development depth was reduced to 55%, and the rest was completed by component manufacturers.

3   Autonomous Driving: The New Battlefield

3.1   New Heights of Industrial Competition

According to the Florida Transportation Authority, "Autonomous driving technology" refers to "the technology that is installed on a vehicle to achieve automatic vehicle travel, in the case where the human factor does not actively manipulate or monitor the vehicle." In recent years, with the rapid development of the Internet, Internet of Things, big data computing, intelligent control, artificial intelligence and other technologies, the trend of autonomous driving has become more and more distinct. Auto makers and emerging technology companies in Europe, America, Japan and China are making great efforts to promote the research and testing of autonomous driving technology. In 2010, the VisLab Laboratory at the University of Parma in Italy launched the VisLab International Autonomous Challenge (VIAC). They organized a fleet of cars to depart from Parma on July 20, four of which were equipped with GPS and GOLD obstacles and lane detection systems. On October 28, after a long journey of 13,000 km, they successfully arrived at the European Pavilion of the Shanghai World Expo. This event made people, especially those in China, begin to notice how autonomous driving is impacting the automotive industry.

The triumph of auto digitalization has irrevocably impacted the car; some even refer to modern cars as "high-performance computer at a speed of 113 km per hour." Peripheral players like Google and Uber now have spared no efforts in the development of autonomous driving technology, not to mention traditional auto makers. Although fully self-driving cars are still in the experimental stage, there is no doubt that other major changes are emerging in the automotive industry.

Digital tools have changed the way people socialize, learn and navigate. Networks, sensors, mobile communications and artificial intelligence have developed rapidly over the past 20 years, and have gradually entered the urban space. These technologies collect and integrate real-time information at all levels. For example, Internet-based interactive visualization system HubCab records 170 million routes taken annually by New York taxis, clearly defining the operation of urban traffic.

Traditional cars are gradually becoming information receptors, from both the riders and the environment. Systems within the car can detect the fatigue and stress level of the driver. Outside the car, radars, camera and LiDAR can "read" the surrounding environment and react accordingly.

These two trends have been synchronized in autonomous driving systems, thanks to the rapid development of artificial intelligence, vehicle interconnection and intelligent transportation infrastructure.

According to IHS and McKinsey, by 2020 autonomous driving technology will be implemented in highways and fleet operations, and relevant laws and regulations and industry standards will also be introduced. By 2025, autonomous driving will increase the industry's worth from 200 billion to 1.9 trillion US dollars. Global sales of autonomous vehicles will reach 300,000–600,000 units, of which about 50,000–100,000 units will be sold to China. By 2035, world sales of self-driving cars will reach 15 million–20 million, and the number of self-driving cars on the road will reach 50–70 million. 25% of the market share will be in China, 29% in North American and 20% in the EU. With such projected growth, it's wonder major OEMs have turned their attention to this field.

At present, many countries are trying to make autonomous vehicle a competitive industry. In September 2016, the US Department of Transportation issued the *Federal Automated Vehicles Policy*, establishing an institutional framework in four areas, from vehicle performance guidance for automated vehicles, model State policy, National Highway Traffic Safety Administration (NHTSA)'s current regulatory tools to modern regulatory tools. After the release of *"Automatic and Connected Driving Strategy"* in 2016, the German Federal Ministry of Transport and Digital Infrastructure (BMVI–Bundesministerium für Verkehr und digitale Infrastruktur) established an interdisciplinary and inter-departmental autonomous driving roundtable system for the technical, regulatory and social issues that autonomous driving may face. In conjunction with the state governments, industry associations, research institutes and consumer associations, they have discussed and reached consensus on some important issues. Germany and France have demonstrated confidence by jointly launching a cross-border road test for self-driving cars. The Japanese Prime Minister has announced that high-tech autonomous driving technology will be part of its economic policy, encouraging the country's auto industry to develop this technology.

In October 2016, China released the *"Technology Roadmap of Intelligent & Connected Vehicles"* to guide the research and development of auto makers and related parties. Some provinces and cities in China have also fully considered the Internet of Vehicles (V2X) and autonomous driving in terms of urban planning and policy support. According to the *Outline of the Shanghai Urban Master Plan (2015–2040)*, Shanghai is preparing space for future new modes of transportation. The city plans to "strengthen the sharing of space resources by various modes of transportation… [to] optimize technical specs on transportation infrastructure and land use, [and to]create conditions for the development of new transportation means such as seaplanes and automated vehicles in the future." Chongqing has made great efforts to promote these vehicle networking applications, through a three-and-a-half-year demo project implemented in two phases. The first phase would be at the comprehensive test site, while the second phase would focus on designated social roads. Simulation and tests will be run on vehicle networking collaborative communication under different communication standards. It will test semi-autonomous and fully autonomous driving vehicles for 5G communication, high-precision digital map, Beidou navigation, cloud computing, under various road situations and multiple traffic scenarios.

3.2 The Future has Come, Along with Fierce Competition

In September 2016, Uber launched a driverless car passenger service on the roads of Pittsburgh, USA, making it the first company to do so in the US. On that day, four Ford Fusion driverless cars, equipped with cameras, LiDAR, radar and other sensors, were sent on the road to transport passengers. To ensure safety, there were two engineers on each driverless car; one sitting in the driver's seat, ready to take over the vehicle when necessary and the other monitoring in the back seat. Pittsburgh has complex roads, narrow streets, steep slopes, and tunnels and bridges. Uber believes that if its driverless car can handle Pittsburgh roads, then it could also handle road conditions in most other cities. Prior to this, Uber's driverless car had been tested on these roads for nearly two years. Test results indicated that performance rates no different from traditional vehicles driven by human drivers.

Despite the trial run, some experts predict that Uber's completely autonomous plan will take years or even decades to achieve. "If a driverless car runs at a speed of about 118 km/h on the road, the situation can get very bad in the event of a breakdown." Prof. Raj Rajkumar at Carnegie Mellon University said, "This technology must be fairly reliable in order to allow the driver to leave the car completely."

Autonomous driving is becoming an investment hotspot, inciting competition from global auto industry and Internet companies. Chinese auto companies have long been in this area, and the R&D process accelerated by foreign auto companies. Changan iconically completed a 2000-km long autonomous car journey from Chongqing to Beijing over five days in April 2016.

The ultimate goal of autonomous driving is to drive unmanned. Companies have already moved forward towards this goal, yet the debates have never stopped. Some do not believe that automated vehicles are a possibility for the foreseeable future. At the SAE 2016 Automotive Electronics Conference, Jeff Owens, Chief Technology Officer (CTO) of Delphi, stated that "Automated driving is right before us, "to wide approval. Ford Vice President Ken Washington agreed with Owens. "Looking of the history of the automotive industry, the speed of maturity of new technologies will always surprise those who are skeptical. It is well known that technology companies are innovating at an amazing speed. Coupled with Moore's Law, the era of driverless cars may arrive sooner than you would think." General Motors has begun to test autonomous vehicles in the United States. Yet skeptics, Jon Lanckner, Vice President of General Motors, joke: "I can assure you that we are not going to test it endlessly."

Phillip Eyler, Vice President of Harman International, argued that the arrival of driverless cars may be faster than expected—however, mass adoption is another issue. This author is agrees. The large-scale popularization of autonomous driving may take some time, but some level of ADAS function could become a new selling point for competition between auto companies. Having these capabilities not only helps car companies build their unique advantages, but also to and influence consumers' choices. GM, Ford, BMW, Mercedes-Benz, Volvo and other mainstream car companies are working hard to build these capabilities.

At the same time, R&D in autonomous driving largely focuses on only the perception and decision-making process, and not the execution process. Emerging companies and traditional high-tech companies, such as Google and Baidu, use their existing knowledge of images and data to build their AI capabilities. The breakthrough of a single technology, computer vision, or deep learning algorithms, still plays an important role.

In the future transportation industry, autonomous driving technology will undoubtedly lead to revolutionary changes; fierce competition among countries and companies has already begun.

3.3 Detroit Versus Silicon Valley: Who Will Come Out on Top?

Rick Snyder, the governor of Michigan, was one of the main driver's of Detroit's bankruptcy a few years ago. Detroit has since embarked on the road to recovery. Since he took office, he has visited China once a year. In his 2016 trip, he looked forward to meeting Chairman of Great Wall Motors Wei Jianjun, because he noticed that the company was scouting potential locations for its North American R&D Center. The two options were Detroit—the old front—and Silicon Valley—the new battlefield of the auto industry.

According to German magazine of "Automobil Produktion", Matthias Müller, CEO of VW Group, Dieter Zetsche, Chairman of Daimler and Harald Kruger, Chairman of BMW discussed the transformation of the automotive industry towards autonomous, electric and customized cars, in a meeting in Munich on November 9, 2016. They agreed that German car companies must prepare accordingly, or risk becoming Silicon Valley suppliers. Dr. Dieter Zetsche said that the German automotive industry faced tremendous changes. Mr. Harald Kruger added that the German car industry could be more confident: "Reviewing the past, any disruption has promoted human progress." Mr. Matthias Müller pointed out that being cautious and avoiding mistakes have led to the success of "Made in Germany", but that "Valley has begun to build the Internet of Vehicles."

Silicon Valley, the city of IT technology, is becoming the auto city that Detroit once was, becoming the favored location for cutting-edge automobile laboratories.

Whether it's a small office "laboratory,"a formal R&D center, or investment fund, car companies vie for at least one house number along the 101 Highway in southern San Francisco. German automakers Mercedes-Benz and Volkswagen set up teams in Silicon Valley in the late 1990s. Ford, Renault-Nissan, Toyota and Honda are relative newcomers, joining only in the last five years. During this period, more than a dozen car companies took root in Silicon Valley, all working towards solving the two major challenges in the automotive industry: the rapid development of autonomous driving technology, and the transition from car "owning" to car "leasing".

Toyota has set up a Silicon Valley R&D center dedicated to artificial intelligence and robotics, led by Gill Pratt, a strongman in the field of artificial intelligence. The company was apparently seeking cooperation with Google, yet the two companies are fiercely competing for the development of a new generation of cars.

Automakers have begun to realize that the driving force for innovation from outside the industry is even stronger than those from within. Establishing laboratories in California can help car companies better cope with this threat from tech companies, as it enables them to poach the best engineers and developers. After all, it is hard to persuade an Apple employee to leave sunny Cupertino and go live in the suburbs of cold Detroit. Furthermore, car companies in the area will can establish long-term relationships with start-ups and universities in the region—especially Stanford University, a center for cutting edge developments in autonomous driving, artificial intelligence and human–computer interaction (HCI).

Jeremy Carlson, an analyst at IHS Automotive Information, believes that Silicon Valley's R&D cycle at odds with the usual "long cycles" of the automotive industry, but is necessary for car companies to adapt to the speed of Silicon Valley. It takes the automotive industry 5–7 years to develop a new vehicle, while it takes tech companies only 6–8 months to develop a new smartphone, and just one hour to design a website. To comply with this faster pace, car companies have begun to cooperate with technology companies to create a travel ecosystem. Ford has set up a 130-person lab in Palo Alto, CA, and partnered with Amazon to integrate Alexa voice assistant into its in-vehicle intelligence system. Mercedes-Benz is working with Google's smart furniture company Nest to integrate its intelligent temperature control system into the car, allowing users to remotely control the temperature at home while driving.

Detroit has fought back as Silicon Valley stepped up. On November 18, 2016, Governor Snyder arrived in Beijing for his sixth visit to China since he took office. He tried to attract the attention of the Chinese auto industry from Silicon Valley to Detroit. As an old car manufacturing center, Michigan was introduced emotionally by Governor Snyder while introducing initiative PlanetM. He said: "PlanetM is Michigan's mobility initiative representing the collective mobility efforts across the state. PlanetM connects you to Michigan's mobility ecosystem—the people, places and resources dedicated to the evolution of transportation mobility."

Mcity is the first real-world automated vehicle testing ground in the city in Michigan. Governor Snyder said: "Mcity is very popular and there is no more space available. We will open a new car test center in the near future, covering ten times the floor area of Mcity." Meanwhile, an intelligent corridor in the southeast Michigan brings together smart car technologies, including construction and road facilities, and will use big data to measure and analyze connected vehicles and autonomous driving test methods. It is foreseeable that neither Silicon Valley or Detroit will surrender this battle ti be the future center of automobile development.

As to the rise of Silicon Valley, Zhao Fuquan, Director of the Institute of Automotive Industry and Technology Strategy of Tsinghua University, commented very well: "who will be the winner in the future development of the automotive industry? Detroit or Silicon Valley? To develop safe, convenient, energy-saving and environmentally-friendly intelligent vehicles, both Detroit and Silicon Valley's capabilities are indispensable. No one alone could afford to build a perfect car in the future. Therefore, competition and cooperation between the two are needed. In fact, this is precisely the embodiment of Internet thinking—competing in collaboration, and profiting from sharing. In the future for a long period of time, cross-industry mergers and acquisitions will occur repeatedly, as the marriage of the two industrial forces of Detroit and Silicon Valley will benefit all of society. In this sense, neither Detroit nor Silicon Valley will come out as "winners", but will, through competition and cooperation, allow the end users of their products to become the biggest winners."

3.4   The "Chasm" to be Crossed

Everyone in technology knows of Geoffrey Moore as the author of "Crossing the Chasm: Marketing and Selling Technology Products to Mainstream Customers" His book has sold over 1 million copies. In Crossing the Chasm, Geoffrey A. Moore has described that in the Technology Adoption Life Cycle—which begins with innovators and moves to early adopters, early majority, late majority, and laggards—there is a vast chasm between the early adopters and the early majority. While early adopters are willing to sacrifice for the advantage of being first, the early majority waits until they know that the technology actually offers improvements in productivity. The challenge for innovators and marketers is to narrow this chasm and ultimately accelerate adoption across every segment.

"Crossing the Chasm" has become a metaphor that is universally used by companies with complex products to explain why they struggle to sell to the mainstream market. As the godfather of insight selling, Mr. Moore believes that early stage, a high-tech technology is only used by enthusiast users, or early adopters. The emergence of enthusiasts of autonomous driving is just the beginning of its era. Only when the majority of society has accepted this new technology, will the era automated vehicles truly arrive.

The change brought by autonomous driving goes far beyond the automotive industry. In its advanced form, driverless driving will completely change the way of travel. In a few years, taxis will be unmanned, the total number of automobiles will be substantially reduced. Yet, utilization rate will be improved, and traffic jams will novelty; "white clouds drifting in the blue sky" will become commonplace. Parking lots will also disappear, replaced by parks, roads and shelters; the chance of traffic accident is almost zero. What an exciting scene!

Chapter 2
Technologies for Autonomous Driving

1 Environment Sensing

In the development of smart car technologies, the perception and understanding on the vehicle's surrounding environment is the basic premise for realizing car intelligence. Only when in the roads, other vehicles, pedestrians and other information around the vehicle are accurately and timely sensed, the driving behavior of the self-driving car will have a reliable decision basis. At present, most smart cars are equipped with various types of sensors, which have different principles, different performances and different functions, as the main sensing devices. These devices can be used to obtain different sensing information and reflect changes in the surrounding environment from different perspectives.

Extended Reading

Selective Attention

Psychophysical studies have shown that selective attention is an important feature of human natural vision. On the one hand, natural vision does not meticulously and accurately identify all objects in the scene. It is often only purposeful and selective to perceive and understand parts of the scene, focusing on the target of interest. On the other hand, the natural vision's sampling frequency and resolution of the scene are not static, but varying with time, that is, selective attention changes with time and is non-uniformly sampled. The division of tasks of different sensors on smart cars reflects the selective attention of humans in the driving process. Vehicle sensors have different ways of configurations. For example, for a visual sensor, it is possible to detect and identify a lane markings, a traffic light, a traffic sign, a peripheral obstacle, etc. by a multi-thread method only by using a monocular camera, or to deploy a plurality of dedicated cameras, each of which performs a single task.

Driver Assistance and Automated Driving systems use sensors to obtain information from the environment. Most commonly used are radar sensors, cameras, LiDAR sensors and ultrasonic sensors to constantly scan the driving environment.

© China Machine Press, Beijing and Springer Nature Singapore Pte Ltd. 2021
Z. Chai et al., *Autonomous Driving Changes the Future*,
https://doi.org/10.1007/978-981-15-6728-5_2

These sensors can detect other traffic participants, such as cars, trucks, motorcycles, bicyclists, persons. Additionally, they can also detect infrastructure, such as lane markings, traffic signs, traffic lights, or other objects such as trees, bushes, buildings.

Different sensors operate using different measurement principles and different signals in the electromagnetic spectrum, i.e. from ultrasound ($\lambda = 4$–7.5 km, $f = 40$–70 kHz) over microwaves ($\lambda = 2.5$–12 mm, $f = 24$–120 GHz) and visible light ($\lambda = 390$–700 nm, $f = 430$–770 THz) to near infrared ($\lambda = 700$–1600 nm, $f = 190$–430 THz) and far infrared light ($\lambda = 7$–11 μm, $f = 27$–42 THz). Some sensors use active dissipation of energy, such as radar sensors, ultrasonic sensors, and LiDAR sensors; cameras on the other hand are passive and make use of environment lighting.

The information perceived by a single sensor is typically not sufficient to support complex autonomous systems. Therefore, sensor data fusion is used in perception systems to combine the information from the various sensors with the goal of generating a uniform and consistent model of the environment, sufficient for the respective function.

1.1 Radar

Radar (Radio Detection and Ranging) sensors were originally developed for military and avionics applications. Radar sensor technology in general is discussed, e.g., here. The first radar sensors for automotive applications were developed in research projects approximately 40 years ago, but the introduction of a radar sensor into commercial automobiles took until 1998. Initially, radar sensors were used for adaptive cruise control. Later, forward collision warning functions were integrated into the system.

Radar sensors operate in the microwave spectrum, mostly in the 76.5 GHz spectrum, which is reserved for automotive radar applications world-wide. A radar sensor emits directed energy through the antenna. The emitted energy Pe is reflected by a target, e.g., a vehicle in front of the sensor, and then detected by the receiver antenna in the radar as the received power Pr:

$$P_r = \left(\frac{P_e G}{4\pi d^2} \right) \cdot \left(\frac{\sigma}{4\pi d^2} \right) \cdot \left(\frac{G\lambda^2}{4\pi} \right) \cdot (\ )$$

The first term of this equation represents the power density of the electromagnetic waves hitting the target. The second term represents the power density of the returning waves. The third term shows the effective aperture or area of the antenna and the last term represents other influences such as rain, snow, or dirt on the antenna. The last term represents other influences.

The amount of reflected energy depends primarily on the reflectivity properties of the target σ as well as on the distance d between the target and the sensors. G is the antenna gain and λ is the wavelength. For example, a metal plate which is exactly perpendicular to the direction of the radar antenna leads to a very strong reflection of the radar signal. Rotating the same metal plate just one degree can lead to significantly reduced reflectivity, which then results in a considerably reduced range in which the plate can still be detected. Generally, different materials also have significantly different reflectivity properties. These reflectivity properties are expressed as the radar cross section (RCS) or σ. In automotive applications, RCS values from 1 to 10,000 m^2 are observed. A large truck might have an RCS value of 10,000 m^2, a passenger vehicle can have a value of 100 m^2, a motorcycle might have 10 m^2 and a person only 1 m^2. Furthermore, it is important to note that radar energy is also reflected off the street surface.

Radar sensors make use of the so-called Doppler effect, which is a frequency shift when sender and receiver—or sender/receiver and reflector—move relative to each other. This frequency change is proportional to the relative velocity of sender/receiver and reflector, see Fig. 1. The frequency shift is positive when sender/receiver and reflector move towards each other, and negative when sender/receiver and reflector move away from each other.

There are many different radar sensor implementations. In automotive applications, predominantly, frequency modulated continuous wave (FMCW) radars are used. FMCW radars use an indirect distance or time of flight measurement, using the frequency difference between the sent and the received signal. The transmitted signal $signal_{sent}$ of a known stable frequency continuous wave varies up and down in frequency over a fixed period of time T_m by a modulated signal, see Fig. 2a.

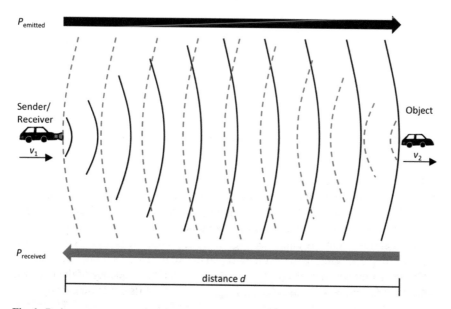

Fig. 1 Radar measurement principle

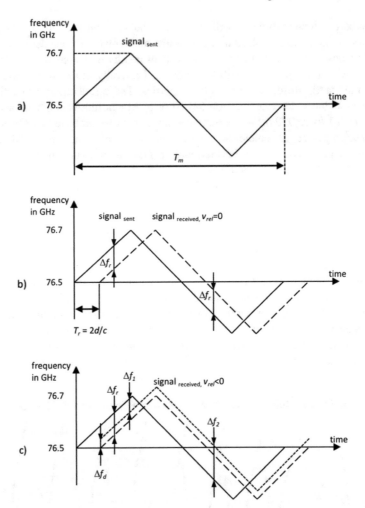

Fig. 2 FMCW radar frequency modulation and frequency shift principle

The returned signal is received after the time $T_r = 2d/c$, where d is the distance between radar and object and c is the speed of light. The frequency difference between the received signal and the transmitted signal Δf_r increases with the distance between sensor and object due to the frequency modulation and, therefore, can serve as an indirect measure for the distance, see Fig. 2b. In addition, in case of moving objects, there is a frequency shift in the returned signal due to the Doppler effect, see Fig. 2c.

Consequently, the distance (or time of flight) shifts the frequency of the returned signal by Δf_r, and relative velocity (or Doppler effect) shifts the frequency by Δf_d. Δf_r is defined as $\Delta f_r = (2 f_{hub})/(c \cdot T_m) \cdot d$, where f_{hub} is the frequency hub of the sender, e.g., 200 MHz. Δf_d is defined as $\Delta f_d = (2 f_{carrier}/c \cdot v_{rel})$, where $f_{carrier}$ is the frequency hub of the sender, e.g., 76.5 GHz, and v_{rel} is the relative velocity between the radar and the object, $v_{rel}, = v_2 - v_1$. Therefore, using two frequency ramps with different frequency modulation results in Δf_1 and Δf_2, where $\Delta f_1 = \Delta f_R + \Delta f_D$ and $\Delta f_2 = \Delta f_R - \Delta f_D$. Two equations with the two unknowns distance d and relative velocity v_{rel} can then be solved for d and v_{rel}. Returns from a target are then mixed with the transmitted signal to result in a signal which will give the distance of the target after demodulation. Automotive FMCW radars typically use several ramps with different slopes to be able to disambiguate multiple objects detected at once. This results in

$$d = (c \cdot T_m) / (2 f_{hub}) \cdot \Delta f_r = (c \cdot T_m) / f_{hub} \cdot (\Delta f_1 + \Delta f_2)$$
and
$$v_{rel} = c / (2 f_{carrier}) \cdot \Delta f_d = c / f_{carrier} \cdot (\Delta f_1 - \Delta f_2).$$

In order to achieve horizontal angular resolution in addition to distance and velocity measurement, several receiver antennas are mounted horizontally with an angular offset relative to each other, for determining the angle between the object and the longitudinal axis of the sensor. This allows determination of the longitudinal and lateral position of the object relative to the sensor.

Temporal tracking means that an interrelation between two (typically consecutive) measurements is created, which is called radar object tracking. In other words, this generates the track of a real object in space over time. Individual radar measurements or so-called detections of one measurement frame are bundled into one object through spatial association. This, in conjunction with the velocity information, is used to create one or more hypotheses where the object will be located in the next measurement frame, typically assuming constant velocity or constant acceleration over the next timeframe. New incoming measurements which fit the previous hypothesis are used as a confirmation and the track is continued. If a previously existing track is not confirmed by new measurements, then it is typically continued for a few frames and a measure indicating object existence probability is decreased. Kalman filters or particle filters are often used for state estimation and object track association.

More recently, automotive radar manufacturers have started to add additional vertical antennas to achieve some limited vertical resolution, which is used to disambiguate objects over the road, such as signs or bridges, from objects on the road.

Different antenna patterns are used for so-called long-range radars with a typical horizontal open angle of around 30° and a range of about 200 m mid-range radars with an opening angle of about 60° and a range of about 100 m, and short-range radars with a range of about 120° and a range of about 30 m. Newer radars can electronically switch between different antenna configurations.

A variety of automotive suppliers have radar sensors in their product portfolio. Among them are Bosch (see Fig. 3), Continental (see Fig. 4) and Delphi (see Fig. 5).

Accuracy of radars regarding range and relative velocity is typically very high, but is decreased in the lateral direction. Since radars detect the peak in the electromagnetic wave, it can be challenging for radars to disambiguate objects very close to each

Fig. 3 Bosch LRR 4 radar
(Photo: Bosch)

Fig. 4 Continental advanced
radar sensor ARS 410
(Photo: Continental)

Fig. 5 Delphi electronically
scanning radar (Photo:
Delphi)

other, such as two bicycles or motorcycles very close together. For the same reasons, automotive radar sensors typically cannot measure the size of objects. Since the radar cross sections of a target and, consequently, the returned energy varies significantly with the angular rotation (e.g., a preceding truck taking a turn), the position of the return on the target can jump from one location to another (e.g., from license plate to wheels or axle). On the other hand, radar waves are almost not affected by adverse weather conditions, such as rain, fog or snow.

1.2 Cameras

Computer vision utilizes digital cameras in which a digital image sensor converts incoming light into multidimensional signals. A camera typically consists of optics to focus the incoming light, an imager with controller plus processor, and memory for image processing, see Fig. 6. The optics of the camera focus the light emitted or reflected by an object from diverging light rays to converging light rays, which results in a focused image of objects on the focal plane. The focal length determines the scale or opening angle of the image and the aperture determines the light intensity. The imager then converts the incoming light into electrical signals, which are then captured by an A/D-converter. The two major types of digital image sensor are CCD and CMOS, with CMOS sensors being most commonly used in automotive applications. The digital signal is stored in memory, processed by a processor, and then handed to an interface controller.

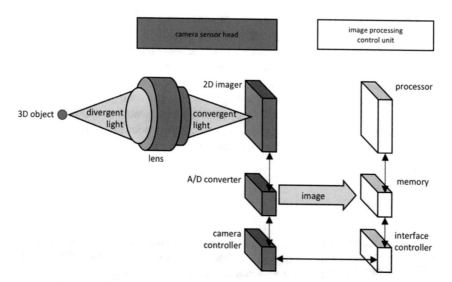

Fig. 6 Principle of an automotive camera

It is characteristic for imaging with cameras that the three-dimensional world is mapped to a two-dimensional image in the imager. Hence the depth dimension is lost during this process, which results in the challenges to be solved by computer vision algorithms.

Light hitting a point on a three-dimensional object is displayed through the camera lens onto a single pixel on the two-dimensional camera imager. The imager then converts light into an electrical signal, which is converted into a digital signal by the A/D-converter and saved into memory. Image processing software is executed by a processor and the resulting image is then saved into memory and transferred by the interface controller onto the bus system.

Image processing is used to extract relevant information for the respective application from the image. Features can be extracted from a single image or a sequence of images. Base features to be extracted from a single image are, for example, edges or corners of objects, which can be described through a characteristic change in the image from pixel to pixel. Mathematically, this is described by the image gradient, which can be extracted by the respective operators.

Reconstruction of the projection of an object point in a sequence of images from one camera or in images of several cameras allows to reconstruct the position of this point in three-dimensional space. Stereoscopy specifically uses images of the same scene from two cameras. From the position of the same object point in the two different images and knowledge of the extrinsic and intrinsic para-meters of the camera, the spatial position of the point in three-dimensions can be derived. So-called motion-stereo uses two images from a single camera in motion, from which the environment can in principle be reconstructed unambiguously if the scene is stationary.

Temporal tracking over time is done by identifying identical object features in consecutive images and tracking them with Kalman filters or particle filters.

Camera imagers typically contain a million or more pixels, typically referred to as megapixel imagers. Due to this fact, automotive cameras are the by far most dense sensing modality, whereas radar and LiDAR measurements are much sparser. On the other hand, cameras depend on external light sources, and need to deal with adverse lighting conditions, shadows, etc. Also, camera components are manufactured in large quantities for many consumer and industrial applications already, resulting in lower cost compared to other sensing technologies.

1.3 LiDAR

LiDAR (also written LIDAR or LiDAR) is a remote sensing technology that measures distance by illuminating a target with a laser and analyzing the reflected light with a detector. The term LiDAR is an acronym of Light Detection and Ranging, and at the same time can be thought of as a portmanteau of "light" and "radar". There are different implementations of LiDAR sensors which differ in the way the laser light is distributed and the reflected light is analyzed. LiDAR sensors are used in automated driving for object detection, 2D and 3D mapping.

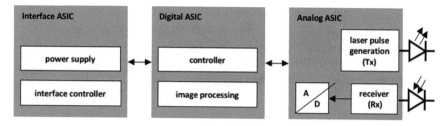

Fig. 7 LiDAR components

The major components in a LiDAR sensor are the laser, the receiver, optics and scanning devices (see Fig. 7). 762 nm or 905 nm lasers are most common for automotive applications, since the components are inexpensive. The disadvantage is that the wavelength is close to visible light (390–700 nm) and hence can be focused and easily absorbed by the eye. Consequently, the maximum power which can be emitted is limited by the need to make the sensor eye-safe in a human environment. As an alternative, 1550 nm lasers can be operated at higher power levels since this wavelength is not focused by the eye, but the technology is less common and, therefore, components are more expensive.

The measurement principle most commonly used with LiDAR sensors is direct time of flight measurement, which means the measurement time t_m is proportional to the distance d:

$$d = (c \cdot t_\mathrm{m})/2$$

One or several short light pulses are emitted by the laser and reflected by an object. The pulse length is typically around 30 ns or less. The time from emission to reception of the pulse is directly proportional to the distance. Since light travels at approximately $c = 300,000$ km per second in air, the time to be measured t_m at a distance of 100 m is only 666 ns.

Since the LiDAR pulse opens in distance and is also partially reflected by translucent objects, such as car windows, multiple reflections can originate from one pulse and most automotive LiDAR sensors have the ability to process multiple reflections of one pulse. This is called multi-target detection. Rain or fog can result in the reflection of some pulses on the water drops, which results in noise in the reflected signal. LiDAR sensors typically adjust emitted power and receiver sensitivity to reduce this effect.

LiDAR receivers commonly use avalanche photodiodes (APD) or positive intrinsic negative (PIN) diodes. Avalanche diodes count individual photons with measurement frequencies up to 100 MHz. PIN diodes are less costly but also less sensitive. Resolution of the distance is achieved by gating, where a time controlled multiplexer saves the signal into separate memory cells or range gates. Each memory cell is equivalent to a distance step of, e.g., 5 cm. Subsequent signal processing then identifies the peak of the signal across several gates. Measurement range depends on the emitted energy, the sensitivity of the receiver and the reflectivity of the target. Typical ranges are up to 250 m for high reflectivity targets and 20 m for very low reflectivity targets.

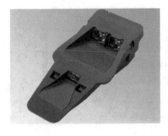

Fig. 8   Continental SRL 1 short range LiDAR (Photo: Continental)

In principle, the Doppler effect could also be used for velocity measurement in lidar sensors. In practice, requirements and cost are prohibitive to detect the Doppler frequency shift in the light spectrum. Differentiation of two or more subsequent distance measurements is used to determine the velocity.

$$v_{rel} = \mathrm{d}d / \mathrm{d}t = \lim_{\Delta t \to 0} \Delta d / \Delta t = (d_2 - d_1) / (t_2 - t_1)$$

There are different implementations of LiDAR sensors that achieve horizontal and vertical resolution. Fixed-beam LiDAR sensors use one or several fixed laser beams. This LiDAR type does not contain any moving parts and, therefore, has low cost and is very robust. Very few LiDAR beams result in little horizontal resolution and sensors are typically designed for highly specific use cases. For example, the Continental SRL 1 Short Range LiDAR (see Fig. 8) is designed specifically for the Volvo city safety function.

Scanning LiDARs rotate or move either the mirror or a platform containing sender and receiver. So-called macro-mirror LiDAR sensors contain a stationary laser diode, typically in the vertical direction. A rotation mirror redirects the laser pulses into horizontal direction. Horizontal resolution can be as low as 0.1°. The sensor's principle is shown in Fig. 9. A common macro-mirror LiDAR sensor is the SICK LMS 291 (see Fig. 10), which was used by several teams in the 2005 DARPA Grand Challenge. In order to achieve some vertical resolution or scanning in multiple distance, the multiple sensors can be mounted in different angles. For example, refer to the 2005 Stanford autonomous vehicle Stanley as depicted in Fig. 11.

The demand for more vertical resolution led to the development of high resolution LiDAR sensors, specifically of the Velodyne HDL-64E (see Fig. 12) for the 2007 DARPA Grand Challenge. This new class of so-called rotating multi-beam LiDARs contains a laser emitter, receiver, and signal processing units on a platform which rotates horizontally typically at 10 Hz. This sensor contains 64 vertical channels or beams with a 27 degree vertical field of view. It is usually mounted on the top of vehicles and scans 360° around the vehicles measuring over 2 million points in 3D space per second (see Fig. 13). In recent years, the integration of rotating LiDAR sensors has made significant advances leading to much smaller sensor sizes, see, e.g., Velodyne (Fig. 14) or Quanergy (Fig. 15).

Fig. 9 Principle of scanning macro-mirror LiDAR. *Source* SICK, Inc.

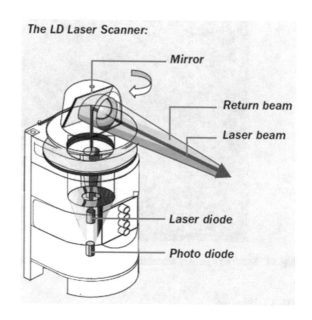

Fig. 10 SICK LMS 291. *S*: SICK, Inc.

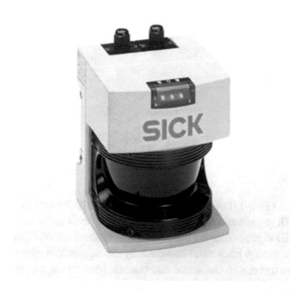

Micro-mirror LiDARs are currently under development. Laser pulses are deflected using a Micro Electro and Mechanical Systems (MEMS) mirror which can be rotated about two axes. These sensors may achieve even higher integration rates resulting in even smaller sensors and potentially higher resolution.

Fig. 11 Stanford's stanley autonomous vehicle. *Source* Stanford University

Fig. 12 Velodyne
HDL-64E. *Source* Velodyne

Flash-LiDAR sensors use a broad LiDAR pulse to illuminate the complete scene. They generate a 3D focal plane array with rows and columns of pixels, similar to 2D digital cameras but with the additional capability of having the 3D "depth" and intensity. Each pixel individually records the time the camera's laser flash pulse takes to travel into the scene and bounce back to the camera's focal plane sensor as well as reflective intensity, resulting in a range and intensity image, see Fig. 17. So far, 3D flash LiDARs have been used successfully in aerospace applications, see Fig. 16, automotive sensors are currently being developed.

Fig. 13   Stanford racing and victor tango together at an intersection. *Source* DARPA

Fig. 14   Velodyne LiDAR product family. *Source* Velodyne

Fig. 15   Quanergy M8.
Source Quanergy

Fig. 16 ASC's TigerEye 3D
flash LiDAR camera. *Source*
ASC

Fig. 17 3D flash LiDAR
image. *Source* ASC

1.4   Ultrasonic Sensors

Automotive ultrasonic sensors use the inverse piezoelectric effect. Through a voltage
applied to a crystal, the crystal is mechanically deformed, resulting in oscillation. A
metal membrane is used to amplify the effect. The oscillation generates a sound in
the range of 40–50 kHz, which is reflected by objects. The same membrane is used
to receive the returned wave, oscillating the membrane in the sensor, resulting in
an electric voltage, ultimately then measured by the sensor. Distance measurement
based on the pulse travel time is technically simple due to the low frequency.

Ultrasonic sensors are manufactured at a very low cost, have a range of a few
meters and are typically used for distance measurement in parking applications.

2 Evolving Digital Maps

In the context of automotive applications, digital maps are used in:

- Navigation,
- Map-supported and map-enabled Advanced Driver Assistance Systems (ADAS), and
- Automated or Autonomous Driving (AD).

Each of those three groups of applications pose different requirements to the digital maps. For example, while address information such as street names and house numbers, is a mandatory part in navigation, it is not used in ADAS and Autonomous Driving. On the other hand, information about road curvature is necessary for some ADAS applications and in Autonomous Driving but it is of no use in navigation. Another example is accuracy; errors in digital maps used in in-car navigation are a nuisance but they may pose a significant safety risk in ADAS and AD.

2.1 Content and Data Models

The first aspects of digital maps to be considered is are *map content* and *map data models*. Content models refer to which real-world entities are part of the map. Data models describe how those entities are represented in the digital map (Fig. 18).

Navigation, ADAS, and AD need different content from the digital map (Fig. 19). For instance, street names are important in navigation but are not part of the ADAS and AD maps. Road curvature is critical for some ADAS applications and required for Autonomous Driving but navigation systems do not use that information. Road geometry, however, is of interest to almost every digital map application.

Fig. 18 Automotive ultrasonic sensor. *Source* Bosch

Fig. 19 Digital map
contents

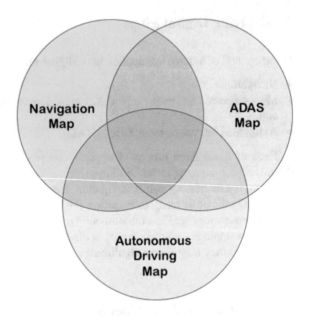

While the majority of automotive OEMs and their suppliers use proprietary data models, most are heavily influenced by the Geographical Data Files (GDF) specifications. The GDF specifications were first released as CEN standard in October 1988. GDF version 5 was released in 2011 as ISO 14825 standard and targets the use of digital maps for a wide array of applications from vehicle navigation systems, pedestrian navigation, and ADAS to highway maintenance systems, road transport informatics as well as telematics.

GDF and similar map data models use three entities:

- *Features* (or objects): points, lines, or areas (*simple* features), or combinations of those (*complex* features),
- *Relationships* between features (such as *belongs-to*, *parent-of*, *applicable-to*), and
- *Attributes* that are properties of either features or relationships.

Depending on which data model is used, one and the same real-world object can be represented differently. In map data models for in-car navigation, for instance, a single road is usually divided into a number of *links* (or road segments) and connecting *nodes* (Fig. 20). For ADAS applications, such a representation is not optimal; the ADASIS version 2 data model is built around a *path* object that is constructed as sequence of connected real-world roads (Fig. 21).

Map data models are inherently very complex. The GDF-inspired relational database schema by HERE consists of 180 tables and a single road segment feature is described with more than 200 attributes. Autonomous Driving needs less complex map data structures, and hence such models are somewhat simpler.

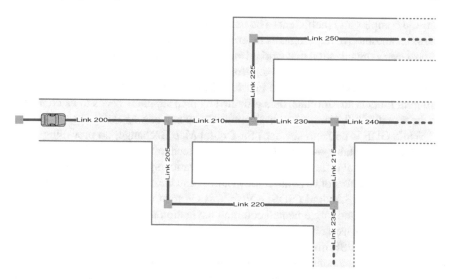

Fig. 20 Road network representation for navigation. *Source* ADASIS v2 API Specification ver. 2.0.1

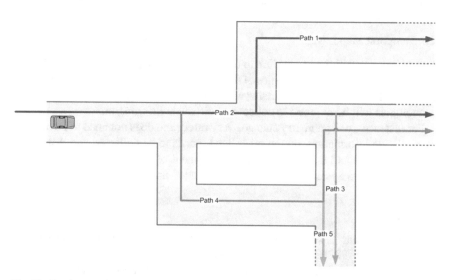

Fig. 21 Road network representation for ADAS. *Source* ADASIS v2 API Specification ver. 2.0.1

(1) Exchange Data Models and Physical Data Models

The GDF data model and format are primarily designed as exchange format— describing the form in which a digital map is delivered by the map provider. As such, it is too complex and inefficient to be used directly by in-car applications. To satisfy the strict automotive requirements in terms of database size and access performance, almost every organization that worked on navigation or ADAS systems designed and developed their own data model and format in which data is stored on the media: Physical Data Model (PDM) and Physical Storage Format (PSF). Because of these different models and formats, data provided by map suppliers needs to be converted to the specific PSF, using a process commonly known as *map data compilation.*

While GDF is de facto standard for Digital Map Exchange, no prevailing PDM and PSF are available—especially because that technology is considered to be one of the fundamental differentiators between navigation system and ADAS suppliers.

In 2009, a number of large automotive OEMs and their Tier-1 suppliers formed the Navigation Data Standard (NDS) e.V. (e.V. is German for *Registered Association*), which designed and implemented common navigation map data model and format. The first systems using NDS are on the market since 2012. With the introduction of the standardized PSF, map data suppliers can deliver data to OEM clients without the need for costly data compilation performed by Tier-1 navigation system providers.

NDS organizes a map database into separate *building blocks* (Fig. 21). First versions of NDS supported only navigation-related building blocks, but newer versions define ADAS-related data building blocks and will be extended to support content for Autonomous Driving.

(2)   Temporal Map Layers

Besides being partitioned according to usage (Fig. 22), map models should take into account the temporal characteristics of data, relationships, and attributions (Fig. 23). Data from different temporal domains will be outdated at different rates and separate technologies will be required to collect and distribute information.

For instance, road geometry changes very rarely and does not need to be updated often. Traffic information, on the other hand, needs to be collected and distributed in near-real-time and that requires the development of technologies such as RDS-TMC (Radio Data System Traffic Message Channel) and TPEG (Transport Protocol Experts Group)

(3)   Accuracy

Another important aspect of the digital map is *data accuracy.* Here we need to distinguish three different accuracy categories.

The first category is *geometrical* accuracy. We separate into two categories.

- *Absolute* geometrical accuracy measures errors between absolute positions of objects in space and positions of the same objects as represented in the map. Navigational maps are expected to have geometrical accuracy of less than 10 m, ADAS maps less than a 1 m and maps for Autonomous Driving less than 20 cm.
- *Relative* geometrical accuracy considers the relative position error between nearby objects in the map. Autonomous Driving maps are expected to have a relative positional error of not more than approximately 20 cm over a distance of 100 m.

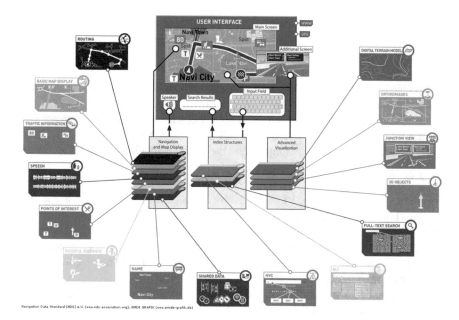

Fig. 22 NDS building blocks. *Source* Navigation data standard (NDS) e.V. (2017)

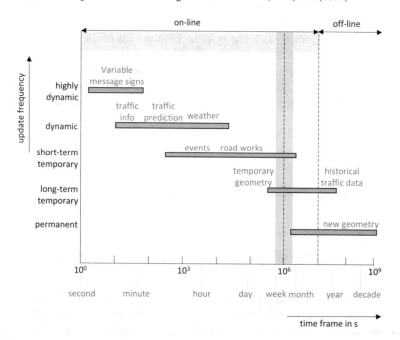

Fig. 23 Temporal map layers. *Source* The ActMAP to FeedMAP Framework Automatic. Detection and Incremental Updating for Advanced In-Vehicle Applications

Accuracy of relationships describes how exact the relationships between objects are captured in the digital map. Errors in this category are particularly critical for the Autonomous Driving applications.

Attribute value accuracy is the last accuracy category. Requirements for accuracy in this category differ depending on the application and the attributes that are used by a particular application and to which extent.

Accuracy of a digital map is not static but it worsens over time due to changes on the ground. Especially safety-critical ADAS and Autonomous Driving applications must take into account that their input from the digital map will contain inaccuracies.

(4) Coverage

Coverage describes which part of the physical space is represented in the digital map available to the application. Oftentimes, in-car navigation and ADAS maps have continental coverage. Simultaneous access to data in such a wide area is necessary to allow for cross-continental destination selection, routing, and guidance functionalities. Depending on the physical storage format and supported features, the navigational map may be between 10 and 40 GB in size.

Apart from being simpler in structure, maps for Autonomous Driving have a much higher level of detail, and they are larger by one or two orders of magnitude. That makes them impractical to be completely stored in the in-vehicle system. ADAS and Autonomous Driving applications, however, need access to a digital map on a very limited area around the ego-vehicle.

Typically, Autonomous Driving systems keep only a small extract of the full country or continental map on board in the vehicle. Depending on the implementation, that may be a single state or county map, or even only an area that is reachable within the next several minutes of driving. However, there must be a mechanism implemented that will fetch another part of the map from a back-end server once the system estimates that the on-board map is not sufficient. That mechanism must rely on high-bandwidth wireless communication channels such as Wi-Fi or LTE.

2.2 Map Data Production and Delivery

(1) Classical Chain

The classical map data production and delivery chain is built from the following links (Fig. 24):

1. Map Data Supplier uses special vehicles to collect map-related data (Fig. 25).
2. Raw collected data is processed to establish proper semantics and relationships between map features. Processing is usually the combination of manual, semi-automatic, and automatic procedures. To minimize cost and improve speed, deep neural networks are applied to reduce or even eliminate manual work. The processed data is provided to clients in exchange format.

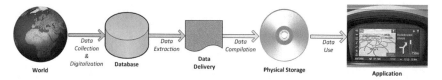

Fig. 24 Classical map data delivery chain

Fig. 25 Map data collection vehicles

3. Data in exchange format is compiled into more efficient physical storage format, usually by a navigation system supplier.
4. The compiled database is distributed to end users on physical media such as CDs, DVDs or memory sticks.

Historically, the primary source of raw map data is limited to a few special vehicles equipped with a range of sophisticated and expensive sensors such as LiDAR sensors, multiple cameras, Differential GPS systems and inertial measurement units. Changes on the road are captured only after one of these data collection vehicle drives by and records the changes, which could be months after a change takes place. In addition, map data providers relatively rarely make the data available in exchange format, usually four times per year. Adding time for data compilation and distribution, even the newest available map data is more than a year old. Besides being slow, the whole process is also expensive.

While outdated map data is troublesome in a navigation system, errors in the digital map pose a safety risk when used in ADAS and Autonomous Driving applications.

(2) Mobile Device as a Map Data Collection Probe

The large number of GPS-equipped consumer devices, primarily personal navigation devices and smart phones, as well as the increased availability and reduced cost of wireless communication opened the door to crowdsourced digital maps. Today, all map suppliers use this technology to a larger or smaller extent. Nevertheless, crowd-sourcing based on mobile devices cannot replace classical data collection because it achieves only low-quality positioning and detects a very limited number of attributes.

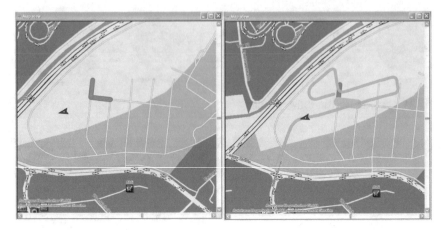

Fig. 26 ActMAP/FeedMAP: Missing road detection. *Source* ActMAP White Paper and Interfaces to the FeedMAP framework

For instance, a missing road or changed road geometry is detected relatively easily by that method (Fig. 26), but presence or absence of a traffic light is much harder to recognize.

(3) Car as a Map Data Collection Probe

Increasing the number of hardware sensors in a car as well as upgrading their sophistication and processing power, significantly boosts both accuracy and richness of information the car can provide to backend data collection and processing. This makes it possible to reduce the number of specialized data collection vehicles or even to eliminate that sourcing method. If digital map data is continuously collected by cars, a map can also be delivered to clients and processed continuously.

The projects ActMAP and FeedMAP (Fig. 27), executed between 2004 and 2008, proved feasibility of such a concept and identified the actor entities, basic algorithms, and processes. Since then, numerous companies and organizations specified, developed, and productized elements in the detection-collection-creation-distribution-use chain. Because of the statistical approach to data creation, standardization of formats and protocols are enablers for higher penetration and efficiency of the system. For instance, one of the primary design goals of the navigation data standard PSF is an 'incremental update' feature, that allows partial and frequent changes in the data that is present in the car. Another example is the *Vehicle Sensor Data Cloud Ingestion Interface specification* that describes the protocol for transferring sensor data from vehicle to back-end.

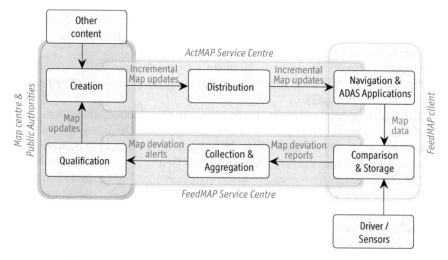

Fig. 27 The ActMAP/FeedMAP loop. *Source* ActMAP white paper and interfaces to the FeedMAP framework

2.3   Digital Map Sub-System Architecture

The purpose of the digital map sub-system is to deliver map data to relevant automotive applications (Fig. 28). It roughly consists of the following modules:

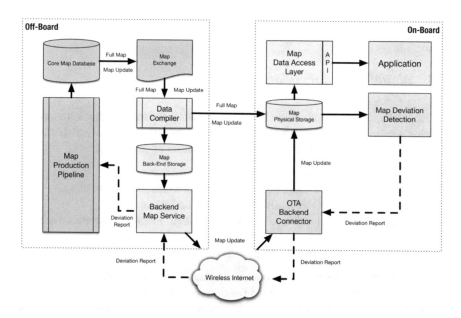

Fig. 28 Generic digital map sub-system architecture

- *Map production pipeline*, usually implemented by a specialized digital map provider such as HERE, Tom-Tom, or AND, collects and processes raw digital map data.
- Proprietary *core map database* is where a complete, potentially world-wide, digital map is stored and maintained. This database is an integral part of the map production pipeline.
- *Map exchange* database is extracted from the core map database and contains parts of it that are provided to clients. Content and accuracy of the data in that database depends on the client's needs. Map exchange database is usually delivered in a non-proprietary map exchange format, such as GDF.
- *Data compilers* are used to transform map data from exchange format to the format more suitable for distribution and use. Two types of compiled digital maps are used; the *backend storage* and the in-car *physical storage* (database).
- *Backend map service* provides a vehicle's complete or partial digital map or a digital map incremental update to the client. This provision is performed using some form of over-the-air communication. Most commonly used are LTE and WiFi but other technologies, such as satellite communication, may be used.
- *OTA backend connector* is an in-car module that connects to the digital map backend service to retrieve complete or partial maps or map updates. In addition, it sends processed or raw sensor information to the backend that can be used in the map production pipeline.
- *Map data access layer* exposes API used by in-car applications to read digital map data.
- *Map deviation detection* identifies issues with the digital map and reports those issues to the backend, where it will be used in the map production pipeline to improve map quality.

2.4 Digital Maps for Autonomous Driving

Conceptually, digital maps can be treated as yet another sensor in the Autonomous Driving applications. Maps have certain advantages over the classical hardware sensors, such as radar, LiDAR, or camera, in detection of static objects:

- Potentially unlimited range in all directions,
- Not influenced by environmental conditions, obstructions, or obstacles,
- Can be used to 'detect' all kinds of static and semi-static objects,
- Semantics of detected object are already provided, including complex relationships,
- Do not use significant part of the processing power.

In other words, maps can be used to "see around the corner", and this information is critical in a number of Autonomous Driving use cases (Fig. 29).

Nevertheless, a digital map's major disadvantage is that the data it provides may be significantly outdated. The digital map industry is targeting that issue by applying

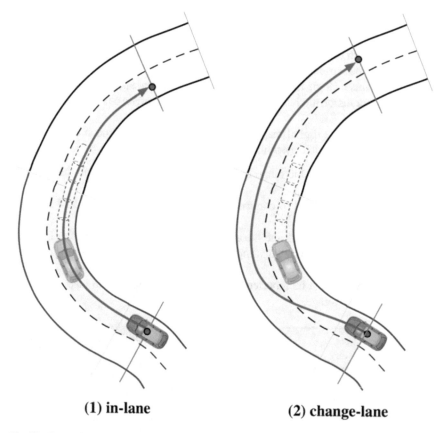

(1) in-lane

(2) change-lane

Fig. 29 Use of digital map in emergency maneuver planning. *Source* Perception Horizon: Approach to Accident Avoidance by Active Intervention

new technologies such as the "car-as-a-probe" concept, incremental, and real-time map updates etc. In time, maps for Autonomous Driving will become more accurate and up-to-date. Nevertheless, one must always keep in mind that change in the ground truth may occur in a short time interval between the last probe checking the location and the ego-vehicle arriving at that point.

In order to reduce the risk of using outdated digital map data, this application may be limited to drivable areas that do not change significantly over time, or when changes do occur they are strictly controlled. Examples of such applications are autonomous highway driving or autonomous valet parking.

2.5 Sensor Fusion for Autonomous Driving

Another approach to use digital maps optimally in Autonomous Driving is to incorporate them to the sensor data fusion process that combines inputs of different sensors and then presents a single model of the environment to the application (Fig. 30).

Such a sensor fusion module is known as *perception horizon* or *environment model* (Fig. 31).

With respect to the digital map, the perception horizon approach offers several advantages to the complete system:

- A digital map extends the range of the perception horizon beyond the range of classical sensors,
- Digital map data is used to confirm or to help interpret the output of classical sensors,
- Fusion of standard sensors and the digital map can detect discrepancies between the sensed environments, indicating either errors in the sensors or errors in the map,
- Those differences can be reported to the backend to correct the map data, or to increase map coverage.

Fig. 30 Perception Horizon visualization. *Source* Gentner (2014) from Single Sensors to the Environment Model: On the Road to Highly Automated Driving. Paper presented at Autonomous Driving Concepts & Technology Minds, Berlin, 27-28 Feb 2014

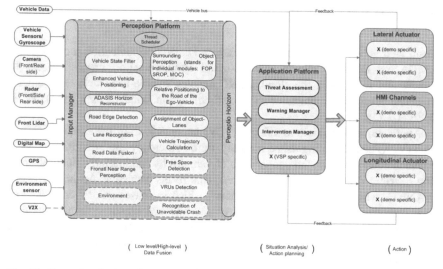

Fig. 31 Perception horizon in automated driving. *Source* Durekovic et al. (2011) perception Horizon: Approach to Accident Avoidance by Active Intervention. Paper presented at the IEEE Intelligent Vehicles Symposium (IV), Baden-Baden, 5–9 June 2011

3 Communication and Vehicle Connectivity: Better Knowing Oneself and Others

3.1 Connected Cars

A connected car is a car that is equipped with access to the internet or with a direct data connection to other traffic participants or infrastructure. This allows the car to share data with other devices both inside and outside the vehicle.

Vehicle connectivity was commercially introduced by General Motors in 1996 through OnStar. The primary purpose was safety and getting emergency help to a vehicle in case of an accident; a cellular telephone call would be routed to a call center, prompting an agent to send help. Soon after, most other automakers followed with similar connectivity features for safety. Also, remote diagnostics were introduced, including vehicle health reports.

Later, the vehicle head-unit, containing the in-car entertainment unit, was connected to the internet. This enabled functions such as traffic information from the internet, internet-based turn-by-turn navigation, internet music or audio playing, smartphone apps, roadside assistance, voice commands, contextual help/offers, parking apps, and so on.

Automated vehicles now communicate map information to and from internet or cloud-based servers, enabling real-time update and download of map information.

3.2 V2V, V2I and V2X

Connected vehicles use cellular network-based data communication such as GSM, CMDA, or DSRC. Cellular network-based communication enables access to servers and services in the cloud. DSRC enables vehicle-to-vehicle and vehicle-to-infrastructure communication. The communication standard used for DSRC is IEEE 802.11p.

Applications of vehicle-to-vehicle (V2V) or vehicle-to-infrastructure (V2I) or both (V2X) can be distinguished in safety, traffic efficiency and infotainment. V2X communication enables the cooperation of vehicles by linking individual information distributed among multiple vehicles. This forms Vehicular Ad hoc Network (VANET), which work like a new 'sensor', which increases the driver's range of awareness to spots which both the driver and onboard sensor systems otherwise cannot see. The vehicle-to-vehicle system electronically extends the driver's horizon and enables entirely new safety functions. Vehicle-to-vehicle communications form a well-suited basis for decentralized active safety applications and, therefore, can reduce accidents, in particular relating to severity. Besides active safety functions, such as Cooperative Forward Collision Warning, Pre-Crash Sensing/Warning or Hazardous Location V2V Notification, it includes active traffic management applications, e.g., Green Light Optimal Speed Advisory or Enhanced Route Guidance and Navigation, and thus helps to improve traffic flow.

One such application, Green Light Optimal Speed Advisory, uses information from the traffic light that can be sent to a vehicle advising its driver the optimum speed to maintain in order to pass through a succession of green lights, which aids in avoiding unnecessary braking for red lights. At red lights, the driver can also receive information about how long it will take for the light to turn green.

Additionally, Emergency Vehicle Warning alerts drivers of a nearby emergency vehicle, which allows them to move into a slower lane or pull over to free up space for the emergency vehicle well in advance. Road Works Warning is similar to the Emergency Vehicle Warning, as construction vehicles and heavy equipment transmit information about where people might be working on the road to approaching drivers. Information such as changed speed limits or altered routes could then be provided to the vehicle.

Finally, Traffic Jam or Accident Ahead Warning alerts the driver to traffic stops, long queues, or accidents. This system is especially useful when the road blockage could be over the crest of a hill, in heavy fog or around a blind corner. Weather information/warning systems issue a warning about local bad weather conditions, such as heavy rain, snowfall or icy roads, e.g., on bridges (Figs. 32 and 33).

Fig. 32 V2X communication on an urban road. *Source* Continental

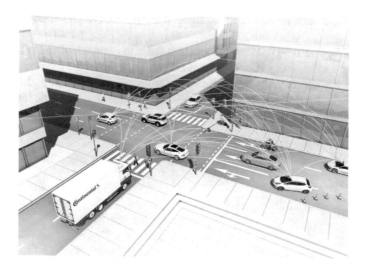

Fig. 33 V2X communication in an intersection. *Source* Continental

4   Algorithms: To Think and to Act

Algorithms for autonomous driving replace the human driver in an automated vehicle. Specifically, there are three primary driving tasks a human driver fulfills:

- Where am I? Determination of the vehicle position. This is called localization.

- Where is what around me? Determination of relevant objects and obstacles in the vehicle environment. This called perception.
- How do I get to my destination? Planning of a route, making driving decisions, and planning and executing the vehicle motion.

4.1 Architecture

Figure 34 shows the typical high-level architecture of a highly or fully automated driving system. The Odometry module integrates information from inertial sensors, wheel speed sensors and steering wheel angle sensors through an estimation algorithm using a vehicle motion model. This results in a locally accurate motion estimate relative to the starting point. This is combined with GPS in the Localization module, which then generates a global estimate of the vehicle position. That global estimate is then consolidated further by matching the position of known landmarks from a pre-recorded map for localization with landmarks that have been detected by the environment sensors of the vehicle, such as cameras, radars and LiDARs. This leads to a robust and accurate estimate of the position and orientation of the vehicle relative to the road and lane, as well as an estimate of the position in global coordinates on the earth.

The Perception system takes information from the sensors of the vehicle such as cameras, radars, and LiDARs, which all constantly scan the environment. The perception module extracts relevant objects from the information and tracks the position of these objects over time. In addition, contextual and semantic information about all objects in the environment is derived in the scene understanding module,

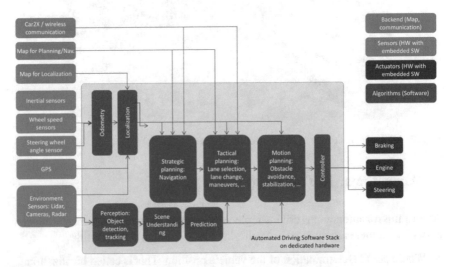

Fig. 34 Simplified algorithmic architecture of an automated driving system

such that the complete traffic scene can be decomposed into roads, lanes, other drivable areas, sidewalks, or other non-drivable areas. All objects can be labeled into the kind of traffic classification they represent, such as cars, trucks, motorcycle, bicycles, etc. This allows for a prediction as to how the current traffic situation will develop in the next few seconds.

The strategic planning module plans a route from the current position to the desired destination, just like a conventional navigation system. The tactical planning or decision making module decides when to take certain driving actions, such as staying in lane, making a lane change, a turn, or an evasive maneuver. The motion planning module then plans the actual motion of the vehicle as a path on the driving surface combined with the velocity on that path. Subsequently, this is fed into the controller, which through feedback control ensures that the trajectory is executed by the actuators, specifically the engine, brake, and the steering system.

4.2 Odometry and Localization

The task of the localization software is to determine the exact location of the vehicle. Specifically, it determines on which road, in which lane, and where exactly within that lane the vehicle is driving. In other words, it must stay localized against the map, which represents the physical infrastructure on which the vehicle is driving. This is called localization relative to the map. In order to be able to determine the route to the desired destination, the vehicle must also determine its absolute position in the world, typically in global coordinates.

Odometry is the use of data from motion sensors to estimate change in position over time, which is used in vehicle automation for estimating the vehicle position relative to a starting location. In a car, typically, wheel speed sensors, inertial sensors such as accelerometers and gyroscopes, and steering wheel angle sensors are used. These sensors are already part of the vehicle stabilization in today's cars' braking systems, e.g., the Electronic Stability System (ESP). Typically, a Kalman filter is used to integrate the sensor data over time. This is very accurate in the short-term but drifts in the long-term, due to integration of sensor errors. Also, odometry only provides an estimation of the vehicle motion relative to a starting point, but not global localization on earth.

Hence, odometry is combined with satellite-based position estimation, which provides a global estimate of the position. This global estimate is correct up to a few meters but it is occasionally unavailable since it relies on a direct line of sight from the receiver to the satellites.

Obviously, vehicle localization must be extremely robust and in real-time. Conventional systems, such as GPS, alone are neither accurate nor reliable enough for vehicle automation purposes. GPS guarantees a "worst case" accuracy of approximately 10 m at a 95% confidence level in space. Real world performance on roads is often further degraded by factors such as atmospheric effects or blockage from buildings, bridges or trees, and receiver quality.

Higher accuracy is attainable by using GPS in combination with augmentation systems. These can enable real-time positioning to within a few centimeters at the expense of higher cost. Differential Global Positioning System (DGPS) is an enhancement to the Global Positioning System that provides improved location accuracy, from the 10-m nominal GPS accuracy to up to 10 cm in case of the best implementations. DGPS provides positional corrections to GPS signals by using a fixed, known position to adjust real-time GPS signals to eliminate pseudo-range errors. Commonly, a set of GPS base stations with an accurately known position receive GPS signals and compare the computed position to the known positions. The difference is the current actual error at the position of the base stations. Errors are then sent via radio or cell phone data communication to a receiver in the vehicle. The GPS position computed in the vehicle is then adjusted accordingly. An important point to note is that DGPS corrections improve the accuracy of position data only. DGPS has no effect on results that are based on other data.

Extended Reading

GPS and DGPS

GPS signals coming down from satellites to the ground travel through two layers of the earth's atmosphere. First, the signals travel through the ionosphere, which is the outer edge of the atmosphere. That bit of the atmosphere is subject to solar radiation, which causes particles to split up and become positively charged. This impacts the electromagnetic signals passing through, including the radio signals from satellites, and can delay them up to ±16 ns. This can introduce an error of up to 5 m to the captured position. The second layer the GPS signals travel through is the troposphere. This is the 'weather' section of the atmosphere, which includes conditions such as clouds, rain, and lightning. This can add delay to the signals of around ±1.5 ns, which introduces a position error of around 0.5 m. It is important to note that the delays are non-deterministic.

DGPS is typically a commercial service, which is associated with a subscription fee and potentially additional fees for transferring the data to the vehicle. DGPS can reduce the error from several meters to less than a meter in the best case but is dependent on the proximity to the base station and the availability of the correctional data service. Hence it is neither sufficient nor reliable enough on its own for vehicle guidance.

In addition to GPS and differential correction service, autonomous vehicles use information stored in the map to further increase precision and robustness of the position and orientation solution. Landmarks or other features are stored in a layer in the map. Sensors on the vehicle recognize those landmarks and detect the distance of the landmark relative to the vehicle. Together with the location of the landmark, which is stored in the map, this allows further refinement and consolidation of the position relative to the map, see, e.g., Figure 35 shows the map used by Google, which contains lane markings, cross walks, and traffic lights in an urban intersection.

Another approach for vehicle localization is to scan the reflectivity of the ground surface and record the result in a map. During driving, the same or a similar sensor

Fig. 35 Localization using map and sensor information to determine the location of the vehicle. *Source* https://youtu.be/tiwVMrTLUWg)

Fig. 36 Perception detecting and classifying objects around the vehicle. *Source* Google

scans the ground surface again and the correlation of that scan with the map is used to derive the vehicle position. This approach has been successfully used in research, but suffers from the large amount of memory required to store the LiDAR reflectivity map.

The first commercial mapping and localization solutions that were launched recently are based on camera-based mapping and localization. Vehicles equipped with cameras and GPS (independent of whether the vehicles are driven manually or autonomously) constantly drive and detect visual landmarks, such as lane markings, signs, and traffic lights. The positions of those landmarks are then uploaded to a map server in the cloud. Many vehicles over time detect and record the same landmarks, which averages out the GPS errors over time. Vehicles requiring precise localization download the landmarks from the required road segments from the server ahead of time and then localize relative to those landmarks.

4.3   Perception

The task of the perception software is to detect and track objects and obstacles relevant for the driving task and to provide a concise and integrated model of the whole environment in order to subsequently generate appropriate driving decisions. This task is also called sensor data fusion. Fused sensors are typically radar, cameras, LiDAR and ultrasonic. Measurements from these sensors are uncertain and normally treated with an estimation framework, which models the dynamic environment and follows the evolution of its environment. Subsequently, the robotic vehicle is able to perform reasoning and make predictions within this environment to accomplish its tasks successfully. Traditionally, LiDAR was primarily used for object detection and tracking.

More recently, computer vision based on machine learning has made tremendous progress. Specifically, convolutional neural networks have been applied successfully for semantic segmentation of scenes. This is one component of a comprehensive environmental model, which is typically broken down into four main parts:

- Free Space: Determining the drivable area and its delimiters,
- Driving Paths: The geometry of the routes within the drivable area,
- Moving Objects: All road users within the drivable area or path,
- Scene Understanding: A semantic description of objects including their type or class. The Cityscapes dataset, see Fig. 37, e.g., uses:

  - Flat objects: road, sidewalk, parking, rail track
  - Humans: person, rider
  - Vehicle: car, truck, bus, motorcycle, bicycle, caravan, trailer
  - Constructions: building, wall, fence, guard rail, bridge, tunnel
  - Objects: pole, pole group, traffic sign, traffic light
  - Nature: vegetation, open terrain
  - Sky
  - Motion class: ground, dynamic, static.

Traditionally, high-level fusion was used to bring information from sensors into one environment model. In this approach, each sensor separately extracts object assumptions and tracks from its measurements. Subsequently, the tracks are associated to each other and the resulting states are estimated.

Fig. 37 Annotated image from the Cityscapes Dataset. *Source* Google

More recently, low-level fusion is predominantly used, driven by the increased ability to transfer information from the sensor to a central processing unit. Then, in order to fuse information from several sensors together into one environment model, Bayesian occupancy filters or particle filters in a grid description are used to describe the scene. Particles in particle filters contain position and velocity. Particles can migrate from one grid cell to another based on their motion model, hence tracking dynamic objects in the driving environment. Similar particles are clustered into groups, which describe objects in the scene.

Figure 36 shows such a description of the vehicle environment, where other vehicles are depicted in purple, a bicyclist is shown in red, and traffic cones are shown in orange (Fig. 37).

4.4 Planning

In order to be able to develop the correct actions of the autonomous vehicle in the given environment, the position and velocity of other vehicles need to be predicted in the future. This is done by taking scene semantics, resulting behavior options, and the resulting interactions between the traffic participants into account. A Google video shows, that a truck is predicted to make a lane change, since the current lane on which the truck is driving is closed due to construction, see Fig. 38. Following the same principle, the behavior of all traffic participants and the resulting trajectories are computed for the respective scene, see Fig. 39.

Then, the decision-making software of the automated vehicle determines how to respond to the scene. Specifically, a motion is planned such that collision with objects is avoided and that it is comfortable for the passengers. The motion planning then

Fig. 38 Pickup truck is predicted to make a lane change due to a lane blocked by construction. *Source* Google

Fig. 39 Prediction of other objects in the traffic scene. *Source* Google

results in a trajectory, which is a path on the street surface and includes a velocity prescribed for each point on the path (Fig. 40).

Fig. 40 Planning the vehicle motion. *Source* Google

4.5   Control

In a final step, the planned trajectory is executed by the vehicle control algorithms. These algorithms provide the interface between the planning algorithms and the vehicle actuation system. Vehicle actuation is provided by:

- The vehicle powertrain for longitudinal acceleration,
- The braking system for longitudinal deceleration,
- The steering system for lateral motion.

Traditionally, the longitudinal controllers used for acceleration and braking were decoupled from the lateral controllers for the steering system, with PID controllers for each path. Then, relevant improvements were done controlling the vehicle motion at the limits of friction.

More recently, new approaches combining lateral and longitudinal control arose, using potential fields or model predictive control. Combined lateral and longitudinal control is relevant in traffic scenarios with highly dynamic changes in the scene.

5   Relevant Fields of Autonomous Driving You Should Know

In September 2016, PwC selected eight core technologies with the most profound impacts and business value in the next five to seven years (three to five years in developed economies) from more than 150 emerging technologies, i.e., Artificial Intelligence, Augmented Reality, Blockchain, Drones, Internet of Things, Robots,

Virtual Reality and 3D Printing. Among the eight core technologies, there is considerable contents related to the applications of autonomous vehicles. The following describes artificial intelligence, Internet of Things, Big Data and Cloud Computing that are highly correlated with autonomous driving technologies.

5.1 Artificial Intelligence

According to Baidu Encyclopedia, "Artificial Intelligence, or AI for short, is a new technological science to research and develop theories, methods, technologies and application systems for simulating, extending and expanding human intelligence. Artificial intelligence is a branch of computer science. It attempts to understand the essence of intelligence and produce a new intelligent machine that can respond in a way similar to human intelligence." As one of the three cutting-edge technologies in the world since the 1970s, namely, space technology, energy technology and artificial intelligence, it is also considered to be one of the three cutting-edge technologies in the twenty-first century, namely, genetic engineering, nanoscience and artificial intelligence. In the past 30 years, artificial intelligence has developed rapidly and has been widely used in many disciplines with fruitful results. At present, artificial intelligence has gradually become an independent subfield, both in theory and in practice.

Artificial intelligence is actually the use of computer technology to complete the replication of human brain functions, and take advantage of computer's high speed to further shorten the time of completion of each function, in order to greatly improve work efficiency. Artificial intelligence has seeded a new wave around the world. Deloitte proposes that the future standard configuration of 80% of the world's top 500 companies is to master cognitive technology; and Science magazine predicts that artificial intelligence could replace 50% of global employment in 2045. In other words, after 30 years, most of our work today will be replaced by artificial intelligence. "The future factory has only two employees—one person and one dog. The person's job is to feed the dog, while the dog's task is to keep people from touching the machine." said Professor Wavren G. Bennis, who has served as an adviser to President Reagan and President Kennedy.

Artificial intelligence technology has a large number of applications in the perception, cognition and decision-making of autonomous driving. In particular, deep learning based algorithms have important applications in improving visual technology, natural language processing, sensor fusion, target recognition, planning and decision-making. Many people believe that the last hurdle for unmanned driving to achieve a real breakthrough is the in-depth development of artificial intelligence, especially deep learning. Therefore, from autonomous driving start-ups, internet companies to major automakers, they are aggressively exploring to construct neural network, using Graphic Processor Unit (GPU), to achieve ultimate driverless driving.

Deep learning, so to speak, a process for the computer to extract the decision basis from the data, is the most popular technology in the field of artificial intelligence.

Deep learning neural networks consist of a series of simple, trainable neurons that work together to learn complex tasks like driving. The biggest difference between deep learning and traditional algorithm-based systems is that after a given model, a deep learning system can automatically learn how to complete a given task, which can be not only to identify images and speech, but also to control a drone to perform a task or to let a vehicle drive automatically. Deep learning simulates to some extent the process by which the human brain learns from the external environment, understanding and even figuring out fuzzy ambiguities. In recent years, an intuitive manifestation of the progress of deep learning technology is the ImageNet competition. In this competition, the entry algorithm tests the accuracy of detection and classification on large-scale data composed of thousands of images and videos. Prior to 2012, the recognition rate of objects in the competition had been increased particularly slowly (less than 70%). After the introduction of deep learning in 2012, the recognition rate jumped to 80% and now exceeds 95%. Deep learning has replaced the position of traditional visual methods in this competition.

Extended Reading

Turing, Father of Artificial Intelligence

Alan Mathison Turing, (June 23, 1912—June 7, 1954), British mathematician and logician. During the Second World War, he assisted the military in cracking Germany's famous cryptosystem Enigma, helping the allied forces to win the war.

In 1936 he published On Computable Numbers, with an application to the Entscheidungsproblem. It is in this paper that Turing introduced an abstract machine, now called a "Turing machine", which moved from one state to another using a precise finite set of rules (given by a finite table) and depending on a single symbol it read from a tape. The "Turing Machine" theoretically proved the feasibility of developing a universal digital computer and laid the theoretical foundation of the entire modern computer.

In 1950 Turing published Computing machinery and intelligence in Mind. It is another remarkable work from his brilliantly inventive mind which seemed to foresee the questions which would arise as computers developed. He studied problems which today lie at the heart of artificial intelligence. It was in this 1950 paper that he proposed the Turing Test which is still today the test people apply in attempting to answer whether a computer can be intelligent... he became involved in discussions on the contrasts and similarities between machines and brains. Turing's view, expressed with great force and wit, was that it was for those who saw an unbridgeable gap between the two to say just where the difference lay.

5.2   Big Data and Cloud Computing

McKinsey defines Big Data as a large-scale data set that is far beyond the capabilities of traditional database software tools in terms of acquisition, storage, management and analysis. IBM first summarized the characteristics of big data into four "V", namely Volume (large data volume), Variety (multiple data types), Value (low value density with high commercial value) and Velocity (fast processing speed). In order to make the definition more accurate and perfect, IBM later added another"V"— Veracity as a new attribute of big data.

Big data is often used to describe a large amount of unstructured and semi-structured data created by a company that spends a lot of time and expense when downloaded to a relational database for analysis. From the category of data, big data refers to information that cannot be processed or analyzed by traditional processes or tools. It defines data sets that exceed the normal processing scope and size and force users to adopt non-traditional processing methods. Big data is normally stored in a non-relational database.

Prof. Viktor Mayer-Schnberger, a big data expert at Oxford University, believes that "the biggest change in the era of big data is to abandon the desire for causality and instead focus on relationship." The strategic significance of big data technology lies in the specialized processing of those meaningful massive data. Using the purchase records of more than 20 kinds of goods that women may buy during pregnancy as data sources to build a model and analyze the behavioral correlation of buyers, Target Supermarket in the US can accurately infer the specific date of delivery of pregnant women, thus its sales department can send corresponding product coupons at different stages of each pregnant customer. The Target example is a typical case of data mining and secondary utilization, and is also a successful case of big data for marketing analysis. There are many technologies for big data, including massively parallel processing (MPP) databases, data mining, distributed file systems, distributed databases, cloud computing platforms, the Internet and scalable storage systems.

From a technical point of view, the relationship between big data and cloud computing is as inseparable as the front and back of a coin. The term Cloud Computing was born in the third quarter of 2007, and only half a year later, the term is more concerned than Grid Computing. The National Institute of Standards and Technology (NIST) describes Cloud Computing as "Cloud computing is a model for enabling convenient, on-demand network access to a shared pool of configurable computing resources (e.g., networks, servers, storage, applications, and services) that can be rapidly provisioned and released with minimal management effort or service provider interaction."

This cloud model promotes availability and is composed of five essential charac-teristics (On-demand self-service, Broad network access, Resource pooling, Rapid elasticity, Measured service); three service models (Cloud Software as a Service (SaaS), Cloud Platform as a Service (PaaS), Cloud Infrastructure as a Service (IaaS)); and, four deployment models (Private cloud, Community cloud, Public cloud, Hybrid cloud). Key enabling technologies include: (1) fast wide-area networks, (2) powerful, inexpensive server computers, and (3) high-performance virtualization for commodity hardware.

This kind of resource pool is called "cloud", which is, based on modern virtual technology, an infrastructure that can dynamically expand network applications at a very low cost, as well as a virtualized computing resource that can be self-maintained and managed. It is generally a large server cluster, including computing servers, storage services, and other broadband resources. It is called "cloud" because in some respects it has certain similar characteristics to the clouds in the sky: large in scale, scalable in size, blurred in boundaries, and unable to fix specific locations but does exist somewhere. The so-called cloud computing is to collect the data in the resource pool, and achieve self-service through automatic management, so that users can automatically call resources when they use, support operations of a variety of programs. They can concentrate on their own business without worrying the details.

Against the backdrop of technological innovation represented by cloud computing, these data, which were difficult to collect and use, can be easily exploited. Through continuous innovation in all walks of life, big data will gradually create more value for humans. Big data must not be processed by a single computer, and a distributed architecture must be used. It features distributed data mining for massive data and must rely on cloud computing for distributed processing, distributed databases, cloud storage and virtualization technologies.

The Cloud Computing model offers the promise of massive cost savings combined with increased IT agility. It is considered critical that government and industry begin adoption of this technology in response to difficult economic constraints. However, cloud computing technology challenges many traditional approaches to datacenter and enterprise application design and management. Cloud computing is currently being used; however, security, interoperability, and portability are cited as major barriers to broader adoption.

The application of big data is still expanding, and has achieved good results in many places. For example, you can use big data technology to find the root cause of faults, problems and defects in time, analyze all the inventory of a supermarket to set prices and clean up the inventory with the goal of maximizing profits, identify VIPs quickly from a large number of customers, and circumvent fraud by click-stream analysis and data mining forbanks and insurance companies and so on.

The big data market is on the eve of the outbreak. IBM has outlined four phases of big data adoption, which include educate, explore, engage and execute. These stages are defined as follows:

- Educate. This phase focuses on knowledge gathering and market observations.
- Explore. After completing the education phase, companies will develop a strategy and roadmap based on business needs and challenges.
- Engage. During the third phase, a business will pilot big data initiatives to validate value and requirements.
- Execute. Companies in the fourth phase have deployed two or more big data initiatives and are continuing to apply advanced analysis.

An important part of the technology chain of connected self-driving cars is data platform technology, including non-relational database architecture, high-efficient data storage and retrieval, big data correlation analysis and deep mining, cloud operating system, information security mechanism, etc. It can be seen that big data and cloud computing are the core technologies.

The auto industry is a rich mine of big data. Hundreds of millions of people in the world use cars to go to work, go home, go shopping and have fun every day. Everyone has his/her own life. It would be very valuable to study certain rules from their habit of using cars. Those data can be aggregated, sorted and packaged, and then sold to anyone who needs them. For example, by collecting GPS data of drivers' mobile phones or their cars, it is possible to analyze which roads are currently in traffic jams, and timely release the road traffic reminders; by collecting GPS location data of cars, it is possible to identify which areas of the city are rich in parking facilities, meaning where there is a relatively active population, and these analytical data can be sold to advertisers. Intel has expected that a car will generate 4000 GB of data every day. Unlike the auto industry, big data analytics is said to have reached 90% of profitability because it doesn't require a lot of capital to invest in factories and equipment!

Automotive big data can be roughly divided into two categories. One type is the data generated by a vehicle itself at work, including the working status of all parts, the working mode and the performance status of the whole vehicle, etc., which can be used by the automaker for different purposes, such as quick analysis of the causes of abnormal state of vehicles and help in establishing more efficient engineering specifications, etc. Another type of data is in close contact with people using cars. A large amount of data can be obtained about the whereabouts of the owners, and their points of interest are summarized. In 2016, just in the US market, Chevrolet collected more than 4000 TB of data from customers' cars.

McKinsey predicts that the automotive big data market will reach a scale of around $4.5 trillion to $7.5 trillion around 2030. Retailers, advertisers, marketers, product planners, financial analysts and even government agencies are eager to get those data to improve their works. In this market, Silicon Valley's emerging companies that have already entered the industry will be able to sell those data to customers over and over again, and the data are constantly being updated, and the business prospects are very promising.

5.3 *Internet of Things*

The internet of things (IoT) is a computing concept that describes the idea of everyday physical objects being connected to the internet and being able to identify themselves to other devices. The IoT is significant because an object that can represent itself digitally becomes something greater than the object by itself. No longer does the object relate just to its user, but it is now connected to surrounding objects and database data.

In 1991, Prof. Kevin Ashton of the Massachusetts Institute of Technology first proposed the concept of the Internet of Things. In 1999, Ashton said it best in this quote from an article in the RFID Journal: "If we had computers that knew everything there was to know about things—using data they gathered without any help from us—we would be able to track and count everything, and greatly reduce waste, loss and cost. We would know when things needed replacing, repairing or recalling, and whether they were fresh or past their best."

The International Telecommunication Union (ITU) cited the concept of the IoT but with modified definition and extended scope in the "ITU Internet Report 2005: The Internet of Things", which has been specially prepared for the World Summit on the Information Society(WSIS) held in Tunis in November 2005. It is no longer just an Internet of Things based on radio frequency identification (RFID) technology. The "ITU Internet Report 2005: The Internet of Things" defines the IoT as follows: A special network, through two-dimensional code reading equipment, radio frequency identification (RFID) device, infrared sensor, global positioning system (GPS), laser scanner and other information sensing equipment, according to the agreed agreement, for any article connected to the Internet for information exchange and communication in order to achieve intelligent identification, positioning, tracking, monitoring and management. According to the definition of the ITU, the IoT mainly addresses the interconnection of Thing to Thing (T2T), Human to Thing (H2T) and Human to Human (H2H). However, unlike the traditional Internet, H2T refers to the connection between general devices and objects, which simplifies the connection between objects. H2H refers to the interconnection between people that does not depend on personal computers (PC). Because the Internet does not consider the problem of connecting anything, the Internet of Things is used to solve this problem. Many scholars often introduce the concept of M2M in the discussion about the Internet of Things, which can be explained as Man to Man, Man to Machine and Machine to Machine. In essence, the interaction between people and machines, machines and machines is mostly to achieve information exchange between people.

In 2009, the Commission of the European Communities issued the "Internet of Things—An action plan for Europe", which described the application prospects of IoT technology and proposed that the EU governments should strengthen the management of the IoT and promote the development of the IoT.

The concept of the Internet of Things in China was introduced in 1999 as a sensor network. As early as 1999, the Chinese Academy of Sciences launched the research and development of sensor network. Compared with other countries, China's IoT technology research and development level is in the forefront, playing a significant role in the world. In 2010, the IoT was officially listed as one of the five emerging strategic industries in China. Since then, the development of the Internet of Things has attracted great attention of the whole society.

The industrial practice of the Internet of Things in the world is mainly concentrated in three major directions.

The first practice direction is called "Smart Dust", which advocates the interconnection of various types of sensor devices and forms a network with intelligent functions. The IoT in the sense of "Smart Dust" belongs to the generalization of industrial buses. Since the mechatronics and industrial informationization, such industrial practice has never stopped in industrial production, but it was called industrial buses rather than the Internet of Things. In this sense, the Internet of Things will move forward steadily, according toits inherent law of science and technology, with the development of various technologies and various local area network communication technologies, and will not accelerate the development speed because of a human movement.

The second practical direction is the well-known logistics network based on RFID technology. This direction advocates the management of logistics and logistics information through the identification of objects, and at the same time forms intelligent information mining through information integration. The EPCglobal Standard, based on the Internet of Things in the sense of RFID, was defined as the core standard of the future Internet of Things when it was launched, however, the inherent limitations of the standard and its only means of implementation through RFID tag make it difficult to really point to the intelligent planet advocated by the Internet of Things. The reason is that the information that people can be informed by the relationship between things is very limited, and the relationship between the things's tates can make people really dig out the universal relations between things, so as to obtain new knowledge and new wisdom.

The third practical direction is called the Internet of Things (IOT) in the sense of data "ubiquitous aggregation". It is believed that the Internet has created a huge ocean of data. It is not only the inevitable requirement for the further development of the Internet, but also the mission of the Internet of Things to fully realize the resource utilization of data by accurately identifying the attributes of each data. "Ubiquitous aggregation" is to realize the omnipresent vast data ocean created by the Internet and realize the aggregation in the sense of mutual understanding. These data represent both the object and the state of the object, and even represent various concepts defined by humans. The "Ubiquitous aggregation" of data will make it extremely convenient for people to arbitrarily search for all kinds of data needed. With the help of various mathematical analysis models, we constantly excavate the complex relationships among the things represented by those data, so as to realize the revolutionary leap of human cognitive ability to the surrounding world.

When the development of the Internet of Things reaches a certain scale, with the help of barcode, two-dimensional code, RFID and other technologies to uniquely identify products, sensors, wearable devices, intelligent sensing, video acquisition, augmented reality and other technologies can achieve real-time information acquisition and analysis, and form a huge network with the Internet. The IoT industry can be divided into five levels according to the level of industry chain: support layer, perception layer, transmission layer, platform layer and application layer. The perception layer and transmission layer of the Internet of Things have attracted many participating companies and become the most competitive fields in the industry.

Generally speaking, the development of the Internet of Things are phased as follows. Firstly, the object attributes are identified, including static and dynamic attributes. Static attributes can be stored directly in tags, and dynamic attributes need to be detected by sensors in real time. Secondly, the identification device needs to read the object attributes and convert the information into a kind of data format suitable for network transmission. Thirdly, the information of the object is transmitted to the information processing center through the network, and the information processing center completes the related calculation about the object communication.

From the initial stage of fragmented, isolated applications, to the new stage of cross-border integration and integrated innovation, the prelude of explosive growth of the Internet of Things has begun. Intelligent hardware, wearable devices and connected vehicles are developing rapidly, which is the confirmation of the acceleration of the T2T and H2H connections. The Internet of Things has deeply integrated informationization and traditional fields, and is also reshaping the way of production organization and transforming the industrial form.

Compared with smart grid, smart security and other fields, Internet of Vehicles (V2X) is not the maturest application of Internet of Things, but with its unique strategic height and huge consumer market, it has won huge attention. With the rapid development of Internet of Vehicles and the wide application of ITS, the practice of transportation industry in the field of IoT has made exciting achievements.

References

Durekovic S et al (2011) Perception Horizon: Approach to Accident Avoidance by Active Intervention. Paper presented at the IEEE Intelligent Vehicles Symposium (IV), Baden-Baden, 5–9 June 2011

Gentner H (2014) From Single Sensors to the Environment Model: On the Road to Highly Automated Driving. Paper presented at Autonomous Driving Concepts & Technology Minds, Berlin, 27–28 Feb 2014

Chapter 3
Applications of Autonomous Driving That You Should Know

1 The First Step Towards Autonomous Driving

1.1 The Starting Point

The dream of self-driving vehicles was already explored in fiction not long after the development of the automobile itself in 1886. In 1915, a magazine depicted a self-driving street car, "a motorist's dream: a car that is controlled by a set of push buttons". Walt Disney presented his vision of futuristic vehicle features in a 1950s TV series, predicting, amongst others, radar sensors, night vision, and traffic guidance systems. In the following decade, he invented Herbie, a vehicle that could think and feel, with "a mind of his own and capable of driving himself". The arguably most advanced fictional robot car was the vehicle KITT from the 1980s TV series Knight Rider. It was able to drive itself and to "see" the environment with the help of a scanner.

The late 1950s marked the starting point for research on self-driving or robotic vehicles. GM and the Radio Corporation of America (RCA) co-tested automated highway prototypes in 1958, using radio control for velocity and steering. GM and RCA developed automated highway prototypes with radio control for speed and steering. Vehicles on that highway were equipped with magnets, and thus able to track a steel cable buried in the road, while the overall traffic flow was managed by control towers. In 1979, Stanford University had its' Cart; initially remote controlled, successfully crossing a room full of chairs in five hours, without any human intervention. Carnegie Mellon University's (CMU) Rover followed soon after, as did Shakey from the Stanford Research Institute, the first general-purpose mobile robot that was able to reason about its own actions.

The Tsukuba Mechanical Engineering Lab in Japan developed the first automated street vehicle with environment perception in 1977. That vehicle was capable of following lane markings for up to 50 m with velocities reaching 30 km/h. The worldwide first actual robotic street vehicles were built at the Bundeswehr University Munich in Germany by Prof. Ernst Dickmanns and his research group in the

© China Machine Press, Beijing and Springer Nature Singapore Pte Ltd. 2021
Z. Chai et al., *Autonomous Driving Changes the Future*,
https://doi.org/10.1007/978-981-15-6728-5_3

1980s. Operating with computer vision and probabilistic algorithms, they drove 20 km on an empty highway at speeds of up to 96 km/h. The publicly funded European EUREKA-project Prometheus (PROgraMme for a European Traffic of Highest Efficiency and Unprecedented Safety) was the continuation of Prof. Dickmanns' research. The project ran from 1987 to 1994, involved many European participants, and received a substantial amount of public funding. Its highlight was the "VAmP" vehicle driving on the Paris Autoroute 1. At velocities reaching 130 km/h, VAmP rode more than 1000 km on-road, and was able to track up to twelve cars simultaneously and automatically pass slower cars using the left lane. In 1995, Prof. Dickmanns' team drove from Munich to Denmark and back, traveling a distance of 1758 km. Computer vision was utilized for both longitudinal and lateral guidance, achieving velocities reaching 175 km/h. The longest part driven autonomously on that trip was 158 km. That same year, CMU conducted the "No hands across America" drive. Their research vehicle covered the entire distance from Washington, D.C., to San Diego, California, with automated steering while longitude was controlled manually. San Diego was also the venue for a technology demonstration of the U.S. Department of Transportation's National Automated Highway System Research Program (NAHS) in 1997. Automated cars, buses, and trucks drove on the I-15 with lateral lane keeping enabled by computer vision and magnets embedded in the road, and longitudinal distance by radar or LiDAR sensors.

1.2 DARPA Competitions

DARPA, the Defense Advanced Research Projects Agency, funded a series of autonomous vehicle competitions. 15 teams qualified for the final competition of the first round in 2004. Their unmanned self-driving vehicles had to attempt a 150-mile course in Nevada's Mojave Desert across dirt roads, flats, and mountain passes. A set of GPS waypoints was distributed to the teams only 24 h before the event. Not a single robot vehicle was able to finish the race, and the best team completed only seven miles.

Extended reading

Revealing DARPA

The Defense Advanced Research Projects Agency (DARPA), an administrative agency under the U.S. Department of Defense, is responsible for developing military technologies. Founded in 1958, DARPA was initially known as the Higher Research Projects Agency (ARPA).

DARPA has been instrumental in the development of many technologies that are still regarded as "cool" today. The first is the INTERNET, formerly the ARPANET, which was developed by ARPA. The second is the Global Positioning System (GPS). Before the launch of GPS navigation satellites, DARPA set up a network, called Transit, that consisted of five satellites. In 1960, Transit began to work to ensure that U.S. Navy ship positions were updated every hour and held to an error-window of no more than 200 meters. The third is a stealth fighter as its concept was also proposed initially by DARPA. DARPA's other little-known achievements also include the development of processing gallium arsenide used as a semiconductor. Although gallium arsenide is significantly more expensive than silicon semiconductors, it can better meet the needs of wireless communication, satellite, radar and military applications thanks to its faster electron transfer speed, which, in return, accelerates the processing for computer chips.

Over a decade, DARPA has provided financial support, coordinated research institutions, automakers, sensor suppliers and semiconductor suppliers to conduct a series of unmanned vehicle challenges. Many of the technologies applied in Google's unmanned vehicles originated from DARPA's Grand Challenge.

The following year, DARPA repeated the event; five teams completed the 132-mile course, providing evidence of the remarkable progress in this field in a very short time span. The first to cross the finish line after six hours and 54 min was the robot car Stanley, which was developed by Stanford University's Sebastian Thrun and his team. It was closely followed by the two Carnegie Mellon University vehicles, provide names HERE AND HERE. The 2007 DARPA Urban Challenge then moved the competition to an urban scenario at an abandoned military base. The teams had to negotiate four-way intersections, blocked roads, or parking lots with other self-driving as well as human-driven vehicles. The first autonomous vehicle traffic jam occurred at this event as did the first minor collision of two robot. The 60-mile race was won by CMU's vehicle "Boss"; Stanford's "Junior" came in second. For the first time, researchers heavily relied on high-resolution LiDAR sensors and maps.

The 2011 Grand Cooperative Driving Challenge (GCDC) took place in the Netherlands and tested cooperative driving behavior in real traffic. Karlsruhe's Institute of Technology's (KIT) team won this competition, followed by Chalmers University. The performance criteria comprised damping to strong oscillating braking, and acceleration maneuvers of the platoon leader to overall traveling time and platoon length.

1.3 Industrial Development

The transition from academic research to industrial development was marked by the Urban Challenge; in 2008 Google started working on self-driving cars and announced its self-driving car program two years later. The program was directed by Sebastian Thrun. Several car manufacturers announced similar projects, such as Daimler, BMW, Audi, Volkswagen, GM, Nissan, Honda, Toyota, Volvo, Ford, Tesla, Hyundai, Jaguar Land Rover, and Faraday Future. Automotive suppliers like Bosch, Delphi, Continental, and Mobileye followed suit. Also, ride share companies, e.g. Uber, IT companies, e.g. Baidu, and chip manufacturers, e.g. Nvidia, Intel joined the field with their own self-driving car projects as did startup companies like Zoox, Cruise Automation, drive.ai, comma.ai, nutonomy, or Nauto. In 2013, a VisLab research vehicle drove autonomously for 13 km in public traffic of Parma, Italy, partially without a safety driver in the driver seat. That same year, Daimler and KIT built the autonomous research vehicle "Bertha," which drove the 100 km long Bertha Benz Memorial Route from Mannheim to Pforzheim, Germany in public traffic. In 2014, Google presented a car that was specifically designed for self-driving and by 2016 Google's fleet had completed over 1.5 million miles. Significant advances in machine learning over recent years enhanced an enormous progress in the area of environment understanding for automated driving, one of the key enablers for developing driverless cars driving in real traffic.

SAE distinguishes three levels of higher automation: Conditional Automation, SAE level 3, does not require constant supervision of the driver anymore. High Automation, SAE level 4, requires the vehicle to be able to reach a safe state in case of failures or emergencies. Full Automation, SAE level 5, does not need a driver anymore at all. The approaches to reach these levels seem to divide *future development* into two distinctively different directions. On the one hand, the established automotive industry mainly follows a more conservative, step-wise approach of gradually increasing the level of automation, starting on highways with traffic flowing at very high speeds but uniformly into one direction without intersections, pedestrians, and bicyclists. Daimler, e.g. predicts that "initially we will drive autonomously on certain classes of roads, starting with the motorway and maybe only under certain weather or lighting conditions. In the beginning the system will also have to be monitored rather than grabbing a book and tuning out completely". On the other hand, Google and startups follow a more progressive approach by developing and testing dedicated driverless vehicles even without steering wheels. However, this approach rather targets *mobility as a service* instead of *end-consumer* owned vehicles. The test cars are, at least initially, limited to low speeds and geographically refined areas.

Lately, automated trucks have gained attention, too. In 2015, Daimler Trucks North America demonstrated a self-driving truck in Nevada. Volvo and other truck manufacturers have also held autonomous freeway driving demonstrations in Europe. The Silicon Valley startup Otto is working on autopilot retrofit kits for trucks since 2016. Another Silicon Valley startup, Peloton Technology, has already been working on truck convoys for fuel efficiency since 2011.

Extended reading

Autonomous Trucks

In 2013, a Japanese research institution showcased its new technology of efficient driving, energy saving and environmental protection. Four trucks, maintaining 4 m spacing between each other, were running at the same speed of 80 km/h. A group of trucks lined up to drive on the road at a fixed interval and at the same speed, like a series of disconnected train cars. This was not due to good driving skills, but due to the integrated intelligent driving system installed in the fleet. Each truck was equipped with an autonomous driving system. Through V2V communication among vehicles, each vehicle can share information such as speed and braking, thus enabling the system to control multiple trucks at the same time. The research institution said that trucks equipped with this system can automatically run in a queue at very short intervals, reducing air resistance, avoiding unnecessary braking, and significantly improving fuel efficiency by more than 15%.

1.4   Challenges

Autonomous driving has developed in over one hundred years from fiction to research, industrial development and first partially automated production systems. Several challenges remain though:

- Technology and functional safety: Algorithms are still under development and being optimized. Suppliers are working on new and better sensors, and companies are testing various different sensors sets. Remaining challenges are, for example, how to deal with unforeseen weather and road conditions? Or with objects the sensors have never seen before? Or temporary construction or changes to a prerecorded map. Automated vehicle technology has made tremendous advances in the past years, but how reliable and safe does the technology need to be in order to be ready for introduction?
- Cyber security: Automated vehicles will likely also be connected vehicles in order to download map and traffic updates, and also connect to other vehicles or infrastructure such as traffic lights and highway signage. But connected vehicles have recently become targets for attacks. Cyber security may never be perfect. How can manufacturers build vehicles that are safe nevertheless?
- Human factors: What is the role of the human inside partially, highly, and fully automated vehicles, respectively? Who is in charge and responsible? How does the handoff transition of responsibility from driver to vehicle and back work? And how will traffic participants and persons outside of the automated vehicle react to it?
- Legality, regulations, liability: The 1949 Geneva and 1968 Vienna United Nations Conventions on Road Traffic require a driver who is "at all times [...] able to control" the vehicle. However, the interpretation of this requirement differs from country to country. Whereas the predominant opinion in the United States is that automated vehicles are legal, for example, the German interpretation is that highly automated vehicles are not [yet legal]. In 2016, the United Nations approved amendments to the Vienna Convention as well as to UN Regulation No. 79, explicitly allowing automated driving technologies transferring driving tasks to the vehicle traffic, provided that these technologies are in conformity with the United Nations vehicle regulations or can be overridden or switched off by the driver. In the meantime, several US states have introduced, passed, or rejected various non-uniform state bills and regulations. Uncertainty also remains in the area of liability: Who will be liable when an AV is involved in an accident? Will liability shift from driver to manufacturer or even supplier as vehicle control shifts from the human operator to the automated vehicle?

- Ethics: Should automated vehicles be as safe or safer than human drivers? How safe is safe enough? What if automated vehicles reduce fatalities significantly, but at the same time cause new, different types of accidents? Arguably, there may never be concrete answers to many ethical questions pointed out by philosophers dealing with the matter.
- Infrastructure: Traffic depends on road infrastructure, and automated vehicles depend on digital and analog infrastructure as well. But how much infrastructure in the form of digital maps, infrastructure based communication systems, dedicated lanes is really required? Can, should, or must automated vehicles rely on the infrastructure? And how often do maps need to be updated?
- Urban planning and road planning: Automated vehicles will have a huge effect on mobility demand and traffic but the concrete changes on the amount of traffic, future traffic patterns, and urbanization are highly uncertain. Will there be more total vehicle miles travelled? But will automated vehicles need less road space than today? And will there be fewer vehicles overall that need less parking?

Significant challenges towards vehicles that can drive themselves in all locations and under all conditions still lie ahead of us. Given the tremendous progress in the field over the past ten years, the next ten years shall be even more exciting.

2　Understanding the Technical Classification of Autonomous Driving

2.1　SAE Autonomous Driving Technology Classification

In 2014, the Society of Automotive Engineers (SAE) released an Information Report on the "Taxonomy and Definitions for Terms Related to On-Road Motor Vehicle Automated Driving Systems". This report provides the most commonly used classification for the levels of automation ranging from no to full automation (see Fig. 1). The highest three levels of automation (specified by the taxonomy as conditional, high, and full automation) are defined in detail with regard to motor vehicles and their operation on public roadways. These three highest levels of automation relate to the cases of driving modes or trips in which solely an automated driving system performs the driving task. Vehicles equipped with some or all of these levels are often also referred to as "autonomous" or "self-driving."

It is important to note that these levels imply no particular order of market introduction. A particular vehicle may have multiple automated driving features such that it could operate at different levels depending upon the feature(s) that are engaged.

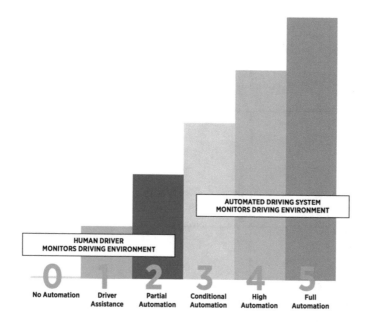

Fig. 1 Summary of SAE's levels of road vehicle automation. *Source* Society of Automotive Engineers (SAE)

2.2 SAE's Levels of Road Vehicle Automation

Detailed explanation of SAE's levels of road vehicle automation is highlighted in Table 1. Level 0 describes "No Automation" or driving when the vehicle is fully controlled by the human driver in all situations of a drive from start to finish. For example, this level obviously exists since the invention of the automobile in 1886.

Level 1 or "Driver Assistance" automates either the longitudinal control of the vehicle, i.e. maintaining a set distance (or more precisely, time gap) to the preceding vehicle, or the lateral control of the vehicle, i.e. keeping the correct position within the lane. These systems were introduced into production in the late 1990s through adaptive cruise control (ACC), which controls distance and relative velocity to preceding vehicles using a radar or LiDAR sensor. Forward Collision Warning (FCW) systems, introduced a few years later, are commonly based on the same sensor. They aim to enhance driver awareness of impending collisions by means of escalating warning levels. Warning levels range from audible alerts to brake jerks. Forward collision prevention systems for low speed applications using low-range, low-resolution, and inexpensive LiDAR sensors entered the market around 2010, e.g. Volvo City Safety. Lateral guidance systems using computer vision to detect lane markers were introduced into production vehicles about ten years ago with systems such as Lane Departure Warning (LDW) or Lane Keeping Support (LKS). These active safety and driver assistance systems partially and/or temporarily control certain aspects of vehicle

Table 1 Detailed description of SAE's levels of road vehicle automation

SAE level	Name	Narrative definition	Execution of steering and Acceleration/Deceleration	Monitoring of driving environment	Fallback performance of dynamic driving task	System capability (driving modes)
Human driver monitors the driving environment						
0	No automation	The full-time performance by the *human driver* of all aspects of the *dynamic driving task*, even when enhanced by warning or intervention systems	Human driver	Human driver	Human driver	n/a
1	Driver assistance	The *driving* mode-specific execution by a driver assistance system of either steering or acceleration/deceleration using information about the driving environment and with the expectation that the *human driver* perform all remaining aspects of the *dynamic driving task*	Human driver and system	Human driver	Human driver	Some driving modes
2	Partial automation	The *driving/node*-specific execution by one or more driver assistance systems of both steering and acceleration/ deceleration using information about the driving environment and with the expectation that the *human driver* perform all remaining aspects of the *dynamic driving task*	System	Human driver	Human driver	Some driving modes

(continued)

Table 1 (continued)

SAE level	Name	Narrative definition	Execution of steering and Acceleration/Deceleration	Monitoring of driving environment	Fallback performance of dynamic driving task	System capability (driving modes)
Automated driving system ("system") monitors the driving environment						
3	Conditional automation	The *driving/node-specific* performance by an *automated driving system* of all aspects of the dynamic driving task with the expectation that the *human driver* will respond appropriately to a *request to intervene*	System	System	Human driver	Some driving modes
4	High automation	The *driving/node-specific* performance by an automated driving system of all aspects of the *dynamic driving task*, even if a *human driver* does not respond appropriately to a *request to intervene*	System	System	System	Some driving modes
5	Full automation	The full-time performance by an *automated driving system* of all aspects of the *dynamic driving task* under all roadway and environmental conditions that can be managed by a *human driver*	System	System	System	All driving modes

Source Society of Automotive Engineers (SAE)

operation. They also include systems that automatically intervene to avoid and/or mitigate emergency situations and then immediately disengage again, but otherwise rely on a human driver to operate the motor vehicle in real time.

Level 2 or "Partial Automation" combines both, the longitudinal control and the lateral control of the vehicle, at the same time. It is important to note that the human driver still remain responsible for monitoring the vehicle environment and the system performance at the same time. Partial automation was introduced for the first time in 2013, when Daimler released a production passenger vehicle combining longitudinal distance keeping with lateral lane centering, but still requiring supervision of the human driver who may need to take over control of the vehicle more or less instantaneously. Shortly after, other automakers released similar systems, e.g. Tesla with the Autopilot system, however, all of them still require constant supervision by the driver. Level 2 systems are commonly a combination of two or more level 1 Driver Assistance systems.

Level 3 or "Conditional Automation" systems also control both, the longitudinal control and the lateral control of the vehicle at the same time but cover *all* aspects of the dynamic driving task. This means that the human driver does not need to supervise the driving environment anymore. It requires a significantly higher system performance and robustness, as the system then needs to cover all, even unlikely, circumstances during a trip. The driver is not responsible for monitoring the driving environment anymore, and consequently may perform other tasks while being driven, such as operating a mobile device or watching a video in the vehicle. However, at this level It is still expected that the human driver will respond appropriately to a request to intervene. Since the human driver is not required to monitor the driving environment anymore though, s/he cannot be expected to respond immediately as s/he will need a certain amount of time to look at the environment and regain awareness of the traffic situation and the environment. The step from level 2 to level 3 is significant. The system robustness requirement increases significantly. Consequently, the sensors of a level 3 system need to be capable of recognizing all driving situations and gain and maintain a situational understanding of the driving scene at all times. But, a level 3 system still requires the human driver to be physically and mentally present as a fallback in case the automated driving system encounters a failure, e.g. of a component. However, since the human driver does not need to monitor the environment constantly anymore, this handover to the driver must not happen instantaneously and the system must possess the capability to overcome the duration until the human driver has gained situational awareness of the driving environment. As of spring 2017, no level 3 system has been introduced to a production vehicle.

Level 4 or "High Automation" builds upon level 3 but removes the requirement of driver take over, e.g. in case of a system failure. A level 4 system must be able to handle failures of system components and must provide a fallback solution. Consequently, a level 4 system must be able to reach a so-called "safe state" in case of a failure, where the vehicle is removed from active traffic and parked, e.g. in a parking lot or in an emergency lane. This requirement generally becomes significantly harder to fulfill in cases of higher velocities on highways, especially in the case of dense, fast-moving traffic. On the other hand, a level 4 system may be much simpler to realize in low speed or parking lot scenarios. Also, a level 4 system may be limited to certain driving modes, e.g. it may only be able to operate in automated mode in a parking lot or on a highway, but not in city streets. As of spring 2017, no level 4 system has been introduced in a production vehicle.

Level 5 or "Full Automation" vehicles can cover all driving modes, i.e. can drive on all streets. Human interaction is only required for setting the destination and starting the system. The automatic system can then drive to any location where it is legal to drive using any legal route.

3 Driver Information Systems

Driver information systems (DIS) provide information and warnings to the driver, without direct access to actuation systems, such as brake, engine, or steering systems. DIS include navigation systems to provide turn-by-turn routing information to the driver, removing the need for dealing with printed maps. Night vision systems enhance the driver's visibility at night. Collision warning systems warn the driver of imminent collisions. Lane departure warning systems warn the drive before unintentionally leaving the traffic lane. Traffic sign detection systems detect signs and provide, e.g. the current speed limit to the driver. Therefore, DIS provides the driver with convenience as well as increased safety.

3.1 Navigation

Global Positioning System was developed by the US Department of Defense. It uses between 24 and 32 Medium Earth Orbit satellites that transmit precise microwave signals. This enables GPS receivers to determine their current location, time and velocity. The GPS satellites are maintained by the US Air Force.

BeiDou Navigation Satellite System (BDS) is a global satellite navigation system developed by China, which is the third mature system after the U.S. Global Positioning System (GPS) and Russia's Global Navigation Satellite System (GLONASS). BDS is composed of three parts: space section, ground section and user section, which can provide high-precision and reliable positioning, navigation and timing services for all kinds of users around the world in an all-weather 24–7 situation, with the ability to convey short messages. It has initial regional navigation, positioning and timing capabilities, with a positioning accuracy within 10 m, and a speed measurement accuracy within 0.2 m/s, and a timing accuracy within 10 ns.

Automotive navigation systems were among the first driver information systems when introduced in the 1980s. Navigation systems obtain the current position of the vehicle from a GPS or similar GNSS receiver and the desired destination from user input. To make up for potential GPS signal loss or disturbance, e.g. in a tunnel or in urban areas, dead reckoning is used for greater reliability, and employs velocity data from wheel speed sensors, turn rate data from a gyroscope, and acceleration data from an accelerometer.

Navigation is based on a map containing streets and information on how those streets are connected. A routing algorithm calculates the route from the current position to the destination based upon the map.

The route is shown in a display, which is typically separated from the instrument cluster. Turn-by-turn directions are usually displayed also in the instrument cluster or on the windshield using a head-up display (HUD). Additionally, directions are often announced audibly.

3.2 Night Vision

Automotive night vision systems use a camera in the infrared spectrum, which is invisible to humans. There are two types of such systems. Active systems use an infrared light source built into the vehicle's headlights to illuminate the road ahead up to approximately 250 m with near-infrared light. The resulting image is then converted into a visible light image. Passive systems capture thermal radiation, which is emitted by the objects using a thermographic camera sensitive in the far-infrared spectrum. This potentially results in a greater range and high contrast for objects emitting a lot of heat, but a generally lower resolution image and poor performance in hot weather environments.

The infrared image is transformed into a visible image. This is displayed either in the instrument cluster, using a high-resolution liquid-crystal display (LCD), in the navigation system display, or on the windshield via a head-up display. More recently, night vision systems with pedestrian, bicyclist, or other object detection have been introduced. These systems visually highlight the respective object in the image.

3.3 Forward Collision Warning

A forward collision warning (FCW) system is a vehicle safety system designed to warn the driver of an upcoming potential collision so that the driver's attention is drawn to the imminent crash. The system uses radar, LiDAR, or camera sensors to detect and track other vehicles in the environment. Typically, the system calculates a time to collision (TTC) by predicting the motions of other vehicles as well as its own. Once the time to collision for one tracked object falls below a certain threshold, a warning is issued to the driver in the form of an audible and visual alert. Some systems also start to simultaneously pre-fill the braking system with pressure, in order to reduce the time for the brakes to engage once the driver presses the brake pedal, activate hazard lights, close windows and the sunroof, and pretension the seat belts. If the driver does not react, the warning is followed by light braking to get the driver's attention. The light braking can be followed by partial autonomous braking.

In 2016, the National Highway Traffic Safety Administration (NHTSA) announced that most of the vehicle manufacturers in the USA had agreed to include automatic emergency braking systems as a standard feature on all new cars sold in the U.S. by 2022. NHTSA projected that these systems could prevent an estimated 28,000 collisions and 12,000 injuries annually.

In 2016, China's Ministry of Transport issued the transportation industry a standard "Safety Specifications for Operating Bus" (JT/T 1094—2016). The standard clearly stipulates that operating buses over 9 m should be equipped with FCW functions, and the grace period to allow for implementation of the standard is thirteen months.

3.4 Lane Departure Warning

Lane departure warning (LDW) systems typically use a camera to detect the lane markings on the road in front of the vehicle using a lane marker detection algorithm. This algorithm utilizes a geometric model of the lane course as well as a kinematic model of the vehicle. Initially, the algorithm detects a potential lane marking by looking for a dark-bright-dark transition in the camera image where a lane marking is to be expected. When this dark-bright-dark transition caused by the white lane marking on the dark road is detected, the algorithm will search for similar transitions in subsequent camera images. Once a sufficient amount of information has been gathered and checked for plausibility, a lane model is then generated.

LDW warns the driver in the event of an unintentional departure from the current traffic lane. Commonly, the time to lane departure is calculated and the driver is warned when a predefined level is reached. The algorithm accounts for lane width, the deviation from the center of the lane, vehicle width, vehicle velocity, and the orientation of the vehicle relative to the lane.

Warning the driver typically takes place in the form of an audible alert, but can also be a visual indicator in the instrument panel or as haptic feedback, such as vibrations in the steering wheel. Commonly, the warning is given when the driver comes close to or crosses a lane without using the turn signal indicator. In the case of an intentional lane change, the driver should use the turn signal to indicate this intention and the system will not issue a warning.

3.5 Traffic Sign Detection

Traffic sign detection systems recognize road signs, such as speed limit signs. The identified information can be displayed in the vehicle's instrument cluster or heads-up display. Typically, a forward-facing camera is used often in combination with speed limit information from a digital map.

Moreover, by detecting the vehicle's speed the system can alert the driver when the speed limit it has recognized on a speed limit sign is exceeded.

4 Advanced Driver Assistance Systems (Level 1)

Advanced driver assistance systems (ADAS) or level 1 systems automate one driving task, either longitudinal or lateral control, whereas the human driver fulfills the other task. Adaptive cruise control (ACC), for example, automates the vehicle's longitudinal control, whereas the human driver still performs the steering task. Lane keeping systems (LKS) maintain the vehicle within a lane by actuating the steering, i.e. the lateral control task, but the human driver maintains the vehicle's speed.

4.1 Adaptive Cruise Control

Just like a conventional cruise control system, adaptive cruise control (ACC) maintains the vehicle at a set constant speed. The difference, however, is that if a car which contains ACC approaches a slower moving vehicle ahead, its velocity is automatically decreased and it then follows the slower vehicle at a set distance. Once the lane ahead is free again, adaptive cruise control accelerates the vehicle back to the previously set speed. The driver can always override the ACC system by pressing the accelerator or the brake pedal. When the accelerator pedal is pressed, the vehicle responds in the common way. And when the accelerator pedal is released again, the vehicle's velocity goes back to the previously set speed. When the brake pedal is pressed, adaptive cruise control is deactivated and the vehicle is under full control of the human driver.

ACC systems typically consist of the following components; a sensor to perceive objects in front of the vehicle, a control unit which processes the sensor information and computes the required actuator signals to perform the functionality, an interface to the engine-management system with torque control, and an interface to the electronic brake control system with active brake-pressure-increase capability, typically based on ESP (electronic stability program).

ACC is activated by the driver, generally by turning on a switch or pressing a button on the steering wheel. ACC turns on when certain preconditions are met, such as the vehicle already has a certain speed, the brake pedal is not pressed, and the vehicle is not too close to another vehicle. Then the vehicle turns on speed control, typically at the current speed. The driver, subsequently, can increase or decrease the set speed to the desired speed. The driver can also adjust the distance or, more specifically, the time gap to preceding vehicles. The gap can typically be set between 1 and 2 s, which is equivalent to a gap of 30–60 m at 65 mph or 100 km/h. Both set speed and set gap are displayed in the instrument cluster. Commonly, it is also indicated if the radar sensor has detected a preceding object in the lane in front of

Fig. 2 Adaptive Cruise Control takes the preceding vehicle as a reference. *Source* Bosch

the vehicle. ACC is deactivated by pressing the brake or by pressing a button on the steering wheel. Most systems contain a button to resume ACC at the previously set speed and time gap.

The radar sensor of the ACC system constantly monitors the environment in front of the vehicle and detects objects in the vehicle's driving environment. Then the relevant control object is selected, which is the object relative to which the velocity and time gap is controlled. This is typically the preceding vehicle in the same lane (Fig. 2). That is simple on a straight road but challenging when one of the vehicles is driving a curve, since the radar sensor can only detect objects but not lane markings. Therefore, a direct association of objects to lanes is not possible. Instead, other vehicles are tracked over time and thus it can be determined if that vehicle is driving in a curve. Similarly, it can be detected if the vehicle is driving a curve through incorporation of steering wheel angle and yaw rate into a motion model. This is called course prediction. A reliable course prediction is required to avoid the reaction of the ACC system to objects in neighboring lanes instead of the own lane.

Adaptive cruise control typically has an upper acceleration limit of approximately 1 m/s^2 and the deceleration limit is typically around 3–4 m/s^2. While this is already significant braking, it is only one third of the maximum possible deceleration on a dry road. The purpose of the limitation of the functional range is to keep the driver aware of the functional limitation of ACC systems. These are, e.g. limited longitudinal and lateral range of the sensor, which may lead to late or no reaction of the system in certain situations, e.g. close lane change maneuvers by other vehicles or missed objects in tight turns. Also, radar sensors cannot distinguish certain small metal objects on the ground from vehicles, hence, standing objects which have never been detected as moving objects are typically ignored. Therefore, an ACC driven vehicle may come to a stop behind a preceding vehicle which has been followed for some time, but may not come to a stop behind a stopped vehicle which has not been seen moving.

Consequently, ACC systems are designed as convenience systems with a limited functional range and require the driver to monitor the driving environment at all times and to be aware and able to take back control from the vehicle at any time.

4.2   Lane Keeping Assist

Lane keeping assist systems (LKS) are a functional extension of lane departure warning systems. LKS similarly detects the lane markings but instead of issuing a warning, the system issues a torque on the steering wheel to actively steer the vehicle back towards the lane. LKS was introduced to the market in 2005.

There are different implementations of this system. Commonly, the driver is required to steer the vehicle and the vehicle merely supports this by issuing a fairly small momentum in the event of lane departure to guide and correct the vehicle back into the lane. Consequently, this is not sufficient in tighter curves when a larger steering momentum is required. More recently, lane center control (LCC) has been introduced, which maintains the vehicle exactly in the center of the lane.

5   Partial Automation

Partially automated driving or level 2 systems combine automation of both longitudinal and lateral control into one system. Typically, this is achieved by combining both adaptive cruise control and lane keeping into one system. The human driver still remains responsible for monitoring the vehicle environment at all times as the vehicle can hand over control to the human driver at any moment. In this regard, a level 2 system is not an autonomous system.

In 2013, partial automation was introduced for the first time, when Daimler released a production passenger vehicle, combining longitudinal distance keeping with lateral lane centering.

Tesla released its Autopilot system in 2015. Tesla Autopilot allows the car to stay in-lane, adjust its speed in accordance with other cars on the road, and change lanes autonomously. Tesla recommends it for highway use only. Autopilot-enabled Tesla vehicles are also capable of scanning for parking spaces and of self-parking, which Tesla calls "Summon." The Summon function gives vehicle owners the ability to call their car from a distance of up to 40 ft, after which the car will drive itself out of a parking space and head toward the owner. Autopilot uses a camera, radar, and 12 ultrasonic sensors.

Audi recently launched its Traffic Jam Assist (TJA), which takes over gas, brake, and steering systems in the speed range of 0–65 km/h (0–40 mph). The system uses two radar sensors and a video camera and follows lane markings and preceding vehicles on the road. The Audi system allows for 15 s intervals of hands-off driving at slower speeds, but the driver is still required to be constantly alert, aware, and able to intervene immediately as needed. Other automakers have launched similar level 2 systems recently.

6 Conditional Automation

As of spring 2017, no production vehicle on the road has surpassed level 2 automation. Several automakers, however, have announced level 3 or conditionally automated driving systems for 2018, e.g. Audi's Traffic Jam Pilot. This system will enable drivers to travel hands free up to 35 mph on certain roads, such as divided highways. For 2020, Audi plans to introduce a level 3 highway pilot system, which offers hands-free driving on limited-access, divided highways with posted speed limits. The vehicle will be able to execute lane changes and pass cars independently.

Level 3 systems do not require the driver to constantly monitor the vehicle environment. However, the driver must remain in the driver's seat and must be able to take over control within a few seconds, but not immediately, as required by level 2 systems. Therefore, Audi plans to implement a "driver availability detection" system, which confirms that the driver is active and available to intervene. If not, it will bring the car to a safe stop.

Bosch outlined the vision for a highway pilot system in a video. The driver would be required to manually drive on urban streets, but once on the highway the car would take over control until the desired exit is reached. While being driven, the driver would be able to write emails or watch videos without having to monitor the driving environment.

7 High and Full Automation

High automation or level 4 systems are similar to level 3 systems, but in principle do not require a driver anymore within their functional range. Level 4 systems, therefore, must be able to handle all expected and unexpected situations on the roads for which the system is enabled. Also, highly automated systems must be able to reach a safe state, i.e. also in the case of system failures since there may not be a driver anymore for handover to take place. This requirement is much harder to fulfill at high velocities compared to low velocities, therefore it is expected that level 4 systems will be introduced first for low-speed or parking applications. As of spring 2017, no level 4 system has been introduced to a production vehicle.

Google is the most prominent company developing level 4 vehicles. Their project started in 2009. In August 2012, Google announced to have completed over 300,000 autonomous driving miles accident-free, typically having about a dozen cars on the road at any given time. In May 2014, Google revealed a new prototype of its driverless car, which had no steering wheel, gas pedal, or brake pedal, and was therefore the first vehicle designed as a level 4 highly automated vehicle. In December 2016, the Google self-driving car project was renamed as Waymo, which is derived from "a new way forward in mobility".

Full Automation or level 5 vehicles can cover all driving modes, meaning they can drive on all streets on which human drivers can drive and where it is legal to drive. Level 5 vehicles will have to be able to deal with all weather and environment conditions in which human drivers can drive. This means, level 5 vehicles are presumably still a long time away from market introduction.

Chapter 4
Electric Vehicles: The Best Carriers for Autonomous Driving

If you were to look back in time, you would realize that all progress has stemmed from some sort of energy revolution. The discovery of fire led to agricultural civilization for mankind; the widespread use of coal catalyzed the industrial revolution in Europe; the discovery and large scale application of petroleum in the twentieth century propelled the manufacturing revolution. Up until now, chemical energy produced by combustion of substances has usually come at the cost of great consumption and waste, yet, like all known discoveries of mankind, resources are always limited.

Therefore to ensure our sustainable development, mankind must find new ways to generate energy that does not over-consume resources. This also motivates mankind to explore green energy. For example, electric energy is adopted as a source of green energy, while the battery is accepted as the energy storage media. Compared with thermal energy, electric energy exhibits natural advantages in terms of control due to its linear physical characteristics. However, there is room for improvement in terms of current energy storage and transmission technology. Along with the rapid development of electronics technology, we have witnessed the emergence of new industries which feature digitized control of electric energy, such as photovoltaic-wind energy storage, peak shaving and valley filling. Similar trends exist in the automotive industry. As we head towards the end of petroleum resources, we must seek new forms of energy sources. Therefore, electrification is inevitable for automobiles.

An electric vehicle has several advantages over a traditional one. Traditional vehicles consist of 30,000+ parts, while electric ones have only 10,000+ parts, giving it a simpler structure. This means that the assembly and maintenance of electric vehicle are easier. As a result, the entry threshold for the vehicle manufacturing industry is lowered to a significant extent. No emission will be generated during the operation of the electric vehicle. In addition, if we were to disregard the environmental harm of the upstream power plant and downstream battery recovery, an electric vehicle generates "zero pollution". Thanks to the absence of the engine, the noise created by an electric vehicle is significantly lower than that of its traditional counterpart—(though the high-frequency noise generated at high motor speed is a pressing issue that requires a solution). Another advantage is that the usage cost of an electric vehicle is as low

© China Machine Press, Beijing and Springer Nature Singapore Pte Ltd. 2021
Z. Chai et al., *Autonomous Driving Changes the Future*,
https://doi.org/10.1007/978-981-15-6728-5_4

as one fifth that of the gasoline-fueled vehicle, while being highly efficient in energy conversion. The energy generated when brakes are applied and when the vehicle is going downhill can be recovered, helping to improve energy utilization efficiency. An electric vehicle can be charged at night using cheaper "off-peak electricity", helping to stabilize the difference between "peak" and "off-peak" electricity in the electric grid. However, consumers have identified some shortcomings, such as long charging time, short mileage, shortage of charging facilities and high prices. Fortunately, we are seeing gradual improvements in these areas.

The electric vehicle is becoming common all over the world. Countries such as the United States, Japan and those in Europe, as well as vehicle companies including Ford, Toyota and Volkswagen, have formulated their own development strategy in this area. Volkswagen has announced its "Strategy2025", in which the company planned to push out more than 30 models of blade electric vehicles, (BEVs) with the target of 3 million units sold by 2025. Benz has raised its target for sustainable development, aiming to develop autonomous vehicles which feature zero emission and interconnectivity. To this end, they have focused on the development of BEVs, plug-in hybrid electric vehicles (PHEVs), and hydrogen-powered vehicles. GM has announced that it would launch nine new-energy vehicle (NEV) models in China by 2020.

At the end of 2016, the global sales of new-energy passenger vehicles (BEVs and PHEVs) reached 774,000 units, among which 48,4000 units had been sold in China. Due to the strong support from the government, China has demonstrated the world's fastest growth in electric vehicles. Only 5 years ago, the Chinese electric vehicle industry was still in its infancy. The E6 BEV launched by BYD was mainly used as taxis, and BAIC Motor had just established BAIC BJEV to launch NEVs based on electrification of traditional vehicles. Tesla has presented us with a brilliant display of its advanced battery management system and model design, thus allowing the ordinary person to fulfil his or her dream of owning an electric vehicle. Subsequently, new emerging forces like NIO and Faraday Future have also entered the field. Participation from both traditional automakers and emerging players had kicked off the rapid development of the electric vehicle industry in China.

In the meantime, technological improvements continue developing within the Chinese electric vehicle industry. For example, notable breakthroughs have been made in battery power and drive motor. Despite being an industry newcomer, NIO won its first Driver's Championship in Formula E of Fédération Internationale del'Automobile in 2015 with its first product—NextEV Formula E (Fig. 1).

1   About Electric Vehicles

Since its inception, an automobile has always been made up of two fundamental sections: mechanical and electrical. The "mechanical" refers to the machines that compose the car, i.e. engine, transmission and structural body. The "electrical section"

Fig. 1 NextEV Formula E electric racing car. *Source* NIO

refers to the electrical parts of the vehicle, which support the vehicle's power output (i.e. engine ignition) and other electric equipment (central control system, onboard computer, ever-increasing car-mounted multimedia equipment, etc.)

1.1 Battery Electric Vehicles

Everything changes when a vehicle becomes electric. The mechanical components of an electric vehicle are significantly fewer than those of a gasoline vehicle. Electric driving and control systems are the core of the electric vehicle, reflecting the biggest difference between an electric vehicle and an internal combustion engine (ICE)-powered vehicle. Just like an ICE vehicle, the electric driving and control system consists of a driving motor, motor control unit, vehicle controller. The power pack consists of the battery pack and battery control unit. Like an ICE-powered vehicle, an electric vehicle requires an uninterrupted power supply. The driving motor converts electric energy from the power supply into mechanical energy, powering the wheels and other working devices directly or indirectly through a transmission. Figure 2 shows the configuration of the chassis in the Tesla Model X. Taking traditional driving habits into account, the electric vehicle has inherited the accelerator pedal, braking pedal and shift stick. In the vehicle, the mechanical displacement of the accelerator pedal and the braking pedal is converted into electrical signals and transmitted to the central control unit, controlling the motor through a power train control module.

The vehicle power pack and onboard equipment can be controlled through a central control unit (CPU), just as in a computer. The control system of an electric vehicle is mainly used to control the vehicle, power battery pack and motor. The output power and torque of the motor vary according to changes in driving conditions. This leads

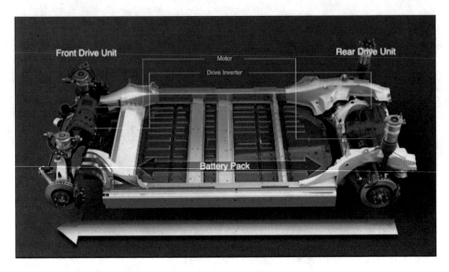

Fig. 2 Typical structure of the chassis of an electric vehicle. *Source* Tesla

to corresponding changes in the voltage and current of the power battery pack. In addition, the control system also includes electronics and other system installations, such as a GPS satellite navigation system, forward collision warning and other ADAS systems.

An electric vehicle features highly electrified characteristics. It is equipped with a complete power driving system network which includes motors, battery and electronic control units. As a result, an electric vehicle is more suitable to mechanical–electrical integration, and the use of an automated control system and management system. This system is poised to be the green and intelligent heart for future automobiles. Compared to its gasoline-fueled counterpart, an EV is therefore more suitable to serve as a carrier for autonomous driving.

1.2 Extended-Range Electric Vehicles

As the energy density of the power battery has yet to reach a sufficiently high level, short driving mileage has become the Achilles' heel of blade electric vehicles. Currently, most blade electric vehicles feature a driving mileage of approximately 200 km. The timely introduction of extended-range electric vehicles has remedied this shortcoming and provided a more functional power supply to the vehicle.

The extended-range electric vehicle features a motor and a traditional ICE, which is mainly used to charge the power battery. While the extended-range electric vehicle relies mainly on the power from the battery, the engine runs only in rare cases to charge when the battery is low. Not only does this design make it possible to charge the battery using EV charging stations, it also retains the advantages of fully electric vehicles, relieving consumers' anxiety over range.

Strictly speaking, most extended-range electric vehicles are variations of blade electric vehicles with a transitional nature. Most extended-range electric vehicles feature only a single driving mode. However, they can be charged both through the power grid and the auxiliary range extender. Around 80% of the time, an extended-range electric vehicle obtains energy through the battery pack, rarely using the auxiliary range extender to supply energy. With the continuous improvement of technology and increasingly strict restrictions applied to carbon emission, blade electric vehicles with a longer driving range will become mainstream.

2 Motor, Motor Control, Battery: The Core of the Electric Vehicle

The motor, motor control and battery are the core components vital to an electric vehicle. According to results of statistical studies conducted by Argonne National Laboratory, the cost of the electric vehicle's powertrain (motor, motor control and transmission) costs up to 15.67% (for passenger car) and 13.69% (for light-van) of the total price of production. This cost is secondary only to the battery and BMS (battery management system). As a key power system component, the development of the battery will eventually determine whether the electric vehicle can replace the ICE Vehicle, and launch a new global energy revolution. Therefore, the motor, motor control and battery (known as the "three electricals") play an important role in the electric vehicle.

2.1 Motor

The driving motor boasts unique features in terms of load requirement, technical performance and operating environment.

Firstly, the driving motor requires high power density. It recovers electric energy through braking, lowering the vehicle's energy consumption.

Secondly, the driving motor features wide speed adjustment and large torque at low speed, allowing for high starting speed, high climbing performance, and high-speed acceleration performance.

Thirdly, the electronic control unit features high control accuracy, a high dynamic response rate as well as high safety and reliability simultaneously.

Currently, the most commonly used driving motors in electric passenger and commercial vehicles are AC asynchronous motor, permanent magnetic motor, and switched reluctance motor. In contrast, hybrid excitation motor, polyphaser motor and other special types of motors are rarely used. The motor types of blade electric vehicles are shown in Fig. 3.

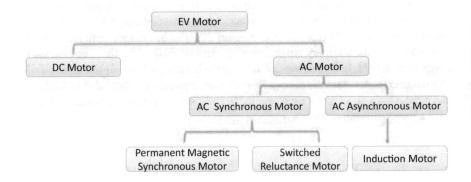

Fig. 3 Motor types of electric vehicles

Although China lags behind the international level in terms of driving motor technology, considerable progress has been made in recent years. These include breakthroughs in general basic technologies, such as magnetic conductive silicon steel, rare earth permanent magnetic material, insulating material, position transducer, integrated design of chips, power electronic system.

2.1.1   AC Asynchronous Motor

AC asynchronous motor, also known as AC induction motor, involves the input of three-phase alternating current to the stator winding, where the exciting current in the stator winding will generate a rotating magnetic field in the metallic stator core. At the same time, an induced current will be generated in the rotor winding, powering the rotor. When the mechanical load is applied to the rotor, the current will increase, increasing the exciting current in the stator winding by property of the electromagnetic induction effect.

Pulse width modulation (PWM) is used by the AC asynchronous motor controller to transform the high voltage DC to three-phase AC. The inverter regulates motor speed, while vector and direct torque control ensure a quick response, appropriate for the variations in load.

The AC asynchronous motor features a simple structure, reliable operation, low maintenance cost, and remarkable speed range and performance. However, its shortcomings include low efficiency, low power factor, and its requirement of a high capacity inverter. Currently, the AC asynchronous motor is mainly used in commercial vehicles and high-speed passenger vehicles with low spatial requirements, such as electric buses, and cargo vans.

Some European and American electric vehicles have adopted AC asynchronous motors to lower their dependency on expensive and hard-to-obtain rare earth resources. Tesla currently uses independently designed AC asynchronous motors in its Model S and Model X lines.

2.1.2   Permanent Magnetic Synchronous Motor

Permanent magnetic motors can be divided into two categories: brushless permanent magnetic synchronous motors (sinusoidal wave) and brushless permanent magnetic DC motors (square wave). Both motors use rotors made of permanent magnets and silicon steel sheets. Stators using three-phase winding allow for the input modulated square-wave voltage to generate a rotating magnetic field to move the magnetic rotor. The permanent magnetic motor features relatively high driving efficiency, high power density and large speed range in its simple structure, as well as a motor with 15% less volume than that of an asynchronous motor with the same power. However, its drawbacks include the potential for demagnetization and back EFM.

Two significant technical barriers for the creation of a permanent magnet motor are the high R&D and design requirements, and the complex manufacturing and assembly process. In addition, the permanent magnet is made from rare earth, and requires the use of costly neodymium iron boron (composing appr. 30% total cost). This is a significant challenge towards the large-scale application of the permanent magnet motor.

At present, permanent magnet motors is widely used in different applications, including within electric passenger vehicles where there are high requirements in terms of space, speed and manipulation performance, as well as in commercial vehicles like buses and trucks, where stronger low-speed torque applies. With the exception of Tesla, almost all new-energy vehicles outside of China have adopted permanent magnet motors. Several vehicle companies manufacture motors for their own vehicles as well, including BMW, Volkswagen, Toyota, and Nissan. On the other hand, component enterprises such as Bosch and Continental excel in this area as well. Almost all Chinese electric vehicles adopt permanent magnet motors. In addition, a kind of brushless permanent magnetic DC motor is also used in certain mini-watt electric vehicles and low-speed electric vehicles.

2.1.3   Switched Reluctance Motor

The stator and rotor cores of the switched reluctance motor are made from layers of pressed silicon steel sheets. Grooves in the lamination are used to form the ridged structure. Adhering to the principle of "minimum reluctance", the stators generate twisted magnetic fields to drive the rotor. The switched reluctance motor features a simple structure and control, high reliability, low cost, good starting and braking performance as well as high efficiency. However, its drawbacks include high noise level, series torque pulsation and high nonlinearity. As the switched reluctance motor has both advantages and drawbacks when used in EVs, it is rarely used in that application at present. BMW i3 adopts the permanent magnetic auxiliary reluctance motor.

2.1.4   DC Motor

In a DC motor, a coil is wound around the main pole of the stator, and direct current is input to generate magnetic field. DC is input to the armature winding of the rotor, which is then placed in the magnetic field to create electromagnetic torque, which acts on the load. Advantages of the DC motor are seamless speed regulation, simple control, mature technology and low cost. On the other hand, the DC motor requires independent brush and commutator, which correlates with limited speed boosting, easily consumed brush and high maintenance cost. The DC motor is mainly used in the driving systems of early electric vehicles, and no longer used for newly developed models.

See Fig. 4 for the comparison of characteristics of the three types of AC motors.

China is rich in rare earth resources. As a result, most electric passenger vehicles use permanent magnet motors, which feature high power performance and small volume. According to statistical data from studies conducted by China Automotive Industry Association, by the first half of 2016, the production and sales volume of blade electric vehicles in China had reached 134,000 and 126,000 respectively, a growth of 160.8 and 161.6% from the previous year. By that same time, the production of blade electric passenger vehicle in the first half of 2016 reached approximately 70,000 of which 65.7% were equipped with permanent magnet motors, a growth of 21% than the last year. Within the same period, market shares of domestic new energy

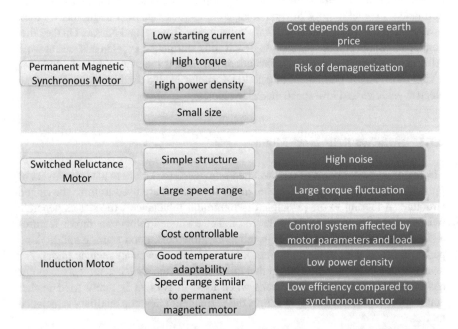

Fig. 4 Comparison of characteristics of the three types of AC motors. *Source* Beijing Huizhi Huizhong Automotive Technology Institute

passenger vehicles equipped with AC asynchronous motors dropped, declining from 35.1% in the first half of 2015 to 32.9% in the first half of 2016. On the other hand, the market share of vehicles with hybrid excitation synchronous motors increased from 0.03 to 1.10%. Although the ratio is low, we can anticipate break through developments in the field of blade electric passenger vehicles in the future.

2.2 Motor Control

In general, motor control in the context of electric vehicles refers to the vehicle electronic control (Fig. 5). Another important part of the electric vehicle control system is the battery management system (BMS). For now, we turn our attention to motor control.

Motor controller technology has been used in the on-board motor of the traditional vehicle. The vehicle air condition compressor and power-assisted steering pump are controlled and adjusted through the medium and low voltage frequencies of the power semiconductor. As the equipment controlling the driving motor of an electric vehicle, the motor controller controls the driving of an electric vehicle by regulating rotational speed, torque and steering, according to input transmitted by the vehicle controller and mechanisms (braking pedal, accelerator pedal and gearshift). At the same time, it also achieves corresponding control against the output of the power battery.

With the increasingly widespread application of microchips in the control of vehicles and assemblies, we can expect a reduction in the cost of all-in-one electric

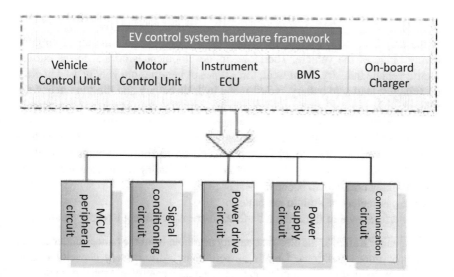

Fig. 5 Hardware framework of the control system of 100% electric vehicle. *Source* NIO

control products, and the gradual consolidation of controls into an integrated "vehicle central control unit".

At present, suppliers capable of R&D and production of electronic control systems can be divided into two categories.

The first category consists of companies whose supply chain for motor electronic control products are partially or entirely produced in-house or by affiliated enterprises. Most of these enterprises comprise the production and R&D base of traditional automakers for new-energy vehicles. For example, enterprises like BYD and JMC New Energy are equipped with independent capabilities to produce and supply motor electronic control products. As a newcomer in the electric vehicle industry, NIO's vehicle parts branch—NIO Power Technology (XPT)has established its own complete design and production capacity for motor electronic control systems.

The second category includes professional auto parts suppliers, such as ZF, Continental, Bosch, Hitachi, Hyundai Mobis and other international giants. This category also includes emerging professional motor electronic control system production enterprises, including Shanghai Edrive Co., Ltd., Shanghai Dajun Technologies Inc., Jing-Jin Electric, and Taiwan Fukuta.

Of the 180,000 blade electric vehicles produced in China between January and July 2016, statistics have shown that 55.4% contained motors supplied by vehicle companies, and 44.6% that were supplied by third-party motor enterprises. They also show that 56.2% contained electronic control units supplied by vehicle companies, and 43.8% that were supplied by third-party electronic control enterprises. The split is approximately equal in both cases.

2.3 Battery

As the only EV power source, the battery plays a decisive role in determining the driving range of the vehicle. Therefore, its development is critical to the current electric vehicle industry. At present, three types of batteries are widely used in electric vehicles: lead acid battery, nickel-metal hydride battery and lithium ion battery. Lithium ion is a newer battery moving into the mainstream, with the best comprehensive performance amongst the rechargeable batteries that are on the market. Lithium ion batteries have the highest energy density among the three, and a relatively long cycle life and service life. In addition, there have been vast improvements in its safety. The lithium ion battery is in its production stage, including large scale automation, and will likely continue to become cheaper.

Main parameters of lithium ion battery.

The performance of lithium ion batteries is measured using five key parameters.

(1) Capacity: Refers to the electric quantity available from the battery under certain discharging conditions. Unit is ampere-hour (Ah). Battery capacity is also further divided into theoretical capacity, actual capacity and rated capacity.

(2) Nominal voltage: Refers to the average voltage during the entire discharging process of the battery under certain discharging conditions.

(3) Internal resistance: Refers to the resistance of current passing through the battery. The internal resistance of the battery is not a constant, but is subject to continuous fluctuation against time during the charging and discharging process. The internal resistance of the battery is the sum of ohm inner resistance and polarization internal resistance. In general, the internal resistance (AC) is used to add a fixed frequency, fixed current (currently 1 kHz, 50 mA) to the battery, then sample the voltage, rectify, filter, and calculate the resistance value through the operation circuit.

(4) End-of-charge voltage: Refers to the maximum working voltage of the battery which is not suitable to continue charging at the time of charging. End-of-charge voltage is related to the characteristics of the materials. Over-charging will lead to lithium precipitation at the negative pole, structural collapse of the anode active materials and oxidization of the electrolyte, which may impact battery performance or even lead to accidents.

(5) End-of-discharge voltage: Refers to the lowest working voltage that the battery should not continue discharging at the time of discharging. Different requirements in terms of the capacity and service life of the battery shall apply based on various types of battery and discharging conditions. Therefore, the specified end-of-discharge voltages are different. In general, when the temperature is low or discharging is completed with high currents, the specified end-of-discharge voltage is lower due to the impact of polarization. On the contrary, during low-current extended discharging or intermittent discharging, the specified end-of-discharge voltage is higher.

Currently, two types of batteries are commonly used in electric vehicles. These are ternary lithium batteries (consisting of nickel and cobalt, with either manganese or aluminum), which are used in most Chinese electric vehicles, and lithium iron phosphate battery, represented by BYD.

The ternary lithium battery features high energy density (over 200 W-h/kg). However, its shortcomings include rapid capacity attenuation, short service life and poor safety performance, as open fires may result from internal short-circuits, or contact with water). The lithium iron phosphate battery features higher output efficiency (standard discharging current of 2–5C, where C refers to the nominal capacity of the battery. The maximum continuous discharging current can be as high as 10C), more charge cycles, longer service life and higher safety performance (no combustion will occur even in case of puncture, short-circuit or exposure to temperatures up to 350 °C). Its drawbacks include low energy density (approximately 150 W-h/kg) and large capacity attenuation at low temperature (the capacity at −10 °C is only about 55% of that at 25 °C).

Cobalt in ternary materials is a relatively rare metal with limited reserves on earth, justifying the high price of batteries containing it. In contrast, raw materials for lithium iron phosphate batteries are abundant and less costly, but also less space-efficient. The energy density of the current lithium iron phosphate battery has reached its upper limit—however there is still much room for improvement for the energy density of ternary batteries. Several manufacturers have started R&D in this direction.

Tesla's early Model S adopted lithium cobalt oxide batteries made by Panasonic of Japan. The size of a single battery was only a little bigger than the common AA battery. Despite its inconspicuous appearance, the battery pack used a series–parallel connection that allowed it to release a remarkable amount of energy. The lithium cobalt oxide battery features mature technology, high energy density, high discharging current and high charging speed. However, it also has relatively poor stability under high temperatures, which leads to higher requirements for cooling and battery safety measures. As such, Tesla had to develop a complex battery protection program and a unique liquid cooling system to ensure the normal operation of the battery pack—no doubt at great cost for the consumer.

At present, lithium-ion batteries are mainly used in blade electric vehicles and plug-in hybrid electric vehicles. All major vehicle manufacturing countries in the world have lent their support to power battery R&D and active technical innovation to further improve safety and energy density of the battery and to lower the cost. Technicians hope to increase the energy density from the current 110–200 to 300–350 W-h/kg through their endeavors.

This optimistic anticipation towards the future development of the electric vehicle industry, battery enterprises all over the world are expanding their production capacity. Global investment in the lithium-ion battery industry in 2011–2014 ranged between $10 and 12 billion. By 2015, the production capacity exceeded 50 billion W-h. Investments from 2014–2017 for Tesla, BYD and CATL alone, were close to 7 billion US dollars. At the same time prices of EV batteries dropped in reverse correlation to the expansion of production capacity. The price of a single battery is now approaching 0.8 RMB/W-h, and that of a battery system is approaching 1.2 RMB/W-h.

Two categories of lithium-ion power battery products are currently mass-produced by various companies. The first category of battery is assembled using small standard cylindrical batteries (most typically model 18,650, but also 21,700, 26,650, 32,650, and 32,700). In general, multiple batteries in the form of series–parallel connection are required to reach the total voltage and capacity required. This may consist of several thousands of single batteries connected in a complex form. The second category consists of batteries with high capacities of tens of ampere-hour or greater. The battery is encapsulated with aluminum-plastic film or welded metal casing. Thanks to the high capacity of the single battery, the number of single batteries in a module or a system can be reduced significantly, and the connection simplified as well (Fig. 6).

At present, Germany, USA and East Asia (China, Japan and Korea) have been recognized as established areas for the R&D and industrialization of batteries. In the long run, East Asia—that is, China, Japan and Korea—will become leaders in both technology and market for small lithium ion batteries for consumer electronics. In addition, the production of lithium-ion power batteries will be concentrated in these countries as well. From the industrial perspective, Japan retains the leading position in terms of technology, but Korea has surpassed Japan for largest market

Fig. 6 The power battery co-produced by Bosch and Samsung SDI

share. However, China ultimately boasts the highest number of battery enterprises, and the largest production capacity.

In China, the development of battery materials has been accelerating, performance steadily improving, and cost significantly reducing. Safety research for single battery, battery pack and battery pack management has also been promoted. An ecosystem for lithium-ion battery sector exists, including key raw materials (the positive pole, negative pole, membrane and electrolytes, etc.), power batteries, system integration, demonstrative applications, recycling, production equipment and basic R&D. Technologies for design formulation, structural design and manufacturing processes of batteries have also been mastered. Additionally, production lines are now transitioning from semi-automation to fully-automated mass manufacturing technology.

Technical levels of power battery cells in China.

The energy density of mass-produced energy mode lithium iron phosphate power battery with square aluminum alloy casing of 10, 20, 86, 120 and 200 Ah is 120~140 kWh/kg.

The energy density of mass-produced 3265 cylindrical lithium iron phosphate power battery of 5 Ah can reach up to 130 kWh/kg.

The energy density of mass-produced, combined lithium manganate battery (with mixed Ni, Co and Mn ternary materials) of 35 Ah can reach up to 135 kWh/kg.

The energy density of mass-produced lithium manganate battery (mixed Ni, Co and Mn ternary materials) of 6.5 Ah can reach up to 81 kWh/kg.

The energy density of mass-produced, 18,650 cylindrical cell-powered battery of 2.6 Ah made of ternary materials can reach up to 200 kWh/kg.

The energy density of mass-produced energy type power battery with capacities of 6, 20, 28, 30 and 45 Ah made of ternary materials can reach approximately 180–220 kWh/kg.

With the rapid development of electric vehicles, Chinese lithium-ion power battery enterprises delivered products with total capacity of 30.5 GWh in 2016, an increase of 79.4% from 17.0 GWh in 2015. Although China's new-energy vehicle market has been considerably impacted by the uncertainty of the subsidy policy in 2016, analysis has shown that the production and sales volume of the entire new-energy vehicle market managed to exceed 500,000 units. Among the 30.5 GWh shipments in 2016, 59.1% were used in purely electric buses, making it the biggest sector consumer of lithium-ion power batteries in 2016. Approximately 25.4% of the batteries were used in blade electric passenger cars. It is anticipated that the ratio of batteries used in the passenger cars will increase gradually, making passenger cars a major consumption segment for lithium-ion power batteries.

55.8% of the total shipments of lithium batteries in 2016 were delivered by the top-three enterprises, with 27% by BYD, 20.5% by CATL and 8.3% by Optimum Nano. This reflects a high degree of industrial concentration. In November 2016, the "Regulations for the Vehicle Power Battery Industry (2017)" (draft for comment) was published in the official website of the Ministry of Industry and Information Technology (MIIT). The guideline strengthens the management of the manufacturing enterprises of vehicle power battery. It is anticipated that the Chinese lithium battery industry will be polarized and integrated, with growth concentrated in the enterprises with extensive production capacities, sufficient cash flow and continuous technology improvement. As such, we shall witness an improvement in R&D and industrialization technologies of lithium power battery sector in China, including the development of system integration technology.

In addition, graphene, which features strong electrical and thermal conductivity, will help to improve the energy density of power batteries. However, the technical progress of the entire system should be considered throughout development, as the factors are interrelated. In addition, design optimizations should be done in parallel with the electric vehicle itself to ensure all targets correlate. For example, the battery should be optimized structurally in consideration with the vehicle's aerodynamic design, to ensure performance during the high temperatures of summer.

Beyond the battery itself, the battery management system (BMS) is also considered a key component to battery application. The BMS monitors and establishes battery status, protects the battery, reports data, and balances the system. The main tasks of BMS in the vehicle include the following:

Protect the cell and battery pack from damage.

Ensure the battery works within a proper voltage and temperature range.

Coordinate the needs of the vehicle with battery capacity, to ensure proper conditions for battery function.

Detection of battery parameters: including total voltage, total current and single cell voltage detection, temperature detection, insulation detection, collision detection, resistance detection, and smoke detection.

Establishment of battery status: including state of charge, state of health (SOH) and state of function (SOP).

Online diagnosis: diagnosis of sensor fault, network fault, battery fault, overcharge, over discharging, over current, insulation fault, etc.

Battery safety protection and warning: including control of temperature control system, high voltage control, and electric leakage protection. Any fault detected will be reported to the vehicle controller and charger by the BMS, and high voltage systems will be disconnected to protect the battery from damage.

Charging control: control of slow charge and boost charge.

Control of battery consistency: single cell voltage information will be collected to maintain battery consistency via equalization (through dissipation or non-dissipation methods).

Thermal management: record temperatures at various areas of the battery pack and determine whether to activate heating or cooling functions.

Network function: including on-line calibration and health check, on-line program downloading. In general, CAN network will be used.

Information storage: Store key data, such as SOC, SOH, AH of charging and discharging, fault code, etc.

A BMS is used by EV manufacturers to regulate a third party's power battery according to its own design standards, and match the function design of the vehicle. Several aspects including chemical safety, structural safety, electrical safety, functional safety related to the power battery are considered when designing the vehicle. Furthermore, various tests including cell impact, cell squeezing, module column press, module flat press, needling, module forward impact, system forward impact and side impact, burning, etc. are conducted to ensure safety. The final objective is to deliver a certified, safe and reliable electric vehicle to the consumer. The above work is particularly important, as 29 fire incidents occurred with new-energy vehicles in China in 2016.

BMS is usually developed independently by an EV manufacturer. In the case of NIO, XPT is equipped with a dedicated production line for the battery pack in addition to its independent supply of motor and electronic control systems. In the future, XPE will supply the battery management system to several models of NIO vehicles, and provide customized battery management system solutions to the electric vehicle manufacturers at home and abroad.

3 Charging and Battery Changing: Supplementing Energy to the Electric Vehicle

In terms of quantity, the electric vehicle has undergone an exceedingly rapid development. However the market share of EVs remains negligible when compared with traditional vehicles. A shortage of charging facilities is a key contributor to this situation, as the construction of infrastructure serves as a limiting factor for the development of electric vehicles. (Charging facilities refer to battery charging and changing stations.)

3.1 Charging

Charging is impossible without charging piles, which functions similarly to gas station fuel dispensers. Charging piles can be installed on the ground, on the wall,

Fig. 7 The charging piles in a charging station. *Photo* Tesla

in public structures such as buildings, shopping malls, or public parking areas, as well as parking lots in communities and stations (Fig. 7). In addition, charging piles should be made available for various electric vehicle models, by voltage level. These piles connect to the AC grid directly, with the output end equipped with charging plugs for electric vehicles. In general, two charging modes are available at these piles: conventional slow charging and boosted charging. Charging piles can be divided into DC charging piles, AC charging piles or DC/AC integrated charging piles.

Charging piles for the electric vehicles are controlled via the embedded processors. User authentication, balance enquiry, billing enquiry and other functions are available to the users via charging cards and self-served operations on the HMI interface. In addition, voice output interface is also provided for voice interaction. Users may choose 4 charging modes from the LCD on the pile, from billed-by-time charging, billed-by-electricity charging, automatic full charging, or billed-by-mileage charging.

At the end of 2015, China issued five standard protocols for charging interfaces and communication. These new standards were officially implemented in January 1, 2016, clearing some obstacles for the construction of charging piles. However, due to low overall quantity of electric vehicles and charging piles and their unbalanced distribution, situations such vehicles not locating charging piles or vice versa, have happened very often. This has resulted in financial losses for nearly all enterprises in the charging pile business. However, we foresee a gradual mitigation of such situations, with the increase of electric vehicles.

Although enterprises have incurred financial losses for charging piles, the future of the electric vehicles remains bright. National policies are supportive, and we still see countless investments pouring into charging infrastructure, and the accelerated construction of charging piles. In 2016, a total of 100,000 public charging piles were newly built throughout China, for a grand total of 150,000. China became the fastest country in building charging piles. The quantity of dedicated charging piles for residents has simultaneously increased. The number of special charging piles for residents has likewise increased, and the installation proportion of matching piles with vehicles has reached 80%. Inter-city boost-charging stations are available on expressways longer than 14,000 km, with an average interval of 48.6 km between the charging stations. The average radius of charging service for the electric vehicles in major cities like Beijing and Shanghai has been reduced to 5 km; public charging network in Shenzhen, Guangzhou, etc. are being planned and implemented rapidly. We are quickly moving towards the target of 5000 m. As stated by the National Energy Administration (NEA, China worked to construct 800,000 new charging piles in 2017, among which 700,000 are dedicated charging piles and 100,000 are public charging piles.

Globally, the construction of charging piles in USA, France, Germany and Japan is witnessing a rapid growth too. Most of the charging piles in USA are private ones. With their eyes fixed on the future, Benz, Qualcomm and TIA are all working on the R&D of wireless charging stations for electric vehicles, in which electric vehicles will be charged in the form of wireless induction through electromagnetic charging coils installed at the roadside and in the vehicles. This is a charging method that saves space and improves charging efficiency significantly.

3.2 Battery Changing

In addition to charging piles, there is another way of charging electric vehicles, namely battery changing. Battery changing has its advantages; namely, the battery can be changed within 3 min and battery changing is not limited by location. Battery changing can be performed through manual and automatic operation, and battery leasing and delivery services are available as well. Moreover, the service life of the battery can be prolonged by professional and delicate control when charging at a station. However, most automakers choose to turn a cold shoulder to battery changing. This is due to many factors, including the involvement of vehicle manufacturers and management by third party battery changing station (the price of the battery is almost half of that of an EV), and the standardized specifications and parameters of the battery system put in time by manufacturers when designing the vehicles. Another issue is the high cost of battery changing stations. Lifan Motors, which launched the concept of "energy station", invested more than 30 million RMB yuan to build the first battery changing station for electric vehicles in Chongqing (whereas standard domestic charging stations only cost tens of thousands RMB). Tesla previously tried battery changing, but ultimately stopped due to its prohibitive cost.

Chapter 5
Infrastructure: Echoing with Autonomous Driving

There is no doubt that autonomous driving technology is not only an upgrade from traditional vehicles, but also a significant feat of engineering. This shift involves not only the transformation of the automobile itself, the reconstruction of social infrastructure, the extensive testing, the promotion of laws and regulations, and even the changes in the behavior of driver and traffic participants. This wide range of issues describes a process by which autonomous driving is gradually accepted by society and the public.

In the autonomous driving programs of the 1990s, the vehicle could only be driven on a dedicated road equipped with signal transmitters. Today, autonomous vehicles use equipment such as radars, LiDAR, and cameras to identify surrounding vehicles, traffic signals, road markings and other information, to guide their own movement. Therefore, clear and intact road markings are especially important for autonomous vehicles. Even if a camera could identify the edges of road markings in broad daylight, it might not be capable of doing so at night or on a rainy day. Some cars are also capable of perceiving posted speed limits, but only if the signs are clearly identifiable. Potholes can also cause other vehicles to swerve, a potential hazard to autonomous vehicles nearby. These potholes also cause damage to the vehicle's tires and suspension, so it makes sense to increase investment in road maintenance.

Elon Musk, CEO of Tesla, spoke to reporters about the impact of infrastructure on the development of autonomous vehicles in October 2015. He said that the lane markings on Interstate 405 near Los Angeles International Airport were unclear. A driverless car cannot drive or change lanes safely if it cannot distinguish the lane. Andrew Ng, Chief Scientist of Baidu, published an article in Wired magazine arguing that for driverless cars to see real-world applications, appropriate changes to the current infrastructure must be made.

© China Machine Press, Beijing and Springer Nature Singapore Pte Ltd. 2021
Z. Chai et al., *Autonomous Driving Changes the Future*,
https://doi.org/10.1007/978-981-15-6728-5_5

Extended Reading

What is "infrastructure"?

The first economist to extensively discuss infrastructure was Paul Rosenstein-Rodan, who divided the total social capital of a country or a region into two categories: Social Overhead Capital (SOC) and Private Capital (PC) in 1943. The former refers to infrastructure. A. Hirschman called the latter "Directly Productive Activities" (DPA). This classification has persisted. Later, H. Chenery and Moises Syrquin clearly divided the social industry sector into two additional parts: the tradable and non-tradable sectors. The former includes the primary product sector and manufacturing and the latter includes social infrastructure and services. These social infrastructures include construction, water, electricity and gas, transportation and telecommunications industries.

McGraw-Hill's 1982 Encyclopedia of Economics provides a more detailed explanation of the infrastructure. According to the book, "infrastructure refers to economic projects that directly or indirectly increase output levels or production efficiency, including transportation systems, power generation facilities, telecommunication facilities, financial facilities, education and sanitation, and an organized government and political system."

Transport infrastructure includes physical networks, terminals and intermodal nodes, information systems and refuelling and electrical supply networks which are necessary for the safe, secure operation of road, rail, civil aviation, inland waterways and shipping.

The future of autonomous driving and transportation rests on upgrading the entire transportation industry. The era of autonomous driving requires a similarly revolutionary renewal and change in transport infrastructure. As unmanned driving is much more dependent on a well-developed road infrastructure, this issue is one on which we must rely on government investment and policy guidance.

1 Transportation Informatization and Intelligent Transportation System

1.1 Transportation Informatization and Autonomous Driving

Since the popularization of urban initiative that prioritizes public transportation, developed countries have integrated modern technology into urban transportation. Telecommunication, Internet and sensors improve the overall reliability, safety, efficiency and convenience of the transportation system, alleviating urban traffic congestion and reducing exhaust pollution.

BMW believes that the significance of this digital trend on traffic and travel is felt across the globe, creating opportunities to make travel easier, safer and more convenient, as well as opportunities to reach new customer groups.

Andrew Ng made an in-depth analysis on the mutual promotion of autonomous driving and transportation systems. He believes that the current driverless cars are not as good as human drivers in the following aspects:

- The driverless cars of today cannot perceive human hand gestures that tell the car whether to go or stop.
- When the sun is positioned behind a traffic light, most cameras are unable to recognize the color of the light due to glare.
- When human drivers see an truck emblazoned with "wide turns," we know how to adjust accordingly. Human drivers may slow down when noticing a child distracted by an ice cream car, because we know that they might run out into the street. Today's computers do not have the same ability to understand such complex situations.

Of course, driverless cars also have some advantages:

- No blind zone, as they maintain 360° observation range.
- No drunk driving, and no distraction during driving. Except for maintenance, they are able to work 24 h a day.
- They have much shorter reaction time than humans.

Because computers perceive the world differently from humans, they also drive cars differently. Artificial intelligence is making great progress. However, it is unrealistic to expect computers to drive like human beings in the near future. At present, our road system is built according to the human driver's defaults. We must to adjust accordingly so that it can support both computer-driven and human-driven automobiles.

Andrew Ng suggests setting up wireless beacons or Apps for driverless cars and guiding them through electronic signals, in lieu of having a human gesturing on the side of the road. Construction plans can be reported in advance, giving driverless cars time to prepare. Emergency service vehicles also need new way of communication, as sirens and flashes have been designed only for human drivers. At the same time, driverless cars should be unique and recognizable, like cars with "new driver" signs, and tell others to treat them differently.

Trains perform badly in terms of obstacle avoidance, but they are safe because they operate in a predictable way. Unmanned vehicles do not have a track to follow, but we should still strive to make their actions predictable. As their behavior may be different from that of manned vehicles, it is crucial for driverless cars to clearly convey their intentions to surrounding vehicles. Therefore, more research on standardized traffic lights and signs is necessary. Well-maintained roads also play a key role to enable predictability. Clear lane markings allow both humans and computers to drive safely on designated lanes. Poorly maintained roads not only make it hard to handle vehicles, but also reduce predictability for both humans and computers.

Andrew Ng further elaborates that multiple traffic signals should be set up inter-sections so that no matter what angle the sun is at, at least one signal can be clearly seen by the camera. Barring that he advocates for the development of a more sensitive camera. A public-private partnership involving auto makers, technology companies and public institutions would be the best way to ensure that a safe and effective transportation system is designed.

Earl Blumenauer, a US congressman, described the challenges of autonomous driving in Wired Magazine. He was especially interested in the impact that upgrading the transport system would have on infrastructure and opportunities. He said that autonomous driving would not only impact the millions of drivers, but also affect car sales, repair and insurance. The development of driverless vehicles might also make downtown parking lots unnecessary: the standard 12-ft-wide (3.7 m) parking spaces would be redundant, as a self-driving car would only need 7 ft (2.1 m) of space. Public areas could be turned into houses, green spaces, or public squares. The new challenge is how to plan for these changes. Automatic driving will require extensive government revenue. Yet unless we start planning now, we will pay the price. Without reliable revenue, the government and parliament would not be able to plan related transportation projects, build new infrastructure, or even maintain the status quo.

Extended Reading

Alternative to fuel tax

The state of Oregon has been exploring alternatives to fuel taxes in attempt to intro-duce a more accurate and fair way for drivers to pay for the roads they use. In a pilot project, one volunteer installed a device in his car that recorded the distance traveled and the amount of oil used. It also generates a bill based on the real-time travel distance. This data collection could become standard in the future, and become a payment mechanism suitable for paying insurance, parking, and tolls. This mech-anism will also enter the market, with fees and charges changing based on the traffic congestion level. During rush hour, users will be subject to a "congestion fee," as it would be unreasonable for drivers to pay the same rate for driving around big cities in rush hours as for driving on rural roads in suburbs on Sundays. Oregon and California have begun pilots of "congestion fees". It can recoup the government's fuel tax losses, allowing it to develop existing infrastructure, and make congestion and travel charges more flexible. This platform will inspire detailed plans to allow autonomous, semi-autonomous and traditional vehicles to coexist peacefully on the road. Most importantly, a "vehicle miles traveled" (VMT)-based alternative to the fuel tax would provide a more stable and detailed source of tax revenue.

The US Department of Transportation (DOT) hopes to establish a model city for policy exploration to enable the future popularization of autonomous vehicles. This requires the joint efforts of public sectors, private sectors, and public communities to generate public interest. Anand Shah, Vice Chairman of American think tank Albright Stonebridge Group, believes that discussions should concern issues such as emission reduction, economic affordability, reduction in public spending, security and potential benefits to consumers. Only by doing so could the United States seize the greatest opportunity of the twenty-first century. Policy development should consider the following five aspects.

1. Recognize that unmanned vehicles hold an obvious environmental advantage.
2. Consumers must benefit from the purchase of unmanned vehicles, which must also be economically sustainable.
3. Transportation policies should help reduce the cost of transport infrastructure for individuals. In 2014, US governments at all levels invested $416 billion in transportation infrastructure. In the future, unmanned vehicles must maximize the use of existing infrastructure and increase existing road usage.
4. Unmanned vehicles must make the community safer. Tens of thousands of people in the USA die each year from traffic accidents—most of them due to driving errors. Studies show that unmanned vehicles can greatly reduce casualties of drivers, passengers and pedestrians. The impact may be even greater.
5. Unmanned vehicles must also bring time savings and asset appreciation to Americans. In 2014, Americans spent a total of 6.9 billion hours in traffic jams. Autonomous vehicles and shared traffic modes would effectively alleviate congestion and reduce waste. More efficient use of vehicles could also increase the value of real estate, as parking lots currently account for one-third of US urban areas.

1.2 Intelligent Transportation System and Autonomous Driving

The popularity of autonomous vehicles will place new demands on Intelligent Transportation System (ITS). ITS is a general term for a transportation system that utilizes traffic information systems, communication networks, positioning systems, and intelligent analysis and route selection to ease road congestion, reduce traffic accidents, and improve the convenience and comfort of traffic users. By using real-time traffic information, travelers can easily understand upcoming traffic environment and make choices accordingly; by eliminating traffic hazards such as road congestion, we could build a traffic control system that reduces environmental pollution. We could also improve driving safety and reduce travel time through the development of intelligent intersections and autonomous driving technologies. In fact, the distance-based fee proposed by US Congressman Earl Blumenauer is in line with ITS design.

As our environments become more connected in general, ITS will play an ever-more important and central role in our cities, towns, suburbs and rural communities, between regions and across borders. The transportation system as a whole can best serve vital needs when it is using technology to its fullest potential and enabling transportation system managers to effectively "connect the dots" of information from various factors that affect transportation operations (e.g., weather, planned special events and response to unanticipated emergencies).

The concept of intelligent transportation was established in 1991 by the Intelligent Transportation Society of America (ITS America). The basic principle is to use existing macro traffic facilities (roads, bridges, tunnels, etc.) to integrate roads and vehicles to solve traffic problems systemically. ITS effectively integrates advanced information technology, data communication technology, sensor technology, electronic control technology and computer technology into the entire transportation management system. Thus establishing a large-scale, real-time, accurate, efficient and comprehensive transportation and management system. The main purposes of ITS is to improve traffic capacity, to improve the safety and speed of traffic, and to save energy in doing so. The ITS America formulated the "Intelligent Transportation System Strategic Plan" in May 1992, which drew a blueprint of intelligent transportation design in the United States for the next 20 years. The plan completely covers all aspects of intelligent transportation and divides ITS into six areas: Advanced Traffic Management System (ATMS), Advanced Traveler Information System (ATIS), Advanced Vehicle Control and Safety System (AVCSS), Commercial Vehicle Operation System (CVO), Advanced Public Transportation System (APTS), Agricultural Transportation System (RTS).

In 2014, the US Department of Transportation Intelligent Transportation Systems Joint Program Office proposed the "Intelligent Transportation Systems (ITS) Strategic Plan 2015–2019", which clearly defined the direction for the development of intelligent transportation in the United States in the next five years. The plan accomplishes the following: (a) Identifies a vision, "Transform the Way Society Moves,"; (b) Outlines technology lifecycle stages and strategic themes articulating outcomes and performance goals that define six program categories; (c) Describes "Realizing Connected Vehicle Implementation" and "Advancing Automation" as the primary technological drivers of current and future ITS work across many sectors; (d) Presents enterprise data, interoperability, ITS deployment interdependent activities critical to achieving the program's vision.

On this basis, the United States has proposed the development of future transportation systems through research, development, education and other means to promote the practical use of information and communication technology. This would ensure the deployment of intelligent transportation equipment and the development of intelligent transportation technology. This proposal put forward the following five development strategy goals:

Firstly, enable safer vehicles and roadways by developing better crash avoidance for all road vehicles, performance measures and other notification mechanisms; commercial motor vehicle safety considerations; and infrastructure-based cooperative safety systems.

Secondly, enhance mobility by exploring methods and management strategies that increase system efficiency and improve individual mobility.

Thirdly, limit environmental impacts by better managing traffic flow, speeds, congestion, and using technology to address other vehicle and roadway operational practices.

Fourthly, promote innovation by fostering technological advancement and innovation across ITS programs, continuously pursuing a visionary/exploratory research agenda, and aligning the pace of technology, development, adoption and deployment to meet future transportation needs.

Finally, support transportation system information sharing through the development of standards and system architecture, and the application of advanced wireless technologies that enable communications among and between vehicles of all types, the infrastructure, and portable devices.

Reading between the lines, we notice that the strategic plan put more focuses on two major development themes: "Realizing Connected Vehicle Implementation" and "Advancing Automation". The United States set up the Mobility Transformation Center (MTC) to conduct large-scale demonstration test of intelligent and connected vehicles (ICV). The USDOT dominates the ICV development, which is expected to introduce the V2V mandatory regulations in 2020. General Motors has planned to install V2V equipment for its cars after 2017.

Extended Reading

Cloud real-time parking map

If a car can be connected to the smart home or smart city through the cloud, connected technology will turn the car into a smart assistant at the wheel. While connected with the surrounding environment, cars also play an important role in interconnected cities. For example, a car equipped with a community parking function also becomes a detector for parking spaces. While the car is driving, it will detect the distance between two cars parked on the roadside. The data is transmitted to a digital street map, to be evaluated by advanced algorithms to verify the parking space. Cloud services can integrate this data to create real-time parking maps, saving drivers a lot of time and money while reducing parking pressure. Currently, Bosch is planning to launch a series of pilot projects in the United States in 2017. Bosch is also conducting a community parking pilot in Stuttgart, Germany in cooperation with Mercedes-Benz.

The European Commission has proposed the "ITS Development Action Plan" and the "Strategic Transport Research and Innovation Agenda (STRIA)" to carry out road coordination, active safety, road safety systems and traffic informationization in the field of transport safety. Since 2002, the EU has invested 180 million euros in research projects for the Cooperative Transport Systems. Based on the results of these projects, the European Telecommunications Standards Institute (ETSI) and Comité Européen de Normalisation (CEN) developed the basic standards for vehicle information connectivity in February 2014, in accordance with the requirements of the European Commission. The standard ensures that vehicles produced by different companies can communicate with each other and with the road infrastructure, or make each other aware of traffic accidents ahead in the course of driving. At present, European countries are carrying out comprehensive application development work, and plan to establish dedicated traffic (mainly road traffic) wireless data communication network throughout Europe, and are developing technologies including Advanced Traveler Information Service (ATIS), Advanced Vehicle Control System (AVCS), Advanced Commercial Vehicle Operation System (ACVO), and Advanced Electronic Toll Collection System (ETC).

Japan is the area with most extensive ITS integration in the world. ITS in Japan started with research and development from the early 1970s, then was framed as a full-scale initiative when the five government ministries and agencies, including National Police Agency, Ministry of International Trade and Industry, Ministry of Transport, Ministry of Posts & Telecommunications, and Ministry of Construction, then involved in ITS affairs formulated the Comprehensive Plan for Intelligent Transport Systems (ITS) in Japan (the Comprehensive ITS Plan) in July 1996. From that point on, ITS was advanced as a national project through the active collaboration of industry, the government, and academia. Japan's ITS communication system includes roadside facilities communication technology, road-to-vehicle communication technology, in-vehicle communication technology and vehicle-to-vehicle communication. The key communication technology used is Dedicated Short-Range Communication (DSRC). This has led to the popularization, with a scale of 40–50 million users till now, of system services such as the Vehicle Information and Communication System (VICS), the Electronic Toll Collection System (ETC), and Advanced Safety Vehicles (ASV), and realized improvements in the safety and convenience of road transportation. This ITS approach makes it possible to use highly advanced roads, and to reduce the burden of driving, walking, and other such road use. It also achieves quantum leaps of improvement in the safety of road transportation, transport efficiency, and comfort, while also contributing significantly to environmental protection by facilitating the smooth flow of traffic through the amelioration of congestion. The ITS initiative does help people realize lives of true prosperity and vigor in these and other similar ways.

In 2015, Singapore government put forward the Smart Nation plan, hoping to enhance the competitiveness of the whole country through the construction of digital infrastructure. Sensors throughout the city will connect people, buildings and vehicles. Urban data will be collected on a unified platform—Virtual Singapore—through a high-speed network. Infrastructure (such as intelligent traffic lights) can be dynamically adjusted after processing and analysis.

With the rapid development in this field, China has built more advanced ITS facilities in Beijing, Shanghai and Guangzhou. Among them, Beijing has established four ITS subsystems: road traffic control, public traffic command and dispatch, expressway management, and emergency management; Guangzhou has established three ITS subsystems, namely, shared traffic information pillar platform, logistics information platform and static traffic management system. China's intelligent transportation system should develop further if we draw lessons from foreign country's mature practices, strengthen the coordination among various regions and functional departments, pay attention to the formulation of unified plans and objectives, realize the seamlessly sharing of traffic information collected by transportation, urban management and public security departments, and give preferential consideration to talents, funds and other resources.

Vehicle connectivity and intelligent transportation system are catalysts to promote the development of autonomous driving technology and popularize its applications. The maturity of autonomous driving and unmanned vehicle technologies will eventually build an urban intelligent driving ecosphere, provide new solutions for future traffic and travel needs, and inject fresh blood into the formation of a new traffic mode. The autonomous driving and intelligent transportation system have powerful capabilities and can provide us with more information about how people use the transportation system. Nowadays, the roads are very crowded, and people spend more and more time on the roads. Any solution that shortens time spent on people's daily journeys may be warmly welcomed.

2 Timely Responses from International Institutions and Governments

2.1 America: Ensuring Leadership

On August 31, 2016, the Michigan Senate Economic Development and International Investment Committee convened to hear testimony and consider Senate Bills 995 till 998, collectively referred to as the SAVE Act, the legislation serves to update Michigan's existing regulatory framework regarding connected and autonomous vehicle (CAV) technology development and deployment and keep the state on the leading edge of a global effort to redefine the future of transportation. All four bills were reported out of committee without opposition.

Testimony focused primarily on the increased safety and enhanced mobility aspects of connected and autonomous vehicle technology, and the economic development impact it will have on Michigan. The bill package positions Michigan at the forefront of the next major transition in the automotive industry. By updating the existing law, the SAVE Act "catches up" with CAV technology and encourages continued development in Michigan without government interference, creating critical jobs. As Michigan's automotive dominance is increasingly under attack from competing states and countries around the world, the bill package is a significant step forward in determining how CAV technology will be defined, incentivized, deployed, and regulated in the twenty-first century. A brief overview of the four bills:

- Allow for an automated motor vehicle to be operated on a street or highway. Previously, automated motor vehicles could only be operated on a street or highway for research or testing purposes.
- Allow for research or testing of automated vehicles/technology/driving systems on a highway or street without a human operator in the vehicle. Previously, an individual had to be present in the vehicle.
- Create the Michigan Council on Future Mobility.
- Establishes a grant-eligible Mobility Research Center at Willow Run in Ypsilanti Township and excludes it from certain provisions that regulate private roads that are open to the general public.

- Exempts a mechanic or vehicle repair facility from liability for damages to an automated motor vehicle, as long as the repairs were made in accordance with the original manufacturer's specifications.

Michigan's cooperative approach is a testament to the type of collaboration that is necessary to lead in CAV technology development and serves as a model for other states—and the National Highway Traffic Safety Administration (NHTSA)—to emulate.

The laws of Michigan have criminalized intrusion into an automotive system as a felony, and even explicitly stipulated that once an autonomous vehicle accident occurs, the automaker and supplier(s) should be responsible. In addition, the NHTSA has taken some power over autonomous driving laws from the states, and related laws differ between states. For example, a person holding a common driver's license is allowed to drive an autonomous vehicle on the road in Florida, while New Jersey requires the person to take a special driving license. In California, car companies or technology companies only need to submit application materials, prepare a $5 million insurance coverage and have a driver/operator training plan if they want to apply to test autonomous driving. As each state goes its own way for drafting laws, companies will always find a suitable state to conduct their autonomous driving projects.

This is just an example of the positive attempt made by the United States and its local governments to legislate on autonomous driving. The factors that restrict the rapid popularization of autonomous driving include technical maturity, cost, and laws and regulations, the last aspect being the biggest obstacle. Auto makers are likely to build autonomous cars with superior performance, but thousands of autonomous cars will eventually be sent to consumers and driven legally on the road. The government's public sector needs to make complex policy adjustments, and the entire policy framework must be adapted to autonomous driving technology. Autonomous driving technology may be able to create a better city, reduce traffic pollution, road congestion and traffic accidents, and make the city safer and more prosperous, but past examples have shown that few policy improvements can keep pace with technological progress.

The Obama administration in the United States was one of the most energetic governments to promote autonomous driving. They saw autonomous driving technology as another opportunity for the country to lead the global scientific and technological revolution, but one that was likely to escape them. For this reason, they spared no efforts in promoting policy changes to adapt to new technological advances. For example, in 2016, the U.S. Department of Transportation announced that it would invest $4 billion to promote commercial development of driverless vehicles. Anthony Fox, then U.S. Transportation Secretary, called for renewed attention to America's crumbling transport infrastructure, while emphasizing that technological advances would radically change the way people travel and the way people think about transportation in the future. Technology would be first used by those who can afford it, then gradually reach the masses over time. In such an era and in such an industry where most of the investment is driven by the government, it is important to consider how autonomous driving technology can benefit people from all sectors of society.

How will autonomous vehicles be integrated into the existing transportation system in the USA? What safety regulations will apply? In September 2016, the NHTSA and the U.S. Department of Transportation issued the Federal Automated Vehicles Policy which set forth a proactive approach to providing safety assurance and facilitating innovation. This is the first regulation on the automated driving, which answers the above questions to some extent. The policy mainly includes the following four parts.

- Guidance for manufacturers, developers, and other organizations outlining a 15-point "Safety Assessment" for the safe design, development, testing, and deployment of highly automated vehicles;
- A model state policy, which clearly distinguishes Federal and State responsibilities and recommends policy areas for states to consider, with a goal of generating a consistent national framework for the testing and operation of automated vehicles, while leaving room for experimentation by states;
- An analysis of current regulatory tools that NHTSA can use to aid the safe development of automated vehicles, such as interpreting current rules to allow for appropriate flexibility in design, providing limited exemptions to allow for testing of nontraditional vehicle designs, and ensuring that unsafe automated vehicles are removed from the road; and
- A discussion of new tools and authorities that the agency could consider seeking in the future to aid the safe and efficient deployment of new lifesaving technologies and ensure that technologies deployed on the road are safe.

The policy was warmly welcomed by the industry. Ford said in a statement: "This policy helps to establish a national regulatory framework to promote the safety of automated vehicles. Ford looks forward to working with the states to achieve this goal." Although this policy is of great significance to the development of US autonomous driving industry, it is still far from perfect. Some have argued that the policy emphasizes too much the federal government's management of safety technology standards, which may weaken the authority of the states; there were also concerns about the possible overregulation. The DOT and NHTSA were also aware of the shortcomings in this policy and indicated that the government would soon communicate with all parties to refine and rationalize various aspects, and might launch a new version within one year.

The current US President Trump's commitment to traffic construction during the campaign is $1 trillion. In a sense, the ability to revitalize America's infrastructure would be one of the key factors for President Trump to "Make America Great Again", which will undoubtedly make the new U.S. DOT Secretary appear more under the spotlights. It is worth mentioning that the Trump administration has appointed Elaine Chao, a Chinese-American woman, as next DOT Secretary. After taking office, Elaine Chao set a "small goal" to carry out Trump's campaign promise to restructure the transport infrastructure of the country.

2.2 Japan is Back: Japan Revitalization Strategy

Japan proposed in the "Japan Revitalization Strategy" (Revised 2015): "In order to achieve the development of technology before level 4 (high automation), it is necessary to conduct empirical tests and continuously verify the effects", and brew legislation for automated vehicles. In the first half of 2016, the Ministry of Economy, Trade and Industry (METI) of Japan established a research team and decided to cooperate with the auto companies in maps, communications, ergonomics and other fields to achieve the goal of testing self-driving cars on public roads by 2020. In May 2016, Japan has developed a roadmap for autonomous driving, indicating that autonomous vehicles (with drivers) will be allowed to travel on highways by 2020. Regarding the responsibility for accidents caused by self-driving cars, the Japanese government is in the process of revising relevant laws and regulations such as the Road Traffic Law and the Road Transport Vehicles Act, and plans to discuss the compensation mechanism for accidents involving self-driving cars.

In May 2016, the Japanese National Police Agency formulated the Guideline for Highway Empirical Testing Related to Autonomous Driving. This policy brings forward issues concerned with road safety and smooth autonomous vehicle tests, and describes its viewpoints on issues related to auto makers, test vehicles, testers (especially test drivers), accident handling measures, and compensation guarantees so as to ensure the road test done reasonably and safely in the following ten aspects. The purpose, the basic system, the basic responsibilities and obligations of the implementer, the security measures consistent with the contents of the highway empirical test, the requirements of the test driver, the requirements of the autonomous driving system related to the test driver, the recording and preservation of all kinds of data of the test vehicle during test, the measures during the occurrence of traffic accidents, ensuring the ability of compensation and informing the relevant agencies in advance.

2.3 Germany: One of the Pioneers

Germany is home to many world-famous auto makers such as Volkswagen, Daimler and BMW. It is also one of the countries in the world that has paid much attention to the automated vehicles and their tests from the very beginning. A consortium of large German auto makers and spare parts suppliers, led by Mercedes-Benz, began to conduct connected vehicle experiments in the Rhine-Maine area as early as 2011, when 120 interconnected cars around Frankfurt were tested for the practicability of the new technology. In 2013, the government approved Bosch to test its autonomous driving technology on a German highway. Later, companies such as Mercedes-Benz and Audi carried out field tests on self-driving cars in the environment of German motorways, urban traffic and rural roads. At the beginning of 2017, the ministries of transport of Germany and France jointly decided for the first time to allocate a specific sector on the cross-border highway connecting the two countries for the testing of autonomous vehicles. The road runs 70 km from Merzig, Saarland in western Germany to Metz in eastern France. It uses 5G wireless network to communicate between test vehicles as well as test vehicles with roadside infrastructure, aiming at "testing true cross-border autonomous driving".

Since the success of the test, Germany's research on automated vehicles has gone one step further. Beginning in June 2014, the German automotive industry launched a discussion on whether or not to install black boxes on automated vehicles. In July 2016, the minister of the Federal Ministry of Transport and Digital Infrastructure said that the country's planned legislation requires the car manufacturer to install a black box for its autonomous car, recording when the autonomous driving mode is activated, when the driver will take part in the driving, and when the autonomous driving system requires the driver to take over the driving, so as to help identify the responsible party for an accident. In October of that year, the minister wrote to Tesla, asking the company to stop publicizing its car's ADAS function using the term of "Autopilot" in order to avoid misleading drivers into neglecting road conditions and causing danger.

2.4 France, Britain and the Netherlands: With Open Mindsets and Firm Paces

In order to achieve industrial revitalization, France launched the strategy of "New Industrial France" in 2013, focusing on the development of 34 industrial sectors, including driverless cars. In February 2014, France announced the development roadmap for driverless vehicles, investing 100 million euros, focusing on the development of driverless cars in three years, and began field test of self-driving cars in 2015. Major French auto makers and spare parts suppliers such as Peugeot Citroen, Renault and Valeo participated in the project. In 2016, in order to ensure this smooth development, France announced that it would realize the vehicle connectivity along thousands of kilometers of roads nationwide and promote the revision of road traffic laws and regulations to meet the requirements of driverless vehicles on the road. In August of that year, the French government formally approved foreign automakers to test autonomous vehicles on the road. Prior to this, only local auto companies were allowed to perform this type of field test. Cross-border field tests with Germany in early 2017 also brought the French auto industry back into the spotlight.

The UK has been studying autonomous vehicles for several years, and Oxford has developed a semi-self-service smart car that allows passengers to drive autonomously when needed. According to the British Science Minister, the car is cheaper than Google's driverless car. In 2015, the British government announced that it would invest £100 million in research and development in the field of smart cars. In February 2016, the first round funds of £20 million was invested in eight projects, one of which involved China's Huawei. At the end of November 2016, the British government announced an additional £100 million for infrastructure testing. The country is also one of the first in the world to encourage the development of smart connected vehicles at the legal and regulatory level. In July 2016, the Britain's Trade Secretary and Transport Secretary publicly stated that the country would remove regulations that restrict the autonomous driving of vehicles, including traffic rules and other policies and regulations that drivers must follow. The Ministry of Science and the Ministry of Transport have begun to study the case. At the same time, the British government will make appropriate amendments to the high-speed traffic laws and regulations to ensure the safe use of the ADAS functions like changing lanes on the highway and remotely controlled car parking. In addition, the UK is discussing and revising the insurance regulations with a view to achieving autonomous driving on the road by 2020. In terms of autonomous driving insurance business operations, Adrian Flux has taken the lead in introducing an insurance policy which contains several exclusive insurance clauses for self-driving cars.

The Netherlands is very supportive and friendly towards self-driving cars. The Netherlands re-examined traffic laws to facilitate a 5-year-period large-scale test of self-driving trucks on roads in 2014. In January 2016, the world's first self-driving shuttle bus was on the road in the Netherlands, making it the first country to drive a self-driving passenger vehicle on the public road. In July 2016, Mercedes-Benz launched a test of self-driving passenger car on the road in the Netherlands. The car successfully completed a 20 km journey, setting a new test record.

2.5   International Institutions: Amending Regulations to Escort Technology

In February 2014, the European Union launched a project named "Adaptive" (Automated Driving Applications & Technologies for Intelligent Vehicles) with more than a dozen European auto makers and spare parts suppliers, aiming to develop partially or fully autonomous driving vehicles on urban roads and highways. The project was expected to last for three and a half years and could receive EU funding of 25 million euros. The "Adaptive" project is headquartered in Wolfsburg, the home of VW Group in Germany. In addition to technical research and development, the "Adaptive" project also studies standards and road traffic laws and regulations that match driverless vehicles.

The 1949 Geneva Convention on Road Traffic and the 1968 United Nations Convention on Road Traffic (the Vienna Convention) require that "drivers shall at all times be able to control their vehicles…", but countries have different interpretations of this. For example, the general view in the United States is that autonomous vehicles are legal, while Germany interprets this to mean highly autonomous cars are illegal. In 2016, the United Nations ratified the amendments to the Vienna Convention, clarifying the autonomous driving techniques that delegate driving responsibility to the vehicle, if in full compliance with United Nations vehicle management regulations or where a driver can manually choose to replace or deactivate the function, can be applied to transportation. At least in the signatory countries, theoretically speaking, this amendment can be followed to give a green light to autonomous driving technology, but it cannot become a nationally enforceable regulation unless the specific signatory country "localizes" this amendment. In other words, autonomous driving test in a country still requires an application and is executed at a given time and venue. Nevertheless, this amendment bill is the first to ratify autonomous driving at the legal level.

Subsequently, it was reported that the UN had set up an expert group to prepare technical standards for autonomous vehicles. This expert group included experts from Japan, South Korea, Germany, France, the United Kingdom and the European Commission. They planned to reach a wide range of agreements for self-driving vehicle's technical standards. USA, which had initially planned to develop autonomous driving regulations separately, officially announced to participate in the autonomous vehicle safety program mission being developed by the United Nations in November 2016. It is understood that this safety program, though not legally binding, would be included in the official UN documents in the future and provide a reference when countries formulate relevant regulations. The first step of this globally unified autonomous vehicle technology standard was to enable the task of overtaking and merging on the highway without the need to operate the steering wheel before 2018. What they are currently doing is the autonomous driving security standard, which is designed to prevent hackers from attacking the communication network used by the autonomous driving system. The World Forum for Harmonization of Vehicle Regulations (WP.29) under UNECE Sustainable Transport Division adopted the Automated Driving Cyber Security Standard in November 2016, which mainly includes measures to prevent hackers from attacking, to warn drivers when detecting attacks and to prevent them from losing control.

In addition, South Korea and Singapore from Asia, and Sweden and Finland from Europe have already begun to modify regulations so that autonomous vehicles can be tested on specific roads, thus paving the way for large-scale applications.

2.6   China: With a Road Map Ready, Standards are also Required

Looking back at China, people's work is not "all the routines." China's overall plan for intelligent and connected vehicles(ICV) began in October 2014, when the Ministry of Industry and Information Technology (MIIT) commissioned the China Association of Automobile Manufacturers (CAAM), the China Society of Automotive Engineering (SAE-China) and the National Automobile Standardization Technical Committee (NASTC) to conduct research. In 2015, the China State Council issued the "Made in China 2025" initiative, including the ICV in the key areas of national intelligent manufacturing development in the next decade. It clearly points out that China must master the overall technology and key technologies of ADAS by 2020 and master the overall technology and key technologies of autonomous driving by 2025. In 2015, the "Standard System Construction Plan for Intelligent & Connected Vehicles" (first edition) was introduced. At the end of October 2016, the "Technology Roadmap of Intelligent & Connected Vehicles" was released to guide the research and development of auto makers and provide support for future policy formulation.

At present, the ICV standard system framework drafted by the MIIT has been formed. The framework includes four main parts: foundation, general norms, product and technology application and related standards. The foundation and general norms refer to the basic standards for connectivity with great commonness; product and technology application involve specific design standards, which are the backbone of the framework, including details of information collection, decision-making and alarms, and vehicle control, while the related standards involve information exchange, communication protocols and connection interfaces. At the same time, the ICV Technical Sub-Committee begins to establish, with a plan to recruit experts from relevant fields to jointly study and formulate relevant technical standards; the development of specific standards for ADAS and information security is also accelerating. Although the standard system framework for ICV has been completed since June 2016, it may take a long time to study and draft specific standard rules due to tedious cross-department coordination involved.

China's auto industry regulator said that it was working with the Ministry of Public Security to formulate laws and regulations on road testing for self-driving cars. Before the release of the regulations, car companies are prohibited from conducting autonomous driving tests on expressways. Under this restriction, it is still a matter of time before the self-driving cars can be measured on the domestic roads. However, through cooperation with developed countries with loose regulations, China's self-driving cars can also be developed, manufactured and tested abroad.

3 Debate Over V2X Communication Standards

With the development of autonomous driving and connected vehicles, the concepts of "Internet of Vehicles" and "ICV" have been repeatedly mentioned. In fact, the Internet of Vehicles, a noun created by the Chinese and a concept derived from the Internet of Things (IoT), is similar in the meanings including Connected Vehicles and Vehicle Networking. In China, the "Internet of Vehicles" has been equated with "Telematics", and a vehicle is regarded as a simple information sending and receiving node. This understanding only sees the role of Internet of Vehicles in the field of providing information services. It is just a one-sided understanding of "Internet of Vehicles".

For a period of time in the future, autonomous vehicles will share the right of way with a human driving a car. The single vehicle intelligence is the necessary foundation, but the future of autonomous driving is not a single vehicle in battle. With the popularity of 5G communication networks, the Internet of Vehicles will enrich the technical connotation and ecology of autonomous driving and amplify its role. The ultimate form of autonomous driving—unmanned driving—will be the perfect combination of single vehicle intelligence and vehicle networking. Although the automated vehicle and vehicle networking are not one and the same, high grade self-driving cars will be connected. The Internet of Vehicles is a kind of Internet of Things. The Internet of Vehicles is a large system network that carries out wireless communication and information exchange between vehicles and X (V2X), while X refers to vehicle, road, pedestrian or Internet, based on the "triple network integration" of the in-vehicle network, the inter-vehicle network and the on-board mobile Internet.

V2X is a security system for bidirectional and multi-directional communications, similar to wifi or cellular mobile network, which can improve traffic safety by enabling important information such as location, speed, barriers, hazards, and various multimedia applications between vehicles, vehicle and pedestrian or vehicle and traffic lights and other infrastructures, and enabling networked interactivity control through big data and cloud computing. This wifi- or cellular data-like connection is V2X communication technology. The V2X complements the ADAS system, such as radars and cameras, providing information about vehicles and obstacles outside the visible range. With V2V, a component of V2X, vehicles can exchange information about position, speed and driving direction with each other, to alert the driver or vehicle of coming danger, avoid collision in the case of low visibility, and provide blind spot assistance. V2I, another component of V2X, can use cameras, radars and traffic infrastructure to achieve similar functions. It can charge automatically, sign in-car and provide traffic information. The comprehensive benefits of V2X technology are to provide a safer, faster and pleasant driving experience, improve traffic flow, help drive to save fuel and improve air quality.

V2X wireless communication guarantees security, efficiency and environmental protection, but it also triggers a series of unique networking problems that require low-latency, reliable data transmission in poor wireless communication environments. Since consumer electronic devices are usually fixed or moving at a slower speed, and cars are moving at high speeds, (often close to or apart from each other) it creates a Doppler effect on wireless communications. In addition, rain, fog, snow and terrain can create additional obstacles to the patency of transmission. For active road safety applications, such as cooperative avoidance of collisions and other serious accidents, delays should vary from 50 to 100 ms. Successful V2X wireless standards need to continue to meet this standard under various environmental conditions.

3.1 DSRC: US-Led, Ten Years of Sharpening

More importantly, if the V2X communication protocols of different brands of vehicles are different, the identification and communication between vehicles, infrastructure and the Internet will not be possible. Therefore, the establishment of V2X communication standard is the most important part of the popularization of vehicle networking. At present, there is no unified communication standard between vehicles in the world. In 1992, the American Society for Testing and Materials (ASTM)

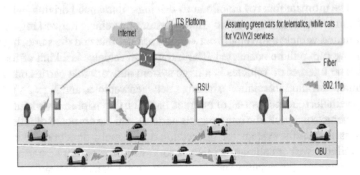

Fig. 1 Schematic diagram of DSRC communication based on 802.11p

first proposed the standard for dedicated short-range wireless communication facilities, namely Dedicated Short Range Communications (DSRC), as shown in Fig. 1. DSRC is an efficient wireless communication technology that provides high-speed data transmission and guarantees low latency and low interference on the communication links. DSRC can ensure a common interoperable safety standard for vehicles, regardless of size, make and model, to help avoid crashes, optimize traffic flow, and reduce congestion. Subsequently, the EU and Japan also launched their own DSRC communication standards, but there are differences in the division, definition and use of frequency bands.

DSRC enables vehicles to communicate with each other and other road users directly, without involving cellular or other infrastructure. Each vehicle sends 10 times per second its location, heading and speed in a secure and anonymous manner. All surrounding vehicles receive the message, and each estimates the risk imposed by the transmitting vehicle. Risks are defined as "safety applications" such as Left Turn Assistance (LTA), Intersection Movement Assistance (IMA) and many others. DSRC was designed for maximal cyber security. The receiving vehicle validates the authenticity of the received messages. The messages aren't linkable to the vehicle, don't expose its identity and therefore don't violate the driver privacy.

It is understood that vehicles equipped with on-board units (OBU) and road-side units (RSU) can realize inter-vehicle communication (V2V) and vehicle-to-road infrastructure communication (V2I) through DSRC technology. DSRC can realize the recognition and two-way communication of moving targets under high-speed motion in a specific small area (usually tens of meters); DSRC can transmit image, voice and data information in real time, realize bidirectional communication of V2I, V2V and V2P. DSRC is widely used in applications such as ETC non-stop toll collection, access control, fleet management, and information service, and has advantages in vehicle identification, driver identification, information interaction between road network and vehicle, and vehicular ad hoc network. Even in adverse weather conditions, DSRC enables high-speed transmission, low latency, fast networking, robust security and high connection reliability.

The DSRC operates at 5.9 GHz and uses the IEEE 802.11p standard at the physical and MAC layers. The 802.11p standard is designed for automotive wireless communications and it is based on 802.11a but with some improvements to make it more suitable for automotive environments. 802.11p changes the bandwidth from 20 to 10 MHz and the output to 27 Mbit/s. By halving the bandwidth and doubling the symbol duration and time guard interval (TGI), the signal stability minimizes its Doppler effect and improves multipath propagation, making it suitable for high-speed vehicle communication in all weather conditions. Designed for automotive communications, 802.11p can also create favorable conditions for extremely short duration messaging. Connections and exchanges of information can be established without the need to join the Basic Service Set (BSS), eliminating the need for acquisition and authentication (authentication at the upper level of the protocol stack), enabling low latency and fast networking.

DSRCs in Europe and the United States share 802.11p at the physical and MAC layers, but differ in their handling of networking, transport, and application layers. Although different methods are used for message routing and channel assignments, both the European ETSI ITS G5 standard and the US WAVE standard can provide stable security and privacy measures, and prioritize communications based on security-related and non-security-related information types. The key to the new DSRC product is the new u-blox THEO-P1 compact V2X transceiver, which achieves a transmission rate of 27 Mbps over a distance of up to 1 km and is suitable for OBU and RSU applications. These transceivers are typically compatible with the US WAVE and the European ETSI ITS G5 standards and can perform in-vehicle hardware encryption acceleration to provide the necessary high-speed encryption for V2X communications.

Thanks to the support of Toyota and General Motors, DSRC standard has become the communication protocol of V2X in North America and Europe. In Europe, DSRC has been used for charging and is planned for V2X communications. The European Telecommunication Standards Association (ETSI) plans to adopt a phased approach, starting with the implementation of the "Day One" deployment of ITS G5, which is now being carried out at the Cooperative ITS (C-ITS) pathway, from Rotterdam, Netherlands through Frankfurt, Germany, to Vienna, Austria. This pathway provides drivers with compatible OBU information about upcoming road works, traffic conditions and road hazards. North America has also conducted extensive research and field testing of the DSRC. The permanent DSRC pilot project is currently underway in New York, Wyoming, and Tampa, Fla., and it has invested $42 million to install DSRC equipment for cars and infrastructure in these areas. It is expected that DSRC will improve pedestrian safety in urban areas, reduce traffic congestion and improve fuel efficiency. The relevant authorities could improve traffic management and infrastructure planning based on the results of the trials. Due to the strong support of NHTSA, DSRC is expected to be a key V2V technology in USA.

3.2 LTE-V: China Program, Freshly Baked

In fact, there have been two camps on the establishment of vehicle-to-vehicle communication standards, one being the DSRC solution and the other the LTE-V solution.
LTE-V (Long Term Evolution-Vehicle) is a communication solution led by Huawei
and Datang Telecom in China. It realizes vehicle communication based on 4.5G
technology and uses LTE cellular network as V2X-based vehicle network proprietary
protocol to meet the needs of active vehicle safety, driving efficiency and multi-scene
vehicle entertainment. LTE-V adopts "Wide Area Cellular Networks (LTE-V-Cell)
+ Short Range Direct Communication (LTE-V-Direct)" (Fig. 2). The former is based
on the expansion of existing cellular technology and supports large bandwidth and
large Coverage communication, mainly carrying the traditional vehicle networking
services, the latter introduces end-to-end direct LTE D2D (Device-to-Device), realizes vehicle-to-vehicle (V2V), vehicle-to-infrastructure (V2I) direct communication,
and carries vehicle active safety services, which mainly meets the requirements of
terminal security with low latency and high reliability. In the future, LTE-V can
evolve to 5G technology smoothly.

Vehicle active safety is the core application scenario of LTE-V. The 27 typical
applications given in the requirements specification issued by 3rd Generation Partnership Project, which is dedicated to the mobile communication standardization,
are mainly based on V2V and V2I's active safety services. LTE-V will help set sail
for autonomous driving: based on LTE-V2X, a real-time, efficient and reliable two-
way information interaction and sharing between vehicles, vehicle and road as well
as and vehicle and pedestrian can be realized, intelligent collaboration achieved,
vehicle active safety realized, and driving efficiency improved. In September 2016,
at the 73rd meeting of 3GPP, the V2V standard of LTE-V was officially frozen in
Release 14, which marked that 3GPP completed the first phase of LTE-V, namely
V2V standardization based on D2D mode. Through in-depth research, more optimized physical layer DMRS (Demodulation Reference Signal), resource scheduling,
interference coordination and other technologies were introduced. V2V is the core

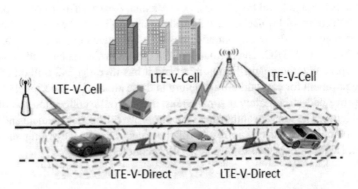

Fig. 2 Schematic diagram of application scenarios of LTE-V-Cell and LTE-V-Direct

of LTE-V, and the freezing of V2V core protocol is of great significance. The second phase will complete the LTE-V standard, including cellular-based V2V, V2I, V2P (vehicle-to-pedestrian) and so on. From the perspective of global development, legislators in Germany, Britain and France stipulate that 5.9G spectrum can be used for LTE-V V2X. Overseas telecom operators such as Vodafone, Deutsche Telekom, Orange, NTT Docomo, and Softbank, auto makers such as Volkswagen, and Land Rover, and auto electronics manufacturers such as Bosch, Harman, Denso, and Continental, all have shown a strong interest in LTE-V. NTT Docomo and Denso have announced that they would jointly study LTE-V. Meanwhile, the research of 5G is further consolidating LTE-V.

Throughout the ITS field, the US-led DSRC standard has lasted for more than 10 years and may steal some thunder. It is also regarded by many auto companies as a core technology of the future internet of vehicles. However, as LTE-V's real-time problem has been completely solved with its fast development, and the industry has gradually come to see that LTE-V stands at the forefront.

In terms of latency, the DSRC-based 802.11p system with short-range communication at 5.9 GHz has a latency of more than 100 ms, while LTE-V is typically 50 ms with less latency jitter. Also, V2X communication requires more system capacity. The system capacity is usually measured by the density of the traffic that can be accommodated, that is, the number of vehicles that the communication system can access in a road per unit length at a given moment. According to the simulation test of Datang Telecom, LTE-V-Direct increases by 257% and 39% respectively compared with unicast and multicast in terms of traffic density. Compared with DSRC, LTE-V has a significant improvement in anti-interference ability due to the intervention of base stations, and an incomparable advantage in vehicle-intensive scenarios (such as in traffic jams). Its value has been recognized by some mainstream telecommunication operators, automotive enterprises and automotive electronics manufacturers. In addition, LTE-V has a better support for high-speed vehicle movement, and LTE-V-Direct has a significantly higher packet transmission rate at each transmission distance. According to the simulation test by Datang Telecom, the coverage of both V2X communication solutions is less than 1 km, and the reliability of LTE-V-Direct data packet transmission rate is improved by 15–25% over DSRC's in the coverage range. At the same time, LTE-V also supports scenarios requiring remote information transmission because of mobile network support. Note that LTE-V is geared towards 5G and is of great significance for autonomous driving. Not only that, LTE-V does not need to rebuild the network because of the extensive coverage of existing mobile communication networks, thus saving a lot of manpower, material resources and time.

Despite having been invented for more than ten years, DSRC has not yet been fully developed due to some factors. By contrast, we can find that LTE-V technology has great value. LTE-V, driven by unified standards, strong policies, communication industry chain and the automotive industry chain seeking innovation, may come to the fore. However, as the LTE-V standardization process and related technology development has not been completed, and DSRC has been developed and deployed by the USA and other countries for many years with rich experience, it is hard to say who wins till now.

Chinese enterprises have accumulated numerous patents in the field of telecommunications, and they have certain advantages in the development of LTE-V. It is expected that a new competitive advantage will be formed in the Internet of Vehicles industry through the combination with vehicles. The Chinese government has approved the 5.9G spectrum dedicated for the LTE-V V2X direct communication test, opening the door to LTE-V industrialization, and making China the first country to plan the LTE-V direct communication dedicated test spectrum worldwide. China Mobile, SAIC, Changan Automobile, Huawei, Datang Mobile and industrial chain related companies are establishing cooperation and jointly researching and promoting LTE-V technology. At the 2016 ICV V2X Technology Summit, Datang Telecom demonstrated its newly launched pre-commercial LTE-V products like the on-board unit (OBU) and the roadside unit (RSU).

In addition to technical accumulation, China's active pilots are of great significance, such as the demo in Hangzhou, Zhejiang. At present, Zhejiang Mobile has planned a total of 34 LTE-V pavement sites, including 33 small and micro stations and 1 macro station. Upon completion, it will become the world's largest LTE-V test network. The entire system will provide drivers with traffic light speed guidance, early warning of pedestrian-vehicle collision, intersection collision avoidance reminder, and lane change assist reminders, according to the signals sent by surrounding vehicles, providing effective auxiliary means for safe driving. This indicates that the test network has initially achieved its intended goal, showing low latency and high reliable access performance, and providing a good information interaction guarantee for vehicle safety assistance.

In February 2017, it was reported that China's formulation of V2X related standards had entered a late stage, and a development schedule for ICV expected to be launched. Around the Technology Roadmap of ICV Development, the MIIT entrusted SAE-China to further refine the specific application technology route. Specifically, "connectivity" is divided into three levels: the first level, or the interaction of the networked auxiliary information, similar to the current information service mode, can be realized through 3G and 4G networks; the second level, or the networked cooperative perception, which completes the information exchange between vehicles and the outside world through V2X, can help vehicles make decisions and control; and the third level, or the networked cooperative control, through V2V, V2I, etc., can achieve coordinated control of the vehicle and the outside world.

Of course, the popular use of V2X still faces many challenges. The industrialization of V2X requires the joint promotion of the government, scientific research institutions, enterprises and other parties. China need strive to lobby the 2019 World Radiocommunication Conference (WRC-19) to accept China's program, promote China's technology to the world and seize the frontier market. As a result, the working frequency of ITS will achieve global unification.

3.3 V2X: Linking Intelligent Transportation to Smart Cities

The value of V2X depends on its popularity. If the V2V coverage of vehicles on the road is not 100%, its significance will be greatly reduced, and V2I will require a lot of infrastructure investment. If V2X is to achieve full coverage, the governments play an irreplaceable role. As with any large-scale infrastructure investment, V2I deployments cannot begin with full coverage at the outset, but start with a partial approach, because such investments must account for business returns. Just like Europe's Cooperative ITS (C-ITS) route from Rotterdam, the Netherlands to Vienna, Austria via Frankfurt, and Germany, China may be able to deploy V2I on several main highways, such as the G2 Beijing-Shanghai Expressway and the Yangtze River Delta city belt expressways to improve the road utilization, and get a better input-output ratio data.

V2X vehicle networking brings upgrades to intelligent transportation, which will bring a variety of application scenarios such as autonomous driving, smart travel and logistics integration.

People-vehicle-road synergy in the context of autonomous driving: With the development of smart cars, vehicles can now achieve assisted driving, and some vehicles can achieve full autonomous driving under test conditions. It is predicted that driverless vehicles will be an important part of the future. Considering that there are also non-intelligent vehicles, it is necessary to set up driverless lanes on highways or urban roads and improve the transportation efficiency by controlling and reducing the safe distance between driverless vehicles to avoid the impact. Some experts think further and suggest that urban roads should be rectified, pedestrians and cars should be diverted thoroughly, intersections should be abolished, and various measures such as one-way lanes and speed limits should be adopted. In order to adapt to the realization of driverless driving, the ITS will manage the people, vehicle and road collaboratively, and realize the efficient operation of the smart city by means of the vehicle networking technology.

Traffic facilitation in the context of smart travel: The emergence of autonomous driving will bring innovation in mobile applications. In an era of intelligent transportation, car rentals will become more and more popular, and private car purchase will become scarce. When a person travels, they can reserve a driverless vehicle through their mobile phone. This method is not only convenient and efficient, but also solves the problem of urban parking. According to the investigation and test abroad, this method can save 80–90% of parking space.

Transportation collaboration in the context of logistics integration: The ITS in the field of logistics, including vehicle distribution, transportation coordination and sharing of real-time information, will develop in a coordinated manner. At present, active safety prevention and control technologies like the real-time tracking of GPS are widely used. Next, it will develop several trends of traffic system safety status identification, emergency response and rapid joint action technologies. In addition, the study of traffic status and active safety assurance technology are also the future development directions.

Intelligent transportation systems (ITS) are central to the Smart City. At this stage, it is mainly reflected in the rapid growth of hardware facilities such as chips, optical fibers and sensors. As smart transportation gradually takes shape and becomes an important part of the smart city, the value of the whole industrial chain will shift to the service side. Mobile communication data, parking data, toll data, and meteorological data are gathered on a smart city platform to form effective big data. These big data are stored, calculated and analyzed by a cloud computing and artificial intelligence platform. Finally, according to the output of the computing platform, the various facilities of the city are automatically managed to provide intellectualized and personalized services for the citizens.

Extended Reading

HD map of infrastructure

If the map is to support autonomous driving, the map data may not be two-dimensional abstract data, but HD map data, including details such as curvature, slope, and height of the road, so that the amount of storage and transmission will increase greatly. If you consider a country, the amount of data and the cost of collection are horrible. If you take the map data needed for autonomous driving, to compare with that of the ordinary navigation map in the past, the growth is not several times, but tens of times, which is an exponential growth.

Fifteen Japanese companies, including Toyota, Honda, Nissan and Mitsubishi Electric, have jointly funded a foundation to draw high-definition 3D maps for driverless vehicles throughout the country, under the authorization of the Ministry of Economy, Trade and Industry of Japan. The project is called "Dynamic Map Planning". Mitsubishi Electric is responsible for the implementation, and has prepared hundreds of test vehicles for this purpose. These vehicles are equipped with LiDARs, cameras and GPS. The captured data is processed to form a digital map with an accuracy of 10 cm. They will start with the drawing of the 30,000 km highway in Japan.

In preparation for the implementation of autonomous driving technology, there is a difference between China and Japan in ideas and practices: one is busy building a backbone communication network, and the other is busy drawing HD maps. China feels that map work could be handled with ease, considering their expertise, while communication network establishment is a race against the clock. Japan sees HD maps as indispensable for any level of autonomous driving, and the government cannot support it. Japanese companies have formed a de facto autonomous driving technology alliance that share not only maps, but also other autonomous driving system technologies.

In the grand picture of the smart city, apart from the tremendous changes in transportation mode, the infrastructure of the entire living environment requires huge funds to transform or upgrade, and the city needs a destructive innovation. This process cannot be done overnight. At the same time, the resources released in the process of transformation, such as land, buildings, and time, will generate new wealth in some form, but the initial amount of funds needed for transformation will certainly create "a psychological shadow that cannot be waved". It is even possible that after the popularization of smart cars, people will no longer be allowed to drive themselves, because they will become obstacles to driverless cars. Future road reconstruction, like lane markings to be replaced by radar or camera recognizable, will also be expensive.

4 The Test Grounds that Have Sprung Up

As the saying goes, "actions speak louder than words": the ever-maturing autonomous driving needs verification through field testing. This requires a venue. A venue for autonomous driving testing and commissioning is an important part of the infrastructure at this stage. More and more geographic "coordinates" (for test or operational purposes, self-use or as platforms) around the world are added to illuminate the journey towards autonomous driving.

The autonomous driving test ground is a test space which reproduces all kinds of road conditions and operating conditions encountered in the use of autonomous vehicles. It is used to verify and test the correctness of the software algorithm of autonomous vehicles. The test road is a kind of road which has been strengthened and modified through centralization, concentration and non-distortion, including normal road surface such as highway, urban road, rural path, and the bad roads which can cause strong bumps of automobiles. In addition, the test site will lays out infrastructure such as GPS base stations, telecommunication base stations, and intelligent traffic lights and so on, to provide a test environment for connected and automated vehicles and technologies. The vehicle test in the test field is more rigorous, scientific, and practical than a laboratory o general driving conditions.

4.1 Mcity, The World's First

The United States has established a small town called Mcity in Ann Arbor, Michigan, in which it operates the world's first purpose-built proving ground for testing the performance and safety of connected and automated vehicles and technologies under controlled and realistic conditions. The functional area of Mcity is illustrated in Fig. 3. The mysterious town controlled by environmental variables has been opened to the public. Figures 3 and 4 show the aerial view and a corner of Mcity. With autonomous driving constantly docking the market, the emergence of Mcity could have a profound impact on the technology.

The Mcity Test Facility sits on a 32-acre site on U-M's North Campus Research Complex, with more than 16 acres of roads and traffic infrastructure. The full-scale outdoor laboratory simulates the broad range of complexities vehicles encounter in urban and suburban environments. There are bridges, underpass, trees, poles, traffic signals, railway crossings, highways, intersections at different angles, blind curve, roundabouts and various buildings in the town. The layout is not much different

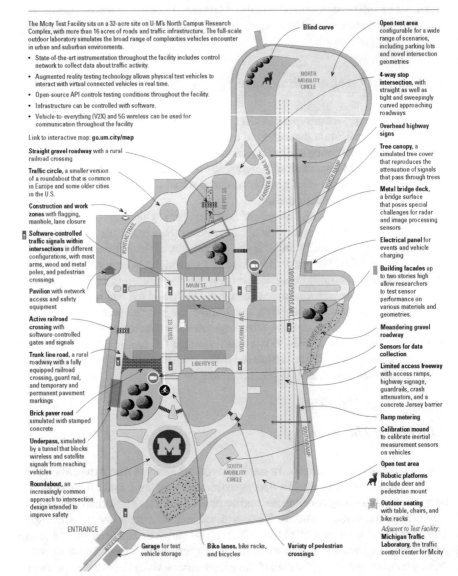

Fig. 3 An aerial view of the Mcity Test Facility. *Source* Mcity

Fig. 3 (continued)

Fig. 4 Building facades up to two stories high line the streets of Mcity's urban downtown area. Photo: Mcity

from that of ordinary American towns. The gap between the trees, the height of the streetlights, and the location of the underground parking lot have been strictly planned to fully simulate the actual road conditions. The calibration device for the vehicle inertial navigation sensor is placed in the open test area. Automated vehicles in Michigan will also undergo a variety of weather tests, from high temperatures and high humidity, biting cold to squalls, which are not always encountered in California's Bay Area. Severe weather is a big challenge for sensors that make automated vehicles possible. For example, hail is not good for LiDAR, while snow can hinder a camera's line of sight and fog can interrupt radar.

Although Mcity is nominally a test center that mimics the construction of ordinary towns, beyond its realistic traffic planning, it also accounts for unforeseen circumstances that may occur in daily life, such as graffiti-covered traffic signs and faded road markings can be seen everywhere. The entire town consists of two basic areas: a high-speed test area for simulating the highway environment; and a low-speed test area for simulating urban and suburban environments.

The design of Mcity Township was mainly carried out by the University of Michigan's Mobility Transformation Center (MTC), which was established in 2013. Mcity has been officially under construction since 2016, with a total startup capital of US$10 million for testing equipment and urban transport facilities, jointly funded by the University of Michigan and the Michigan Department of Transportation. MTC has already reached a cooperation agreement with 15 OEMs and suppliers including Ford, GM, Toyota and Delphi. In the coming three years, each company would invest US$1 million to develop and test autonomous driving and V2V/V2I technologies. The other 33 project affiliates each has pledged $150,000 in financial support. By doing so, universities, technology companies, and auto makers form industry alliances. While improving core technologies, R&D costs are also reduced. China's Changan Automobile and Guangzhou Automobile Group are also members of the MTC.

The University of Michigan has issued an official statement: Mcity will be open for free, with priority to project members, and teachers and students at the university. Mcity will build an experience learning base, TechLab, to expose students interested in practical applications to early technology and learn from it to develop practical technology in order to enable students to promote the innovation of ICV into the market. This kind of learning opportunity is available only at the University of Michigan.

It is worth noting that through the demo test in Mcity, the conclusion that the Internet of Vehicles technology can reduce traffic accidents by 80% has directly promoted the US government's announcement that it will force the installation of vehicle-to-vehicle communication system to improve driving safety. It is expected that the relevant compulsory standards would be implemented in 2020. At the same time, the development of this demo project has determined the United States' world leadership in the field of ICV development and standards drafting, thus has greatly promoted the development of its smart cars and related industries.

4.2 Willow Run Proving Ground—"American Center for Mobility"

Michigan's Willow Run test facility is only 15 km away from Mcity and covers an area of $1.2 km^2$, 10 times the size of the Mcity. Willow Run has a deep history of innovation and was initially built by Henry Ford as an advanced aircraft manufacturing facility, also known as Air Force Plant 31, to produce B-24 bombers during World War II. At the time it was noted as the largest factory under one roof in the world, offered equal pay for equal work and laid the groundwork for sweeping social change. It was later redeveloped by General Motors in the 1950s as a powertrain plant that operated until 2010. The Michigan State Government later spent $20 million to buy the site from GM and is now remodeling it into a large state-level proving ground for autonomous driving.

The Willow Run Proving Ground embodies Michigan's ambition to establish a national leadership in autonomous driving. With its special climatic conditions, abundant facilities and space to expand, and a strong atmosphere of auto industry, Willow Run is alluring to auto makers and technology companies. Google is very excited about testing in the area. Google is currently testing driverless cars in Mountain View, California and Austin, Texas. The tech giant hopes that driverless cars will be able to adapt to different road conditions, traffic scenarios and driving conditions. Michigan is not only rainy but also snowy in winter. If Google wants to work further with car companies, Michigan is a good choice. It is also known that Google also has plans to establish a R&D center in Michigan.

The Michigan State Government plans to link Mcity and Willow Run Proving Ground to create the world's largest test center for advanced autonomous vehicles. Among them, Mcity can be used for early research, and the product development stage can be carried out in Willow Run. Willow Run Proving Ground has been named "American Center for Mobility", operating as a global center for testing and validation, education, product and standards development related to connected and automated vehicles. The facility is open to private industry, government, standards bodies, and academia.

Extended Reading

Testing in highways

According to reports, in October 2016, HERE, which was acquired by BMW, Audi and Daimler for $3.1 billion the year before, announced that it would assist the Iowa Department of Transport in building a nearly 30 km long driverless highway on State Highway 380. Sources from HERE said that the Iowa DOT would use its open positioning platform, including real-time and predictive traffic maps, to create exclusive roads for autonomous vehicles, enabling them to communicate better with infrastructure and other vehicles. This is an infrastructure upgrade, and many experts believe that if unmanned vehicles are to make full use of their potential, such work will become a must. A large number of cameras and sensors will be installed in fixed *traffic devices, such as traffic lights and streetlights. Although companies such as Google and Uber have their own driverless car teams and have field tests in San Francisco, Austin, Pittsburgh and other places, HERE is the first company to explore unmanned freight vans on the road. In many ways, freight transportation is the best way to assess the commercial value of autonomous driving technology, because it requires careful review of driving routes, strict observation to timetables and close monitoring of surrounding objects and human drivers.*

At the beginning of 2015, the German government planned to start the test section on an expressway for unmanned vehicles. The test section was located on the A9 Highway in Germany. The test section would be used by vehicles equipped with ADAS or an autonomous driving system. The German government also uses Hamburg as an administrative division to promote autonomous driving testing and has launched a local ICV project. VW Group and the Hamburg Municipal Government will make every effort to promote the relevant cooperation plan.

4.3   Construction of China's Testing Grounds: Aiming High and Going All Out

In developed countries and regions such as the USA, Europe and Japan, cooperative assisted driving technologies based on V2I/V2V technology are developing applied technologies and going through large-scale field testing at proving grounds. Experts and scholars in China have also suggested that China to catch up. Prof. Li Keqiang of Tsinghua University urges that China should promote the construction of ICV demonstration zones and application demos, including standard construction. The ICV demonstration zones will greatly promote the maturity of relevant technologies and industrial development and provide an important basis for the Chinese government to make relevant decisions. At the same time, the establishment of the demonstration zones will also help establish the leading position of local enterprises and research institutions in the field of ICV in China, which also plays an important role in promoting regional economic development.

Previously, Tsinghua University, Tongji University and other universities cooperated with Changan Automobile to carry out the application research of cooperative vehicle infrastructure system (CVIS) technology under the national "863 Project", and small-scale demo tests were made. Various auto makers have also been conducting preliminary research. With the support of the MIIT, Shanghai International Automobile City, SAE-China, Tsinghua University, Tongji University and Shanghai Automotive Industry Corp (SAIC) began to build the National ICV (Shanghai) Demonstration Base, the first of this kind testing facilities in the country, in early 2015, aiming to promote the maturity and application of ICV technologies. The base's first phase—closed testing zone—was put into operation in Anting, Shanghai in June 2016.

It is understood that a GPS differential base station, 2 LTE-V communication base stations, 16 sets of DSRC RSUs, 4 sets of LTE-V RSUs, 6 intelligent traffic lights and 40 various types of cameras have been installed in the closed testing zone. The road identifies the centimeter-level positioning of the Beidou system and the full coverage of Wi-Fi, and completes the simulated traffic scenes such as tunnels, avenues, refueling/charging stations, underground parking lots, intersections, T-junctions, and circular roundabouts, which can provide a comprehensive test and meet functional requirements for autonomous driving vehicles. It can undergo closed tests in four areas of safety, efficiency, information services and new energy vehicle applications where a variety of test function scenarios, V2X communication technologies such as DSRC and LTE-V are adopted. At present, 200 vehicles from SAIC, Volvo, GM, Ford, Changan and many other automakers, have been tested there. Typical application scenarios include forward collision warning, emergency braking warning, lane change assist, blind spot warning, and intersection collision warning.

News released from the launching ceremony said that, by the end of 2016, 50 demo scenarios would be launched in the closed testing zone, simulating different road types such as expressways, urban road and country paths, natural environments such as rain, fog, ice, daytime, nighttime and artificial lighting, as well as various traffic scenarios such as pedestrians, non-motor vehicles and disturbing vehicles.

In addition to the first phase closed testing zone, the second and third phases of the plan are also in progress. Construction of the 27-km^2 Phase II open-road testing zone has begun. It was expected that by the end of 2017, the demo area would form 36 closed traffic scenarios around Boyuan Road and Moyu South Road in the core area of the Auto City, enable real-time measurements from 1000 vehicles within its 73 km of road, and finally build the first fully functional public service platform for ICV tests and demos. The third phase of the project was planned to be built between the end of 2018 and the end of 2019, and a pilot project of typical urban comprehensive demonstration zone would set sail, with its scope gradually expanded to 100-km^2 in the whole town of Anting, accommodating 5000 vehicles trial run. Thus a regional ICV test demonstration public service platform, based on the concept of smart city, would be built, an ICV industrial cluster would be initially formed to eventually become a regional ICV standardization industry base in China.

Shenzhen Chains Tech Co. and the University of Michigan introduced the Mcity project to Shenzhen, China in October 2016 and established a $27 million partnership program. Chains Tech funded the University of Michigan's autonomous driving technology research and established a Smart Car Joint Research Center, targeting the landing of "Urbanized Social Laboratory" that features a technological innovation transformation model, in Shenzhen. At the beginning of 2017, Chains Tech transplanted a similar concept to Zhangzhou, Fujian to create the "world's first city-state autonomous driving social laboratory" and claimed to set up a smart car industry fund of 10 billion RMB yuan. Similar projects and huge amounts of money are dazzling, and people simply cannot interpret what they indeed intend to do.

Before and after the G20 Summit in early September 2016, the MIIT and Zhejiang Province jointly led the construction of LTE-V Demonstration Zone in Yunqi Town, Hangzhou, which was built by Zhejiang Mobile, Huawei, SAIC, Alibaba, West Lake

Electronics, Zhejiang Supcon and other heavyweight manufacturers in the industrial chain. Subsequently, Chongqing Demonstration Zone Industrial Park, led by China Automotive Research Center and Changan Automobile, opened in Chongqing on November 15th, applauded by relevant units like China Automotive Engineering & Research Institute, China Mobile, China Unicom, Huawei, Datang, and Tsinghua University. At the same time, relevant organizations in Beijing, Changchun, Wuhan and other cities are also actively introducing LTE-V for testing and verification.

Extended Reading

Overview of the global autonomous driving testing grounds

Based on relevant reports from China Automotive News and other materials, the partial summary of the global autonomous driving testing grounds is shown in Table 1. From this table, we can see that China has taken the lead in terms of the

Table 1 Summary of partial global autonomous driving testing grounds (till December 2017)

Name/date in operation	Location	Investors	Unique features
Beijing Yizhuang Intelligent Vehicles and Smart Mobility Demonstration Zone (China)/January 2016	Beijing Yizhuang Economic and Technological Development Zone	Intelligent Vehicles and Smart Mobility Industry Joint Innovation Center/MIIT, Beijing Gov't, Hebei Gov't	The first large-scale intelligent vehicle test area located in Beijing, Tianjin and Hebei
National ICV (Shanghai) Demonstration Base (China)/June 2016	Jiading District, Shanghai/5 million m^2, covering the whole area of Anting Town and Hongqiao Transportation Hub by 2020	Shanghai International Automobile City Group	Promote the integration of CBUs and parts, electronic communications, scientific research, information services and demo applications, focusing on building six functional services through the construction of a third-party public service platform

(continued)

Table 1 (continued)

Name/date in operation	Location	Investors	Unique features
Panjin, Liaoning (China)/September 2016	Panjin, Liaoning Prov.		China's first unmanned commercial operation project
Wuhan Driverless Vehicle Test Park/December 2016	Wuhan Sino-French Ecological Demonstration Park	Dongfeng Renault	Allow visitors to visit and experience autonomous driving in the park
Wuhu Driverless Vehicle Operation Area, Anhui Province (China)	Near Chery Factory, Wuhu City, Anhui Province		An exclusive operation area built to serve Baidu's production plan for driverless cars. Plans to extend to the whole Wuhu Municipality in the future
Shenzhen Driverless Demonstration Base (Town) (China)/Second Half of 2017 (Planning)	Longgang District, Shenzhen or Shenzhen-Shantou Cooperation Zone, Guangdong	Chains Industrial Fund 10 billion yuan	Integrate production, education, research and urban construction, and explore new forms of society after the arrival of autonomous driving
China Intelligent Vehicle Integrated Technology Research and Test Center (China)/2018 (Planning)	Changshu, Jiangsu	China Intelligent Vehicle Integrated Technology Research and Test Center	The third-party intelligent vehicle evaluation organization composed of scientific research institutes and technology companies

(continued)

Table 1 (continued)

Name/date in operation	Location	Investors	Unique features
Zhangzhou City-state Autonomous Driving Social Laboratory (China)	Zhangzhou Development Zone, Fujian Province	Zhangzhou Development Zone, Fujian Yuanchuangli Intelligent Automobile Research Institute and Chains Industrial Fund	600,000 m^2 of closed test site, 2 million m^2 of Industrial Park experimental site and 56 km^2 of large-scale experimental site under actual traffic situations
Mcity (USA)/20 June 2015	Ann Arbor, MI/130,000 m^2	Mobility Transformation Center(MTC)/Univ. of Michigan, Ann Arbor, DOT, MI., and many partners/over $20 million	The world-renowned large-scale ICV testing site brings together senior talents from the automotive and science and technology fields. Its test scenarios are rich, and many related enterprises participate actively
Willow Run Proving Ground (USA)/2018 (Planning)	Michigan, only 15 km from Mcity/More than 1.2 million m^2	The Michigan State Gov't/$20 million in financial assistance and needs another $60 million to start the project	Close to Mcity, the area is ten times larger. Vehicles can conduct highway and long-distance driving tests
GoMentum Station (USA)	U.S. Navy Concord Base/20 million m^2	U.S. Navy	Near Silicon Valley, CA, where the U.S. Navy is responsible for security and high confidentiality. It is reported that Honda and the Apple team are testing here

(continued)

Table 1 (continued)

Name/date in operation	Location	Investors	Unique features
Castle Air Force Base (USA)/2011	Atwater, CA/240,000 m²	Google	The test site "customized" by Google Inc., which is "financially strong", is not open to other companies
Part of Pittsburgh City (USA)/September 2016	Pittsburgh, Pennsylvania	Uber, Volvo, Pittsburgh Gov't	The first US city which has turned on a "green light" for autonomous driving commercial operations (The Uber driverless fleet also operates in San Francisco, USA)
VW Driverless Special Test Zone (Germany)/September 2016	Hamburg, Germany	Hamburg Gov't and VW	
Autonomous Driving Transnational Test Area/2017	Merzig, Germany, to Metz, France, 70 km long	1 million euros as project fund from German Gov't	Test the ability to communicate for V2V and V2I, using 5G wireless networks
AstaZero Comprehensive Safety Technology Test Site (Sweden)/2014	Gothenburg, Sweden/240,000 m²	5 institutions and companies including Swedish SP Inst. of Technology, Charms Polytechnic Univ., Swedish Innovation Authority, EU Cohesion and Regional Development Fund/ 500 million Swedish kronor	Volvo and Scania conducted relevant technical tests here in the world's first test center for automotive safety
MIRA Proving Ground (UK)	3.5 km²	HORIBA of Japan/10 million pounds aid from UK Gov't	Simulating urban road environment

(continued)

Table 1 (continued)

Name/date in operation	Location	Investors	Unique features
Tsukuba Science City (Japan)/2019 (Planning)	Japan Automobile Research Institute (JARI)/150,000 m^2, Ibaraki Prefecture, Japan	Ministry of Land, Infrastructure and Transport, Ministry of Economy, Trade and Industry, JAMA, etc./34 billion yen	Japan's first large automated vehicle test site
Chiba Toyosuna Park (Japan)/August 2016	Chiba Prefecture, Japan		Japan's first unmanned bus line (The trial operation of EZ10 is also carried out in parts of the Netherlands, France, the United Arab Emirates and other countries)
Pangyo Autonomous Driving Road (Korea)/2017 (planning)	Gyeonggi-do, Korea/Two-way four-lane, about two kilometers long	Gyeonggi-do	
K-City (Korea)/2018 (planning)	Hwaseong, Gyeonggi-do, South Korea/360,000 m^2	Ministry of Land, Infrastructure and Transport/10 billion won	K-City has established five distinctive spaces with local transportation environment including the school area, bicycle lanes, automatic parking areas, highway and suburban road
A Pilot Program of Driverless Taxi Operation at One North (Singapore)/August 2016	One North, a business park in downtown Singapore/6.5 million m^2		One of the few autonomous driving projects in the city-state in Asia

number of plans. The boom in China's autonomous driving test fields has become overwhelming, but the USA is the undisputed leader in terms of research and test results. Considering China's huge market capacity and numerous automobile enterprises, there are reasons for such a big number in China. However, if the management and coordination is done poorly, analysis is not done carefully, and the development is carried out unorderly without thoughtful review of objective conditions, it will likely waste valuable funds, which may be detrimental to the overall development of China's autonomous driving sector. Only with both emotional enthusiasm and rational thinking will this project succeed.

Chapter 6
Top Ten Challenges Facing Autonomous Driving

In his 1980 book Stratified Flows, Yi Jiaxun, a Chinese-American hydromechanical scholar, used the phrase "the spring breeze rises, wrinkling blue waters" to preface his book's fourth chapter, entitled "Fluid Dynamics Stability". The phrase was written in calligraphy by his good friend, Feng Yuanzhang, a foreign academic of the Chinese Academy of Sciences, to a surprisingly refreshing effect. The instability caused by wind blowing across the water surface is the well-known "wind-generated wave" phenomenon in fluid dynamics. In some aspects, we can see the origins and impact of autonomous driving on society as being quite similar to these "wind-generated waves."

Despite the continuous development and maturation of technology under the staunch support of various enterprises, numerous challenges nevertheless impede the practical use of autonomous vehicles. Even the most basic legal issues become a philosophical quandary; if an accident happens in a driverless car, who is held responsible? How would insurance companies settle claims? Social acceptance is another important issue. What kind of paradigm shifts would drivers have to experience, in order to let go of the steering wheel? Are pedestrians ready to entrust their safety to machines? Can a hacker mar the reputation of autonomous driving by manipulating a few lines of code and causing accidents? What changes will autonomous driving bring to our lives? How will it affect our travel experiences? Will the era of dedicated car-sharing come? What about issues of personal privacy? These are just a few of the myriad questions facing the future of the automotive industry.

Technology always improves, as does human society; though these issues will be addressed over time, the process of change dictates the gradual implementation of roadblocks to success.

© China Machine Press, Beijing and Springer Nature Singapore Pte Ltd. 2021
Z. Chai et al., *Autonomous Driving Changes the Future*,
https://doi.org/10.1007/978-981-15-6728-5_6

1   Machine-Controlled World and Nowhere-To-Place Privacy

When science fiction author Isaac Asimov (1920–1992) put forth his "Three Laws Of Robotics" in 1942, there were no robots in the world. Nor was there such thing as robotic studies, or companies dedicated to the field. It wasn't until 1959, when engineers Joseph F. Engelberger and George Devol created the world's first industrial robot, declaring that robots have changed from science fiction to reality. Despite the continuous advancement of robotics, Asimov's "Three Laws of Robotics" was held fast as the "golden rules" of robotics. However, as the development, manufacture, and application of robotics has grown more complex, so has the applicability of these "Three Laws of Robotics."

In a broad sense, a robot can be thought of as an intelligent machine. This of course includes cars, especially autonomous ones, "car robots" if you will. The "Three Laws of Robotics" naturally applies to these autonomous vehicles. The challenges encountered in the development of robots will naturally also be reflected in the development of autonomous vehicles; in fact, they compose only the tip of the iceberg.

In Asimov's works, his has begun to worry that the world of machine control may pose a huge threat to humanity itself, thus, he formulated the "Three Laws of Robotics" as proscriptive rules governing what robots can and cannot do, according to a simple moral code. Asimov's Three Laws are as follows:

(1)   A robot may not injure a human being or allow a human to come to harm.
(2)   A robot must obey orders, unless they conflict with law number one.
(3)   A robot must protect its own existence, as long as those actions do not conflict with either the first or second law.

Asimov's "Three Laws of Robotics" has since become one of the guiding principles of modern robotics. He later added another rule to precede the others. Rule Zero states that "A robot may not harm humanity, or, by inaction, allow humanity to come to harm."

In "Sally," (a short story within the collection "I, Robot,") Asimov portrays a tempestuous battle of wits and courage between man and robot. The titular hero(ine) is an autonomous vehicle, living on a farm with fifty other retired autonomous vehicles and her owner, Jake. Sally's powerful motor catches the attention of a businessman named Gellhorn. Coveting her to scrap for parts, he attempts to hijack her. Though Gellhorn attempts to manually override her controls, Sally remained autonomous and in control. Undeterred, Gellhorn takes her owner as hostage aboard an autonomous bus. The bus is then surrounded by the autonomous vehicles, who communicate with the bus to open a door and let their owner free. The bus drives off with the hijacker, and Sally and Jake go home. The next day, Gellhorn's body is found in a ditch, having been run over.

The conclusion of this story has become somewhat of a storybook cliche; the heroes have prevailed, and the villain has been defeated. However, what is certainly surprising is the level of intelligence that the autonomous vehicles display. Not only did they identify "good humans" from "bad humans," but independently perceived a dangerous situation, planned a reaction, and executed it. But what about the moral conundrums that these vehicles must work through? The villain's death is poetic justice, but by killing him, hasn't the bus violated the very first law of robotics? Taking this further, what if the autonomous vehicle's AI was defective or buggy, or if it simply broke trying to uncover the perpetually gray morality of "good" and "bad"? In another situation, the vehicle might have ended up executing an innocent, based on misguided AI.

The ramping prevalence of autonomous driving and incandescent optimism of its proponents naturally inspires a certain level of enthusiasm within society, and not least some fear of the uncertainty. If machines were to become capable of controlling the world, who could guarantees the safety of humans? For some, autonomous vehicles are the prelude to a world in which the safety of humanity—or even its mere existence—is not guaranteed.

In early 2016, the White House held a series of national seminars to discuss the impact of AI, publishing a report on the topic in October that year. In the same month, the Obama administration released a new roadmap for robotics in the United States. The roadmap called for a better policy framework to safely integrate new technologies into daily life, particularly when it came to autonomous vehicles and commercial drones. Current progress in autonomous vehicle development has far exceeded the predictions put forth in the 2013 edition of that same roadmap. We have also crucially recognized that the accident rate for autonomous vehicles is far less than that of human drivers (in which a fatal accident is likely to occur once for every 3.3 million km driven); yet there are still several obstacles that need to be overcome. Autonomous vehicles need to become more like industrial robots—capable of running for years without human intervention and repair. With the continuing advancement of technology and deep learning in AI, autonomous vehicles will also grow increasingly intelligent and powerful, potentially gaining the capability to rebel against human control. In this regard, humans must remain dominant over autonomous vehicles, and unconstrained by them. The construction of autonomous vehicles should also be standardized with regards to method and techniques. Finally, governments need to formulate policies and regulations to ensure that these vehicles can safely share roads with human drivers.

Autonomous driving is also often negatively associated with a lack of privacy. Fully autonomous vehicles, whether used by a single person or rented to a hundred, will undoubtedly continue to collect large amounts of personal, and sometimes sensitive data. Most of this data is relevant to the life of the owner: for example, where they went on a certain day, how long they stayed, in which supermarket they frequent go shopping, their body height and preferred temperature settings, and so on. In this way, automakers can easily access the personal information of owner and their family, such as the ages of any children, or their daily schedules. In the future, cars may be equipped with black boxes—not unlike airplanes—to record data and allow authorities to monitor the owner's every move. This realization might rightfully unsettle any user of those cars.

This future is fast approaching. In July 2016, German Transport Minister Alexander Dobrindt announced planned legislation that would to require automakers to install black boxes for all vehicles equipped with autonomous driving, to help establish responsibility in the event of an accident. The US NHTSA has established similar regulations. Since September 2014, all automakers have been required to report vehicle drive data in at least fifteen aspects, including driving speed, braking parameters, throttle opening, etc. This data must be made public, and has since been used to establish relevant standards. The US Department of Transportation also launched the Federal Autonomous Vehicles Policy on September 20, 2016, making the US the first country in the world to issue an overall autonomous driving policy. These federal guidelines have pushed autonomous vehicle manufacturers to share data about their failures with each other and with the government. This move has met with resistance from the tech and automotive industries, in no small part because this "invaluable" data may include sensitive information, from competitor data, to confidential business information and even personal information.

How data will be shared between the companies racing against each other to build the first autonomous car is yet to be determined, but what's clear is that tech companies aren't happy about the idea. David Strickland, who represents Uber, Google, and Lyft through the Self-Driving Coalition for Safer Streets, told reporters today that "the devil is in the details" when it comes to data sharing, and that's going to be a sticking point for all private companies involved, especially in a space as closely competitive as autonomous vehicles. Strickland indicated that the industry would likely push back against the data sharing requirements. "There is competitor data, there's confidential business information, there's a number of aspects which have to be respected. But on the other hand, safety is a number one priority, and figuring out the right context and space that we can ensure that while protecting the data rights and, frankly, the property of all the innovators and manufacturers should be properly balanced and that's going to take some time," he said.

In an interview, US Secretary of Transportation Anthony Foxx once used this analogy: "When I drive through a pothole, the car behind me can see that, and they can avoid the pothole as they're driving. But if it were an autonomous car driving through this pothole, could it communicate and share data with the myriad of other vehicles that follow—not just one from the same model or from the same manufacturer, but between all types of autonomous vehicles, regardless of the manufacturer? That solution is going to take a lot of time, as well as a lot of data, and there is a problem of information legitimacy."

Consumers' opinions differ. Today, car manufacturers have collected and stored location and driving data from millions of cars on the road. As the era of autonomous driving approaches, consumers are concerned that data collection will create more security risks. Consumer regulators in California say that the more sensitive data is, the more profitable it is for large companies that earn most of their revenue from advertising, like Google. Yet, we are left in the dark about which companies are to have access and utilize this data. Regulators say that once blackbox technology is widely used in cars, these corporations will have extensive information about the vehicle's location, destination, driving speed, and so on, and the individual will have no idea how it is used. The American Automobile Association (AAA), the largest auto club in the United States, requires its partnered auto companies to adopt the Consumer Rights Act, which sets stricter privacy standards. Another 2014 California legislative proposal aims to provide a framework for how car companies use such data in the future. The proposed "Consumer Vehicle Information Choice and Control Act (SB-994)" required car buyers to be informed that car manufacturers were collecting vehicle information, and accessing and controlling any shared information. However, this bill ultimately failed, in no small part due to the opposition of car companies.

Crucially, the issue is not whether vehicle data should be shared or not, but how to do so as to protect the copyright of each corporation and the privacy of the vehicle owner. In the aviation sector, all flight information is shared between all commercial flights anonymously. In this sense, information shared by an aircraft does not show specific carrier information, but does record the surrounding flight environment to help other pilots better deal with flight safety issues. Similarly, in smartphone applications, Apple's approach is also worth study by automobile manufacturers; for private and sensitive information, Apple only retains local AI algorithms, and not the user's archives; non-private information stored in the Cloud must also undergo extensive measures to ensure that even vendors themselves cannot trace or identify the user through their personal information.

Every once in a while, news breaks that smartphones are collecting user information, inciting public condemnation and vitriol. The increasing amount of data being collected by each vehicle positions the industry at a similar risk of data and privacy breach and scandal, a concern which is only becoming more prevalent over time. Manufacturers should be able to collect data, to improve their products and the overall experience of autonomous driving; however, it is their duty to do so in a way that protects the privacy of their customers. In the era of big data, all enterprises experience a sense of "data hunger"—but this data should be to the user's benefit, and not at the expense of their privacy. Tesla has made a commendable attempt to maintain their user privacy: for each journey, the data for the first and last five minutes—which show exact addresses and locations of the user—is not recorded. This surveillance shutoff of course excludes functions necessary for normal, safe vehicular function. Despite the debates held between government and automaker on the extent to which vehicle data needs to be shared, both parties concur that "safety is the first priority." After all, what good is a car that shouldn't be driven? As former President Obama said in an op-ed, "Make no mistake: If a self-driving car isn't safe, we have the authority to pull it off the road." However, it is necessary to find a proper balance between the government's safety regulations and the data and property rights of the manufacturers. This dilemma will likely require discussion.

2 Hackers are not Just Technical Problems

The term "hacker" has existed since the time of William Shakespeare, when it was used to define someone who struck their target with irregular blows. However, in the machine era, this term refers to a breed of computer-savvy technicians, developers, and voyeuristic criminals.

With the continuous development of autonomous driving technology, the safety of automobiles has become a significant subject of public scrutiny. Beyond questions to the safety and reliability of intelligent technology, the threat of hacking has now become a palpable concern for consumers and lawmakers. In 2014, the car-hacking project Open Garages, (which compiles and shares open source code for hands-on auto enthusiasts to fine-tune their cars) compiled a "Car Hacker's Handbook," detailing the vulnerabilities in the inner workings of computer systems and embedded software in modern vehicles. It was met with considerable enthusiasm, and was downloaded more than 300,000 times in its first week, causing the site to crash twice. Though the handbook is not meant to aid in crime of any sort, its popularity certainly illustrates a growing interest in the inner workings of each car.

In the context of autonomous driving, the Chinese term "安全" (an quan) has two meanings. It first means "safety." In this sense, it refers to safety functions such as a 360° sensor with no blind spots, multi-sensor integration, an accurate sensing algorithm, real-time control feedback, multi-layer redundancy of hardware and software, its resistance to large temperature variations, anti-vibration and dustproof functionalities, and so on. The auto industry also follows the ISO26262 standard for functional safety. The second meaning is "security." Technically speaking, autonomous vehicles are more vulnerable to attack than traditional vehicles, simply because a hacker has more venues to exploit. Once the central processor is compromised, the entire vehicle is shackled; hackers can also attack environment-aware nodes like radars and cameras to raise false alarms to cause security incidents; the networking features on connected cars also expose the vehicle to the Internet, increasing accessibility and creating vulnerabilities.

Autonomous vehicles are highly complex and interconnected devices that are difficult to protect against cyber attacks. These attacks often come in two varieties: indirect and direct attacks. Indirect attacks come from hackers who manipulate data streams over cellular networks, traffic and infrastructure signals, GPS information, and other sources to implant error messages via the vehicle's connection with these sources. Security researchers have confirmed the possibility of this type of attack. However, direct attacks are even more dangerous. These attacks refer to invasions of the autonomous vehicle system itself, and could result in accidents, kidnapping, or even murder. Such an incident would instantly destroy public confidence in autonomous vehicles, and with it, the fruits of years of research and development.

Perhaps, one day, you might suddenly receive a text message from a hacker asking you to pay $20 to unlock your car door. The more costly, time-consuming alternative is to hire a tow truck to move your car to a shop, where you can pay a service professional to unlock it. It's no stretch to think that one would just pay the vehicular kidnapper and go on one's merry way. Andreas Mai, director of Cisco's Smart Connected Vehicles Division, admits that this is just one example of the numerous crimes that hackers could potentially commit. As of 2014, a Cisco team had spent three years identifying threats and analyzing the vulnerabilities in vehicles by a number of different automakers.

In July 2013 at the DEF CON hacking conference in the US, two US security researchers demonstrated how the computer networks of the Toyota Prius and Ford Escape could be directly attacked by hackers, releasing a 100-page technical report detailing the process. Using laptops patched into vehicular systems, the two were able to force a Prius to "brake suddenly at 80 miles an hour, jerk its steering wheel, and accelerate the engine", and forcibly "disable(d) the brakes of a Ford Escape traveling at very slow speeds." Though these researchers had to directly connect to the vehicle networks in order to carry out their attacks, the extent of their control was alarming. It is worth mentioning that the vulnerabilities in the Toyota and Ford drive systems uncovered at that time have since become a focal point for both automakers, resulting in their eventual cooperation in further developing their on-board computer security technology.

In 2015, American hackers Charlie Miller and Chris Valasek used remote methods (the Internet of Things) to invade Cherokee's Uconnect car system, resulting in the recall of millions of cars, and shocking the entire industry. As one researcher puts it, "The real problem is not the vulnerability of a car, but the misconception that people have long relied on a closed system and their over trust in security."

Extended Reading

Hackers captured Cherokee

At the 2015 Black Hat hacking convention, Charlie Miller and Chris Valasek showed that they had broken into Cherokee's Uconnect system through Sprint's wireless connection, found a way to get back to the CAN bus by reconfiguring the V850 controller, and took control of the car computer, controlling functions such as braking, acceleration and so on through their laptops. At late July 2015, the two engineers remotely commandeered a Chrysler Cherokee on the St. Louis Highway in a car hacking experiment, leaving the "Wired" magazine reporter behind the driver's wheel helpless. Miller and Valasek turned on the car's wiper, maximized the stereo and AC systems, and cut the car's transmission on the highway, essentially making it inoperable—all having never left Miller's basement some ten miles away. The two American car hackers gained notoriety for their remote Cherokee takeover. Within few days, Chrysler's parent company Fiat Chrysler recalled 1.4 million vehicles with the same risk of being hacked. Assuming the eventual popularity of autonomous driving, it would be very frightening for hundreds of thousands of autonomous cars to lose control at the same time, whether in a city or a country, due to a hacker's intrusion.

After the hackers captured the Cherokee, the NHTSA spent five months conducting a comprehensive investigation, arguing that though most brands, (including Volkswagen, Audi and Bentley) use similar radio systems, relying on their security systems to deter hackers. Two US senators, Ed Markey and Richard Blumenthal, have proposed the "Security and Privacy in Your Car" Act to prevent future disasters on the road. They proposed that all cars produced in the US must be equipped with "reasonable measures" to prevent hacker attacks, specifically that system entrance should be protected, and interior networks isolated so as to avoid tampering with important software responsible for braking and car control. Under this act, cars must also undergo security vulnerability assessments in accordance with "best security practices," and subjected to third-party penetration tests. The bill also provides for privacy and transparency to better protect the privacy of car owners.

In September 2016, China's Tencent Keen Security Lab announced the first-ever successful remote invasion of a Tesla vehicle. Without any physical contact with the car, or accessing the Tesla mobile app, researchers at Keen Lab were able to break into the vehicle's computer system, targeting the electronic gateway and exploiting multiple high-risk security vulnerabilities through the internet to control the CAN bus and send directives to the vehicle.

After the successful takeover, researchers were able to shift the Tesla from park and drive at will, from thousands of miles away. They were able to force the car door open without a key, misalign its rearview mirror suddenly while changing lanes, open the trunk of the car while it drove, and even toy with crucial braking and steering functions to make them malfunction (for example, disabling the brake booster or power steering). In previous car-hacking cases, a failsafe mechanism in the car would prevent certain orders from being executed while the car was in motion at high speeds. In previous cases, this prevented hackers from braking at high speeds. However, Keen Labs seems to have worked around this problem.

The research results of Tencent Keen Security Lab have affected the sales of a number of Tesla models. After receiving the vulnerability report submitted by Keen Lab, Tesla officials quickly repaired the bug and requested its users to download the patch into their cars via Over-the-Air (OTA) Technology. Tesla hopes to protect users as early as possible in order to achieve "zero casualties" and "zero recall." Shortly afterwards, Tesla's official website publicly thanked Tencent Keen Security Lab. The lab became the only team inducted into that year's "Tesla Security Researcher Hall of Fame," for their achievement of "first remote physical contact-free intrusion into Tesla cars in the world."

Regulators as well have begun to grow concerned about car hacking. In addition to the introduction of strict data protection laws, the European Commission has taken measures to design and develop a modular system to protect the safety of in-vehicle network connections. In attempt to discover deeper vulnerabilities through reverse development, DARPA in the US has launched a cyber-security grand challenge for the creation of a system that can automatically defend a network by generating security patches. Additionally, the US Department of Transportation launched the Federal Autonomous Vehicles Policy in 2016, which proposed 15 safety assessment guidelines for self-driving cars, as well as provisions to prevent hacking. The

UN is reportedly drafting autonomous driving safety standards for use by various automakers to similar effects. Some experts believe that governments also need to participate in this regulation; however these regulations might be inadequate to test and ascertain the security of complex autonomous driving systems.

Though hacking appears to be a technical problem, vehicular safety affects both the owner's property rights and personal safety, becoming both an economic and social issue. A cyber attack on two cars is an isolated crime; however, the connected future of automobiles makes it entirely likely for this crime to scale. The economic, political and even military effects of these attacks on society could be catastrophic. Given the increasing topicality of cyber-terrorism, political, religious, and ethnic division in areas across the world, the hazardous potential of car hacking cannot be underestimated. This is especially true in the case of fully-realized Level-5 automation, which would leave the vehicle's occupants trapped in their incontrollable, speeding vehicle, at the mercy of hackers and inertia.

With the continuous improvement of vehicle automation, the risk of cyber attack also increases. At autonomy levels one and two, the vehicle can only passively detect risks, serving a secondary assisting function to drivers who retain control over its path. Between levels two and three, these controls are given to the vehicle itself, with are controlled by the system. Yet, the human driver is still liable for the system's choices. Beyond this level, it is crucial for network security to be ensured. Penetration testing and vulnerability scanning prevents direct attacks on the vehicles via the Internet. From autonomy levels three to five, analysis and control of vehicles are entirely controlled by the system, necessitating additional security measures. Though vehicles with Level-4 autonomy no longer need to be monitored, there are additional testing requirements to ensure control security. Additional security extension analysis for autonomy levels three to five is designed account for the required aftermarket modules that accompany the development of autonomous driving. The security measures necessary for vehicles each level of autonomy thus changes in terms of monitoring, control and responsibility.

How do we address these security risks? To do so, we must first carefully define the security requirements of autonomous driving systems, conduct risk analysis, establish a software and hardware platform with a trusted computing foundation, isolate key modules via partitioning and virtualization, and ensure end-to-end data encryption. Secondly, the security design methodology must be evaluated at every step, from design confirmation to implementation. Is it secure against attacks? How can it be updated or patched? Is there enough coverage to protect against areas of failure in the system? In case the system falls, is there any way to regain control or terminate all systems? Thirdly, the future needs a security information market. Through this market, researchers or white-hat hackers can sell security flaws to host manufacturers or technology providers. The current state of this field is somewhat chaotic, making it the perfect time for individual companies and corporations to move towards regulation, and establish norms. The business opportunities in this future are sizeable.

For each hurdle we cross, another rises in its place. Through cooperation between the auto industry, the IT industry and consumers in society, we will eventually be able to build a strong "firewall" to ensure the security of autonomous vehicles, and prevent hackers from penetrating into our cars.

3 Ethical Disputes Caused by Autonomous Driving

The now-famous "Trolley Problem" was first proposed by the philosopher Philippa Foot in his paper "The Problem of Abortion and the Doctrine of Double Effect." The paper was published in 1967 to criticize then-mainstream theories of ethical philosophy, particularly utilitarianism. The "Trolley Problem" has since become one of the most well-known thought experiments in ethics.

There are several versions of this experiment, one of which goes like this: Five innocent persons are tied to a tram track. A runaway tram hurtles towards them and is just about to crush them. You have the choice to pull a lever and divert the tram turn to another track, where only one person is strapped. You have two choices: don't pull the lever, and five people die under your nose; pull the lever, and become responsible for a single person's death. What would you do?

Utilitarianism believes that in pursuit of the greatest benefit for the most people, the minority should be sacrificed to save the majority. Therefore, one person should be sacrificed to save five people. Kantianism believes that morality should be based on the necessary obligations and responsibilities. If it is a moral obligation not to kill people, then you should not kill the single person by pulling the lever, even if it means sacrificing five others. Interestingly, in a survey, 68.2% agreed that you should pull the lever. However, pretend that you were the single person tied to the tracks. Would you still pull the lever?

Though Google's autonomous vehicle development program had a record of more than 2.2 million km safely driven, its program was suddenly halted on February 14, 2016 after one of Google's self-driving cars crashed into a bus in California. In a statement, Google said: "We clearly bear some responsibility, because if our car hadn't moved, there wouldn't have been a collision. That said, our test driver believed the bus was going to slow or stop to allow us to merge into the traffic, and that there would be sufficient space to do that." This is the first accident caused by a self-driving car. Because driverless cars obey traffic rules strictly, it is easy to collide with vehicles driven by less attentive human drivers when confusion and congestion occur. There are some similarities between Google's incident and the "Trolley Problem". Just as the tram could kill a single person instead of five, Google's self-driving car crashed into the bus to avoid the sandbags. This compels reflection on the ethical issues surrounding autonomous driving and the technology it is based on. The era of autonomous driving is now, as it the era in which to think about these problems.

In fact, a professional organization has begun to develop a version of "Trolley Problem" specific to autonomous driving. This version goes like this. Suppose an autonomous car is carrying you on a one-way street, with a roadblock on the left and a stone wall on the right. Suddenly a group of children are run across the street. It is too late to hit the emergency brake. What decision does the AI make? Does it save the children and swerve the car to hit the stone wall, or does it prioritize the protection of the passenger—you? The organization surveyed more than 2000 people and found that 76% of respondents replied that it is more ethical to sacrifice one passenger for ten pedestrians. However, 81% of respondents replied that they wanted to buy an autonomous vehicle that would ensure the passengers' safety at all costs. The results of such a survey appear contradictory, but reflect the different standpoints of people when they are outside and inside the car. People put their lives in the first; such is the simple nature of human beings. When a person is disadvantaged or threatened, their sense of self-preservation will inevitably urge them to protect their lives.

Yet, the pragmatic approach does not solve all ethical conundrums. Complications arise when placing this sterile hypothetical in the complexities of the real world. Taking the previous scene, let's imagine that the five people crossing the road are not innocent children, but criminals in the middle of committing a robbery. The one passenger in the car is a scientist, whose discovery has cured cancer. Would you still wish for the autonomous car to (pragmatically) maximize the number of lives saved?

When applied in reality, a car must consider several factors when making its decisions. This includes the magnitude of potential collision in relation to the targets, and the quantities of passengers on board. Unlike the instinctive, split-second decision made by a human in a similar situation, autonomous vehicles would rely on well-designed risk management strategies when making the decision—risk being defined as the product of the degree of misfortune and the likelihood of the event. Could the solution be to allocate a "value" or kinship index to everyone, to facilitate the vehicle's risk management procedure? This would certainly complicate these calculations. However, as long as there are rules to follow, computer-controlled cars are more than capable of embracing complexity. Therefore, in this case, we assume that the "value" of the scientist's life is greater than those of the five robbers. As a result, those five robbers must die.

A similar algorithm has already been employed. Google was granted an application patent for risk management in 2014. Google's patent describes a scenario where a car may change lanes to better observe traffic lights; it may also choose to stay in the current lane to avoid a very low probability of collision (such as a collision due to a sensor reading error), and bear the loss of incomplete reading of the signal data. The outcome of each potential event includes its probability of occurrence and positive and negative values (pros and cons). The vehicle would multiply the magnitude of each event by its probability of occurrence, and then add them up to assign a value. If the pros outweigh the cons to a certain margin, the vehicle will take the action under consideration.

The International Humanist and Ethical Union points out the potential risks of this strategy. Accidents such as brake failure, sudden and unpredictable behavior by other drivers, cyclists, pedestrians and animals are unavoidable; therefore, it is very likely that these cars will need to make difficult choices. However, would you, the consumer, feel comfortable in a world where automakers will assess the value of your life, a number that will determine whether cars should hit you or another person?

Though it is tempting to conclude that this philosophical approach is "simple," it conjures a host of complications and dilemmas. When autonomous vehicles become reality, this problem will rise again and again. The AI algorithms of each vehicle, representing the opinions of each automaker, answer the question: to pull the lever, or not?

In October 2016, Mercedes Benz executive Christopher Hugo proposed his response on the dilemma of whether an autonomous car should prioritize the lives of its passengers or passers-by. "If you have the ability to save a person, save him first at least." Hugo said at the Paris Motor Show that year, "If you have the confidence that you can save your own passengers, do it first." Of course, this is not to say that Mercedes-Benz's self-driving cars will run over pedestrians without consideration—however, it is one answer to the "Trolley Problem". Mercedes-Benz subsequently denied this statement, saying that "it is obvious that programmers or autonomous driving systems are not qualified to determine the value of human life." An official statement from Mercedes-Benz also attempted to debunk the reality of "Trolley Problem," stating that the company "is committed to avoiding the dilemma completely by implementing risk-proof driving strategies".

Mercedes-Benz's denial after expressing its opinions indicates that automakers are in a dilemma on this sensitive issue, with pressure from all sides. In contrast, Google is more frank on its stance. Sebastian Thrun, the founder of Google X, said that the company's self-driving cars would choose to collide with the smaller of the two objects. A Google patent dated in 2014 related to lateral lane positioning also follows similar logic, describing that autonomous vehicles should be kept away from trucks in one lane and closer to cars in another lane because it is safer to collide with smaller objects. Of course, colliding with smaller objects is an ethical choice that protects the interests of the passengers and minimizes the damage they suffer. Yet, in doing so, it passes the danger to pedestrians or other passengers in small cars. As Patrick Lin, a professor of philosophy at California Polytechnic State University, has asked, "what if the smaller object is a baby cart or a child?".

In March 2016, Google Self-Driving Car Director Chris Urmson described a more "worldly-wise" rule of thumb to the Los Angeles Times, "Our self-driving cars will try to avoid colliding with any road user without any protective measures: cyclists and pedestrians, and they will avoid bumping into a moving object." Compared with the strategy of crashing into a smaller object, this method pays more attention to practicality and tries to protect the most vulnerable people in accidents. In accordance with this principle, autonomous vehicles must leave more space for motorcycle riders without helmets, because unprotected riders are more likely to be killed in a crash than motorcycle riders wearing full protection devices. This doesn't seem fair—why should safety-conscious cyclists be punished for their virtues?

Another difference in ethics between robots and people is that even if a programmer's intention is good, the ethical standards of a robot may be contrast with the original intention. Imagine that the computer program of autonomous vehicles assigns different buffer intervals to pedestrians in different districts by analyzing the number of civil cases of traffic accidents in each residential area. Although this is a reasonable, well-intentioned and effective way to control vehicle behavior, it may have adverse consequences. For example, the real reason for the smaller number of claimed pedestrian's collisions could very well be that people in the area have low incomes and no money for lawsuits. Computer programs, then, inadvertently disadvantage the poor, providing them with fewer buffer intervals and increasing their risk of being hit while walking.

Ford said it has been working with several well-known universities and industry partners to address the ethical issues of autonomous vehicles. At the same time, Ford also warned of excessive philosophical thinking. Wayne Bahr, global head of Ford's automotive safety, said: "We are trying to solve this problem from a professional perspective and avoid being caught in unrealistic assumptions. In discussions about the ethics of autonomous driving, a common problem is that the underlying assumptions about the ability of an autonomous vehicle are wrong. For example, valuing one person's life as higher than another is based on the assumption that autonomous vehicles can identify who has higher value."

The question now is, how should the decision-making standards for software design of automakers be formulated? How can the industry discuss it openly and reach a consensus? This is not only an ethical issue, but also a political one, which must be solved through the formulation of policies and regulations by the government. NHTSA issued a report in September 2016 stating: "Manufacturers and other entities, working cooperatively with regulators and other stakeholders (e.g., drivers, passengers and vulnerable road users), should address these situations to ensure that such ethical judgments and decisions are made consciously and intentionally." In any case, no matter how enterprises, governments and experts argue, the attitude of the users remains consistent. As the results of the survey show, users only want to buy cars which are able to protect their safety at any cost.

For the technology's actuators, there are other ethical issues that need to be considered, such as potential conflicts of interest and opposition. There will inevitably be individuals in favor of or against such "statistically safer" technology, especially when it is prone to low-level mistakes. Most people do not have a good sense of "safer," or notice a lack of accidents; but they are much more sensitive when accidents happen, and could likely oppose in that case.

Tesla is a courageous spearhead of this technology, and is attempting to give its cars autonomous driving functions through OTA. However, the initiative can do better in terms of due diligence, failing to distinguish between autopilot and autonomy in marketing materials, failing to stress the unreliability beta version software, and overly lax in issuing notifications to pull a driver's attention back on the road. Sure enough, the industry is still at odds about Tesla's approach. In an international conference held in Beijing in April 2016, when asked about Tesla's system update methods, Dr. Peter Mertens, Volvo's senior vice president of research and development, euphemistically described it as "more risky." He argued that autonomous driving, unlike AlphaGo, involves ethical and legal aspects. To ensure consumers' autonomy and privacy, the deep learning of the system must always be under control.

When considering the supervision of these safety measures, the electronic control system of cars is quite predictable. For example, the safety of electronic stability control systems can be easily determined, based on the clarity of the physical models and accuracy of its sensors. The car's electronic control unit (ECU) will thus reflect a situation close to Ground Truth. However, according to a study by German safety certification body TÜV Rheinland this may no longer be applicable to self-driving cars, especially when ADAS begin to move towards autonomy.

Compared with the data required for the Electronic Stability Program (ESP) of a feature car ("steering angle" and "wheel rotating speed"), the data used by a smart car needs to undergo another high-order interpretation process. The data interpretation and subsequent prediction and action planning are accomplished by software or AI algorithms, which are also continuously optimized through machine learning to improve daily. These AI algorithms deal with numerous variables, from understanding driving intentions to predicting driving behaviors. Taking intelligent collision avoidance as an example, planning errors or wrongly interpreted sensor readings have fatal consequences to passengers, especially at high speeds. In the worst case, ADAS could even cause traffic accidents. These scenarios have given rise to new problems, such as whether is it possible to gauge the role of vehicle error in accidents, who is responsible, and to what extent.

With the help of AI algorithms, software engineers consciously and unconsciously allow autonomous vehicles to use its own judgment in circumventing the law. Google admits that if it is dangerous to drive at a slower speed, it should be allowed to exceed posted speed limits in order to keep up with traffic. In other cases, such as getting passengers to a medical facility, most people would also support speeding. Chris Gerdes and Sarah Thornton of Stanford University oppose mandatory inclusion of laws in software coding. Drivers seem to view most laws as flexible, allowing instances of speeding if it brings positive rewards. No one wants to be held back by a slow-moving cyclist, just because their car refuses to slightly cross the double yellow line to pass.

Semi-autonomous driving systems are also unavoidably accompanied by ethical issues. During an outdoor test for automatic collision avoidance safety system in November 2013, a Mazda CX-5 SUV crashed into a side guardrail, injuring two people on board. In contrast to this "False-Negative" incident, Toyota had a "False-Positive" incident in 2012, which activated the emergency braking system. When the system detected scattered waves reflected between the vehicle body and side lane, it mistakenly judged that there was an obstacle ahead and hit the brakes. In addition to smart cars with ADAS capabilities, fully autonomous vehicles without human interventions will also require safety oversight, and give rise to more ethical dilemmas.

Five major tech companies—Amazon, Facebook, Google, IBM, and Microsoft—established partnerships to help study ethical guidelines for designing and deploying AI systems in September 2016. In the context of ever-increasing international attention to this field, Carnegie Mellon University established a research center focused on the AI ethics in November 2016. This has led a series of academic, governmental and private studies to explore new technologies, some of which had only exist in science fictions until recently. Over the past decade, faster computer chips, cheaper sensors, and massive amounts of data have helped researchers and driven the development of technologies such as computer vision, speech recognition, and robotics.

Subra Suresh, President of Carnegie Mellon University, believes that it is necessary to add ethical norms to AI technology with the development of technology. Although the concept of the "Terminator" robot seems distant, the US military is already working on weapons that can determine to killing autonomously. "We are at a unique point in time where technology is far ahead of the society's ability to limit it." Suresh pointed out. At the same time, he also questioned the general optimism regarding AI, especially in the field of autonomous driving.

Prof. Sam Peltzman of University of Chicago pointed out in 1975 that seat belts and airbags actually led to more traffic accidents. Autonomous vehicles may bring about similar, unexpected problems in fields such as safety, ethics, and law. We should maintain awareness, but also should not give up eating for fear of choking. Both automakers and consumers must have the courage and patience in the early stage of new technology development. The path ahead is long, but we will eventually reach the finish line.

4 Challenges of Crossing the Road "Chinese-Style"

On October 11, 2012, a netizen by the username "This is super interesting" posted a message on the microblog "Crossing the road Chinese-style: jaywalking based on safety in numbers, regardless of the traffic lights." The microblog also includes a photo of pedestrian crossing the road (Fig. 1). Although there are no traffic lights, several pedestrians jaywalk in a group. Among them are old people pushing a baby cart, as well as electric bikes and wagons. This post resonated with netizens, and was shared by nearly 100,000 users just in one day. As one netizen commented, "it's vivid."

The so-called "Chinese-style" crossing is not unique to China, and is actually a symptom of poor urban traffic management. This phenomenon reflects problems including lack of management, failure of governance, and a weak awareness of traffic rules. The key issue is that all road traffic participants compete disorderly for Right of Way (ROW).

The ROW principle is the principle of going one's own way, and serves as the essential basic principle for traffic participants. Modern transport facilities provide separate routes for all traffic participants, giving pedestrians, non-motor vehicles and motor vehicles their own prescribed routes. The struggle for ROW is to balance the rights and interests of all parties in sharing the road, so that all parties benefit. Pedestrians, non-motor vehicles and motor vehicles all have corresponding ROWs. They should have their own roads, abide by rules and regulations, and neither party can infringe on the traffic space of the other party. Only in this way can traffic order be improved, accidental hidden dangers be reduced, and road safety and smoothness be realized. However, in the context of limited road resources, parties often compete and deny others' "ROW". Pedestrians do not look at traffic lights, do not walk the zebra crossing, or even cross the road over traffic barriers; motor vehicles rampage on the sidewalks and non-motorized lanes, even running red lights. Some pedestrians ignore

Fig. 1 "Chinese-style" crossing a road. *Photo* Microblog of "This is super interesting"

lights, while some drivers skirt the rules, by changing lanes without queuing, and dodging cameras while violating traffic rules. Some drivers drive crazily to provoke other cars that have blocked them, with no consideration for the safety of others. They may also provoke other parties into violating traffic rules, and so on. This kind of disorder damages other people's rights and interests, artificially creates unnecessary traffic congestion, and also causes huge safety hazards.

There have also been interesting debates between automakers and consumers about ROW. Many enterprises that develop autonomous driving technology say that in order to ensure the safety of pedestrians, the autonomous vehicle will take priority to avoid pedestrians. Though this conclusion appears to be a given, a second thought reveals an problem: giving priority to pedestrians means that an autonomous car needs to stop whenever approaching a pedestrian, thus becoming unable to move forward when driving on a crowded road—especially an urban road with "Chinese-style" crossing. This would greatly reduce the efficiency of self-driving cars. When this strategy was put forward, some consumers came out to say that they would not such a vehicle. Did these consumers overreact? After all, were they expecting the self-driving car to crash into the pedestrians it comes across? In this sense, autonomous driving AI has not evolved to the point of human behavior—knowing when to stop and when to scoot forward. Therefore, the best solution is for pedestrians, non-motorized vehicles and motor vehicles to abide by their respective ROWs and go along their own paths.

Beyond law-challenging "Chinese-style" crossings, and the mix of traffic users (including pedestrians, bicycles or electric bikes, low-speed three-wheeled vehicles, etc.), other factors such as unreasonable road designs, traffic signs, and complex road systems have also posed new challenges to the application of autonomous driving. Conventional functions such as stopping are less daunting, as the parameters are common have clear solutions. The real difficulty of autonomous driving is how to deal with traffic irregularities, such as a driver running a red light, crossing a guardrail, or another unexpected situation.

The development of autonomous driving is highly effected by geographic region, as driving conditions in Europe and America are different. This difference is even more pronounced when compared with China. Regional characteristics often contribute to the high failure rate of functions already mature in Europe and America (such as changing lanes and crossing intersections) when used in China.

Although autonomous vehicles have been tested on closed testing grounds in China, accidents may occur if they have not passed the tests on the actual roads in China. In this sense, localization of data processing and autonomous driving decision-making algorithms cannot be avoided. A number of foreign automakers and international parts suppliers have extensively discussed this widely-recognized issue.

It is precisely because China's complex traffic conditions and "Chinese-style" crossing chaos that traffic management must be reimagined. Due to its software, autonomous vehicles are more disciplined than their human drivers, acting only on actions supported by a factual, comprehensive analysis of the situation through its sensors and V2X. Its path is optimized with regard to actual road and traffic conditions, to deliver passengers safely to the destination, balancing the interests of its passengers as well as other travelers, and reducing meaningless delays and jams. With autonomous vehicles, Chinese road users can save time, labor, and worry. Why not implement it?

"Surveys show that Chinese consumers have the highest willingness to accept autonomous driving." Wang Xiaojing, Chief Engineer of Research Institute of Highway Ministry of Transport, supports the above judgment from yet another angle. He argues that because China's car ownership rate is lower than that of Europe and America, the development of car sharing in the future might increase the acceptance of autonomous driving in the country.

Multinational companies have also been rushing for a slice of China's huge auto market pie. Håkan Samuelsson, President and CEO of Volvo Car Group, said in an international forum in April 2016, "China's traffic congestion has led to high accident rates, and intense air pollution, so autonomous driving technology is needed to improve transportation efficiency and commute times through scientifically planned routes." He also had confidence that China's complicated road conditions would greatly aid and polish the autonomous driving system. In response to the concerns of translating existing technology to China's particular situation, another Volvo executive commented, "Don't exaggerate the differences between cities. There are two kinds of road conditions: high-speed and urban. We just have to fine-tune to China's situation, and open up road tests as soon as possible."

5   Autonomous Vehicles Interaction with Human Beings

Volkswagen "The Wizard of Oz" study placed volunteers in a human-driven car, while tricking them into thinking that they were inside an autonomous car. Though the volunteers were uneasy at first, they were eventually able to calm down and enjoy the journey in about ten minutes. This test shows that it is relatively easy to establish trust between humans and autonomous vehicles. With trust, a "close" relationship can begin between the two.

The relationship between an autonomous vehicle and humans is a fascinating topic worth exploring. Are autonomous cars servants of mankind? As it provides a service to human, the vehicle must go wherever we go. Is it a student? The vehicle must learn and consider a myriad of human driving habits, even if driving by itself. It also must remind people to maintain control in case of emergency, just like a student doing homework has to ask her teacher for help on a difficult problem. Is it also a friend to mankind? Humans enabled the vehicle to make its own decisions on how to operate better and faster. In the sci-fi story written by Asimov, the relationship between the robot car Sally and her master is more like that between pets and people. Perhaps there are even more types of relationships between autonomous vehicles and humans. However, the communication and interaction between the two is nonetheless becoming more important than ever.

Let's first discuss the uncertainties affecting driving behavior, including factors in the environment, vehicles and other people.

(1) Uncertainties in driving environments include: weather patterns, such as season, temperature, illumination, changes in wind, frost, rain, snow and haze; road geography, such as width, curvature, slope, access ramp, intersection, road quality, and material,; traffic signals such as signs, traffic signals, posted signs, construction; and real-time traffic conditions such as changes in traffic flow, surrounding random obstacles, environmental noise.

(2) Uncertainties in the vehicle itself include: physical properties of the vehicle such as body weight, size, etc.; dynamic properties such as steering wheel torque, engine power, traction, braking force; and wear and tear on the vehicle.

(3) Uncertainties of the driver's behavior include: the driver's personality, current mood, knowledge, experience and skills, familiarity with the road, proficiency and habits.

No two drivers behave in exactly the same way. No two cars on the road are in the exact same condition, and no two intersections are exactly the same. The uncertainty of the driving environment, the driven vehicles and the drivers' behavior highlights the challenge and importance of interactions between humans and autonomous vehicles.

There are still many hurdles facing the successful implementation of autonomous driving in the real world, including the issue of how to return control to human beings in case of emergency. The Organization for Economic Co-operation and Development (OECD) released a report on autonomous driving on May 28, 2015, hosted by economist Philippe Crist. As cars become more and more self-driving, alternative control between machines and humans "will become a thorny problem," Crist said. This is one of the key reasons why automakers are reluctant to launch a completely unmanned car soon. Such vehicles may lead to new types of traffic accidents.

In the era of intelligent driving, people and cars will undoubtedly form a new type of relationship. Decoupling the lateral steering wheel and the longitudinal throttle and brake is the key to the success of autonomous vehicles. With advancements in input modes, such as touch, voice, gesture and brain-computer interface, the UI system will become an important indicator for evaluating an autonomous vehicle. Maybe autonomous vehicles of the future won't look much different from today's vehicles, merely boasting a dual driving mode that makes driving as simple as operating an elevator. By simply inputting the destination to the car and pressing "go," it will automatically deliver you there. With the driver's attention off the road, it's possible them to watch movies, drink tea, nap, or work. However, even an elevator has emergency stop and fire safety, in case of unexpected situations—so should be the case for a self-driving cars. The controversy surrounding the human driver's control and driving style essentially boils down to what kind relationship that humans should have with autonomous vehicles. Does an autonomous vehicle require a steering wheel? Or should we cut off the most direct point of interaction between the automobile, and the human inside?

Traditional automakers may not wish to surrender control of the steering wheel so quickly, choosing instead to bring new technologies to real life through phased upgrades, subtly influencing the future—warming the frog in the waters, rather than dropping it in a boiling pot. When a driver is tired from or when stuck in traffic, autonomous driving technology can give them a chance to breathe, or provide a more comfortable experience (for example, collapsing the steering wheel to free up elbow space). When the driver is ready or needs to take back control, they can do so at any time.

The business model that automakers and auto parts companies are striving towards involves automobiles that drive automatically, but retain a physical steering wheel. Obviously, they are still keen to provide ADAS, meaning we need to conduct research in both technical and social science aspects. When a car tells its driver to take over the steering wheel, how can it be done safely? How does a self-driving car shake a human driver from their sleep, pull them back the reality from a video, or interrupt his phone call to catch their attention? It's not uncommon for a person absorbed in loud music or an engrossing film to miss voice or light cues.

Ultimately, it is most effect to "wake" the driver through vibration. In just 0.8 s, a passenger focused on writing email can switch back to "driver" mode. At the same time, how can the vehicle communicate where to direct their attention? Previous studies have shown that human perception of horizontal vibration along a tactile belt can be accurate to 10 degrees. Therefore, a vibration along the belt in the direction of trouble would be an effective communication. However, even if the vibration warning were effective, it may be confusing when paired with other complex sound and visual signals. To avoid this, a reasonable suggestion would be to use different warning modes at different stages of the takeover process. For example, a vibration is used to arouse people's attention, while a visual aid is used to inform the direction of danger. In this way, the tactile sensation enables a quick reaction, while the visual sense enables accurate information.

Another view argues that enhanced assistive technology is an unsafe approach: people must either drive on their own or not at all. The Transportation Institute of Virginia Tech conducted an experiment in which dozens of volunteers were recruited to sit in an autonomous car. The test required the testees to vigilantly observe the road and surrounding conditions and control the car if necessary. However, most testees quickly relaxed, hands and feet moving out of position. There is a similar issue in aviation. Aircrafts can already travel between runways autonomously. The pilots need only take control of the landing and takeoff. However, this technology is not widely used because the pilots could be distracted if not controlling the aircraft. Even if relevant road regulations declare that drivers must keep their eyes on the road, must not use phones, or must put their hands on the steering wheel, the human psychology dictates that people cannot be fully focused while in an autonomous vehicle driving. When their attention is suddenly drawn, they are wholly unprepared.

Internet giants who share views include Google, Apple and Baidu, who represent a forward-looking, radical force in the development of autonomous vehicles. They believe that autonomous vehicles should not have steering wheels. These companies are keen to maximize the online time of "drivers." Once people no longer need to spend time driving, Google's business model can take advantage of people's daily commute time to increase its advertising revenue. Google does not want a people behind the wheel to drive; therefore, it pushes fully autonomous vehicles.

However, this model also poses a problem for Tesla and many other automakers, especially high-end luxury brand manufacturers. The attraction of luxury brands to customers is often the driving experience. If a car does not have a steering wheel, the "driver" is just a passenger—the driver loses the most direct sense of communication with the car. That's why Tesla, Jaguar, Land Rover and other automakers use autonomous driving technology only as a supplement to their current unmanned vehicles, and still allow the driver to control the vehicle. Porsche even claimed that it would not use autonomous driving technology. Oliver Blume, CEO of Porsche, said in an interview that if driverless technology were to replace human driving, no one would buy a Porsche, whose main draw is the driving experience.

The debate continues, and it too early to draw conclusions about which mode will ultimately win. Perhaps only when the era of completely unmanned driving arrives will we know. At that stage, there will be a mix of both human and autonomous drivers, complicating road traffic. At present, some research institutes have been discussing whether so-called "effective avoidance" in autonomous driving technology will cause chain reaction, leading to more severe traffic accidents and personal injury when the road is filled with autonomous, semi-autonomous and manual vehicles. What strategy does one use when surrounded by human drivers? How does one maximize the safety of people and property? Autonomous vehicles can avoid collisions, but what about human-driven vehicles? How can a semi-autonomous vehicle be controlled? This requires is a comprehensive strategy for mixed traffic coordination. When properly established, it can be taught to and recognized by consumers, speeding the introduction of autonomous vehicles into the market, and promoting the early maturity of future traffic patterns.

When human drivers encounter a blocked intersection, most of the time they wave to other drivers or pedestrians to let them go. In this case, it is impossible for an autonomous vehicle to interact with others like a human driver. The same problem occurs when a car is merging or changing lanes in a highway. Human beings can communicate their wishes through interaction in many ways, while interactions between robots can be exhausting or frustrating.

In the fall of 2015, a New York Times reporter personally experienced the two demonstrations of Google's driverless car, to understand both the car's cautiousness and the unpleasantness of conflicts between the car and the people. In one demo, the driverless car had to make a sharp turn to avoid a car parked irregularly in a residential area. If the sensor was not so sensitive, a collision would have probably occurred. In another demo, the passengers felt a "violent shock" At that time, the driverless car "saw" a red light and began to slow. The laser system on the roof also sensed that a car on the opposite side was driving towards the intersection at a speed exceeding the safe range. The autonomous car immediately swerved right to avoid collision. However, the human driver in the opposite car had stopped in time at the red light. Courtney Hohne, a spokesman for the Google's driverless car program, said that the current test sought to "eliminate" this contradiction between automotive software and people. For example, the program allows the car to creep forward at intersections like a human driver, to clarify its own sequential position during release.

John Lee, a professor and automation and safety driving expert at the University of Wisconsin, points out that at such times, human drivers often solve problems by "making eye contact and, at a glance, deciding who will go first". However, where is the eye of an autonomous vehicle? To this end, American startup Drive.ai is seeking to install an eye for the autonomous vehicles through the use of AI "human–computer interactions." As Drive.ai has judged, it is necessary for autonomous vehicles to disclose their perception and decision to human drivers. Autonomous vehicles need to communicate unambiguously with humans both inside and outside of the vehicle.

The company is trying to use deep learning techniques to understand human language, and improve the ability of computer vision systems to recognize objects. Drive.ai's cars don't talk to pedestrians or cyclists, but instead visually communicate with them through the information display (Fig. 2). This information is not limited to traffic signs that are currently in use, and may be also be expanded to include text messages and sound signals.

Fig. 2 A car is visually communicating with pedestrians through the information display. *Source* Drive.ai

Drive.ai's CEO remarks, "there is no human factor in most discussions around autonomous vehicles, which is a strange thing, because this is the first time a robotic system has the opportunity to really appear in the real world and interact with humans." With the constant progress of artificial intelligence and robot technology, the interaction and mutual learning between autonomous vehicles and human beings will be important for the development of related science and technology in the future.

6 Autonomous Driving and Sharing Economy

John Krafcik, CEO of Google's driverless car program, said in an interview in August 2016: "People's ideas change, for example, from buying a car to renting a car. Fully self-driving cars will become more expensive. Society will optimize the use of these assets. A car is only used 4% of the time a day, and the remaining 96% of the time is parked in the parking lot. Each car has at least three or four exclusive parking spaces in the US. This is a shame for the city." Krafcik's argument advocates the socialized optimization of autonomous vehicles, which caters to the trend of the "sharing" economy.

According to Adam Jonas, an analyst at Morgan Stanley, "in the future, only wealthy people may have cars, and only robots will drive." It sounds incredible at first, but it couldn't be more true. Jonas divided the development of the automotive society into four phases (Fig. 3) and used four quadrants to describe it vividly. The first quadrant is the current social transportation mode, when everyone owns or rents their personal car, and drives it themselves. At this point, autonomous driving technology has just begun to develop. The second quadrant occurs a few years later, when online car-booking services, like Uber, Lyft, and Sidecar will become increasingly popular, and decreasing the demand for private cars and increasing the number of drivers. This stage is the first step towards autonomous driving. The third quadrant

Fig. 3 Four development phases of the automotive society. *Sources* Morgan Stanley

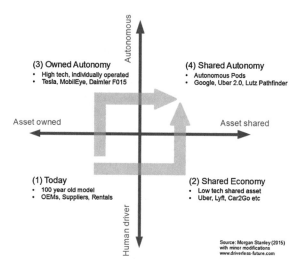

occurs more than a decade later. By this time, the rich will abandon the cars that require human drivers. Finally, by the fourth quadrant, most people will commute via self-driving cars. These cars may be owned by companies or individuals. Regulations may restrict human drivers to drive only on certain designated roads, and public transport systems may no longer use human drivers. According to Jonas's prediction, car sharing will undoubtedly become a contributing factor to the continuous improvement of automotive society.

The future belongs in the future, while we grow more curious about what is happening today. Didi announced on August 1, 2016 that it reached a strategic agreement with Uber to acquire all the assets of Uber China, including its brand, business and data operated in mainland China. After the two parties reached a strategic agreement, Didi and Uber Global reached a mutual agreement towards shareholding. Uber Global now holds a 5.89% stake in Didi, which is equivalent to 17.7% of the economic rights. The remaining Chinese shareholders of Uber China receive a total of 2.3% of the economic rights. Meanwhile founder and chairman of Didi Cheng Wei joined Uber's global board of directors, and founder of Uber Travis Kalanick joined Didi's board of directors. Didi, with more than 15 million drivers and 300 million registered users, thus joined hands with Uber, another world-class sharing economy giant, to build an ecosphere that links people, cars, transportation and lifestyles through interconnectivity and open sharing.

Coincidentally, China's National Development and Reform Commission and the Ministry of Transport jointly issued the "Implementation Plan for Promoting "Internet +" Convenient Transportation to Promote Intelligent Transportation Development" (referred to as "Implementation Plan") on August 5th. This plan promoted the convergence and development of transportation enterprises and Internet companies, giving full play to their respective advantages, encouraging the integration of online and offline resources, realizing the deep integration of information resources, capital, technology and business, and organically integrating with upstream and downstream industrial chains. It is foreseeable that the "Implementation Plan" will promote the development of a sharing economy.

The essence of the sharing economy is to enable idle offline goods or services so that the supply-side can provide goods or services at a lower price. The supply-side transfers the right to use the goods or provides services within a specific time. The demand-side does not directly hold the ownership of the goods, but uses them by leasing, borrowing and sharing. Once-popular DVD and book rentals are all part of the sharing economy. It is based on this concept that the first time-sharing car rental was established in the US in 1999, and then rapidly developed in Europe. Now, Uber and Didi have launched the carpooling function, upgrading the sharing of cars into the sharing of seats, reducing the vacancy rate of shared cars, reducing travel cost by about 30%, and further improving travel efficiency.

The implementation of time-sharing car rental programs has grown to impact even the automotive industry, and major automakers will certainly not let go of this gold opportunity. BMW invested in the MyCityWay car sharing project for $5 million in 2011. In the following five years, a total of 14 projects were invested in 2016, totaling $1.6 billion. Mercedes-Benz has not only started car sharing projects around the world since 2009, but also invested in 500 smart cars and launched Car2go projects in Chongqing, China in 2016. Toyota has invested in Uber and entered into a partnership with the Japanese Taxi Association to explore autonomous driving, car sharing and other mobile connectivity technologies. Volkswagen has invested $300 million to the taxi company "Gett" to build its own online digital platform, aiming to achieve 80 million active users by 2025, and invested in the "Gofun" car rental in China and signed a strategic cooperation agreement with Didi. It hopes to transform itself into a travel service company targeting annual revenue of billions of euros in 2025. Volvo has also established a dedicated business unit focused on car sharing, based on Sunfleet, a car-sharing service that launched in 1998. Afraid of being left behind, other major automakers are also in the process of launching their own time-sharing platform, such as Ford's Ford2Go and Audi's Audi on Demand.

Detroit News reported on September 29, 2015 that GM added a new position with the primary responsibility of overseeing GM's strategic decisions across of the world, combining projects of urban mobility, car sharing, carpooling services, and autonomous driving. In the process, they also acquired the Maven car-sharing brand founded by Lyft for $500 million. General Motors said in July 2016 that with the help of the Maven platform, the first driverless car would be launched for all consumers in the next few years, and an autonomous electric taxi fleet was scheduled to be tested in 2017. If the plan is successful, it will have a major impact on the traditional automobile manufacturing industry and rental services. Thanks to the popularity of car sharing, and the ease and low cost of online bookings, car rental companies have been forced to invest more in technology to maintain competitiveness. The sardonic complaints of car hirers should be enough to alert them—"Car rental companies now still need 20 min to do paperwork, which feels very twentieth century."

Ever-ostentatious Tesla claims that by simply clicking on its mobile App, anyone can join their driverless car to Tesla's sharing service fleet, with the revenue earned from this service used to offset the monthly car loan charges. Surprisingly, just as the partnership between Tesla and Uber seems to be on the horizon in 2015, Travis Kalanick, CEO of Uber, hinted that if Tesla could achieve full automation by 2020, Uber would buy every car produced by Tesla. However, in the blink of an eye, Elon Musk rejected Kalanick's proposal, making the tantalizing offer above part of the development blueprint. This shows that automakers see new business opportunities and are unwilling to pass them over, instead going into battle stripped to the waist.

In China, the progress of SAIC in their car time-sharing project is representative. In mid-2016, SAIC's "e-Enjoy Daily" (Fig. 4) completed a merger with Jiading International Automobile City EVCard. They planned to develop the operating vehicles from 2800 to 5000 within the year, and expand their network to 2000 points. SAIC also plans to expand its business to 100 major domestic cities and several foreign cities by 2020 with 300,000 electric vehicles. At the end of 2016, a service called "TOGO Tour" appeared on the streets of Shenzhen. This service allowed users to rent special vehicles parked in major public parking spaces at any time through their mobile phones. After arriving at the destination, they can park in any legal parking space to complete the whole process. The cost was calculated using time and mileage. From here, we can already glimpse the new development strategy and ambitions of these car companies. More importantly, these companies have detected new business opportunities: expanding from product providers to service providers, presenting transportation solutions for ordinary citizens, and becoming enterprises capable of "hardware" and "software" solutions.

At present, car sharing services face an insurmountable obstacle. For the dedicated platform, if its fleet is not autonomous, then there must be a costly hired driver. Uber data shows that the company's unit cost is $2.8/mile, 80% of which comes from the driver. When the vehicle enters full, driverless automation, its rental cost could be

Fig. 4 "e-Enjoy Daily" time-sharing car rental fleet. *Sources* SAIC

reduced to $0.53/mile. In the era of autonomous driving, the cost of hiring drivers will be greatly reduced, also reducing travel costs. Autonomous driving and the sharing economy naturally complement each other. The two combine to reduce travel costs by 70%.

"Autonomous driving" within a "sharing economy" can also cultivate user habits through shared vehicles, while reducing operating costs. Driverless cars are more expensive, and consumers' purchasing power limits the number of driverless cars entering the market in large quantities. However, corporations are less sensitive to this cost. In theory, vehicles can run 24 h a day and recover the cost faster. Therefore, the sharing economy can also accelerate the rate at which autonomous vehicles enter the market.

At present, the current system of travel rooted in the past century is undergoing major changes, to be replaced by a new ecosystem of interconnected individual travel. Automobile companies are turning away from large-scale transportation products in order to provide consumers with the most economical and convenient travel solutions, without the trouble of paying for cars, applying for licenses, parking and after-sales service concerns. According to eMarketer, a sharing economic research website, about 15 million Americans used shared travel or taxi services in 2016. By 2020, this number will rise to more than 20 million. After car sharing has been fully developed, buying a car will not be as convenient as renting a car, and 99% of people will not need to own their own private cars. The Boston Consulting Group (BSG) research report believes that the widespread use of driverless cars and robo-taxis in the urban area can reduce the number of cars on the street by 60%, emissions by 80% or more, and road traffic accidents by 90%.

The economic model of sharing everything and the technological trend of autonomous driving are recognized as the most important factor in promoting a fundamental shift in the relationship between cars and people. Car companies actively participate in the development of autonomous vehicles and the sharing economy, in order to adapt to social progress and future traffic development. This will inevitably bring about changes in the automotive industry chain.

As society shifts to car-sharing, demand for cars will become short-term and single, as its singular use will define the car's performance clearly and simply. Traditional understanding of vehicle function will no longer be applicable. Attributes based on identity (such as personalization) will gradually disappear. Designers won't have to scratch their heads about how to balance various contradictory needs, because users won't mind if the seat has a memory function or it doesn't have hi-fi sound. Instead, they can design vehicles with seats suited for different body sizes, and at different rental rates, according to consumer data. People will be able to choose different types of cars according to the different modes of travel, and completely break through the limitation of one person, one car.

At the same time, unrealistic and luxury functions can be omitted in vehicles used for sharing businesses. Influenced by the sharing concept, the auto parts and assemblies of the future will be standardized and generalized. Therefore, the structure of auto parts suppliers will change greatly, and the total number reduced by 30–40%.

Corresponding auto after-sales services, including dealers and repairers, will face a business reduction, intensifying the resultant corporate shuffling.

Though of course, the purpose of vehicle-sharing is to simplify, it also has complex aspects. Through their mobile devices, users are able to customize and personalize their experience within the car. In a more thoroughly-conceived autonomous vehicle, the car itself can be a simple capsule or a box, while the interior is completely personalized to resemble the design of one's own home. Unmanned driving has the potential to change car design even more profoundly. The structure of traditional automobiles is based on human driving needs, such as the visual field and layout of the controls. After autonomous driving technology matures, this foundation will no longer exist—the windshield can be completely turned into a screen, and people can ride in any orientation and sitting position.

In fact, the design changes initiated by the combination of autonomous vehicles and sharing have already begun to take shape abroad. South Korea's Ulsan University of Science and Technology (UNIST) is working with the country's electric car manufacturer Power Plaza to develop a small electric car with integrated an input interface for two to three people. The folding design of the car allows it to be used when standing and sitting, making it suitable for smart cities with complex traffic conditions.

Smart cities and intelligent transportation are being constructed in contemporary society. Car sharing and autonomous driving are the elements that promote each other's development. When autonomous driving technology is improving day by day, there is huge potential for development in car sharing.

7 Responsibility Assignment of Traffic Accidents

The popularity of autonomous driving technology can also cause legal problems. The issue of liability for traffic accidents will probably make lawyers argue endlessly, as there is no explicit provision in any laws on this issue. The responsibility for car accidents is currently mainly borne by the driver. If AI is involved in making driving decisions, this premise can be subverted. Once a traffic accident occurs, especially when human lives are involved, machines can potentially become killers. Legal issues are most problematic during the period when humans and machines drive on the road at the same time.

In March 2016, there was a slight collision between a Google's driverless car and a bus. According to Google, The vehicle "detected sandbags near a storm drain blocking its path, so it needed to come to a stop," Google said. "After waiting for some other vehicles to pass, our vehicle, still in autonomous mode, began angling back toward the center of the lane at around 2 mph—and made contact with the side of a passing bus traveling at 15 mph. Our car had detected the approaching bus, but predicted that it would yield to us because we were ahead of it." In fact, for human drivers, such misjudgments occur every day. "Obviously we need to take some responsibility, because if our vehicles were not moving, there would be no collision," Google admitted in a notice.

After Tesla's autopilot car had its third accident in China, the company immediately spoke up and attempted to wipe the slate clean. "When the road conditions became very uncertain, the car warned the driver to put his hand on the steering wheel, but the driver did not do this, then the car hit the roadside stakes." However, the car owner said that he did not hear the warning that danger was imminent, and measures needed to be taken before the accident. Not to mention that the car system uses English and he only speaks Mandarin. After that, the owner received a "careless driving" traffic accident ticket, which made him feel depressed for a few days. Unfortunately, some of the self-driving cars at this stage are still occasionally making industry headlines due to accidents, which show that this occurrence is not uncommon. One of China's limited edition 390 imported Volvo S90 sedans with ADAS functions has its first collision in Guangdong Province only two months after its launch. The car, which has not yet been licensed, crashed into an engineering vehicle at 115 km/h and set off a fire. Fortunately, the driver survived, but he was severely injured.

The accidents of Tesla and Volvo have exposed some safety hazards of current autonomous vehicles. The biggest controversy is how to divide responsibility after an accident occurs in an autonomous vehicle. However, autonomous vehicles are not clearly legislated, and are basically blank in terms of driving norms and responsibility determination. The lack of legislation has become an obstacle to the development of autonomous driving in the future. The most worrying thing in the auto industry is that vehicles are commercialized and used on the road without adequate testing.

Extended Reading

Trial of a driverless car accident

Dr. Noah J. Goodall of the Virginia Transportation Research Council in the US virtualized a trial of a driverless car accident in 2034 to see how the prosecution and the defense bothered.

It's 2034. A drunken man walking along a sidewalk at night trips and falls directly in front of a driverless car, which strikes him square on, killing him instantly. Had a human been at the wheel, the death would have been considered an accident because the pedestrian was clearly at fault and no reasonable person could have swerved in time. But the "reasonable person" legal standard for driver negligence disappeared back in the 2020s, when the proliferation of driverless cars reduced crash rates by 90 percent. Now the standard is that of the reasonable robot. The victim's family sues the vehicle manufacturer on that ground, claiming that, although the car didn't have time to brake, it could have swerved around the pedestrian, crossing the double yellow line and colliding with the empty driverless vehicle in the next lane. A reconstruction of the crash using data from the vehicle's own sensors confirms this. The plaintiff's attorney, deposing the car's lead software designer, asks: "Why didn't the car swerve?".

If it were a human driver, no court will ask "why the driver can't do something in a critical moment of traffic accidents". This question has no practical significance for responsibility determination—the driver loses the ability to think when panicked and can only react by instinct. However, if the car is controlled and driven by a robot, "why not do something?" is a question of legal effectiveness. The law's provisions on human ethics are not perfect, so software engineers are afraid to make various assumptions. One of the most important assumptions is that people with good judgment know when to implement the legal spirit and ignore the legal provisions. What software engineers must do now is to give good judgment to cars and other autonomous devices, that is, robots.

What is interesting is that car companies have also reacted differently to the issue of affiliation of rights and liabilities. At a news conference in Washington in 2015, Håkan Samuelsson, CEO of Volvo, promised that "when a car is in a driverless mode, Volvo will take full responsibility for the traffic accident." Volvo has become the first automaker in the world that made this commitment. An industry expert said in an earlier interview that if a self-driving car were to bear too many safety responsibilities, it would induce car companies to adopt a conservative driving strategy, which may not be conducive to the development of autonomous driving technology. Peter Mertens, Vice President of Volvo, had a different view: "taking full responsibility is good for auto companies, drivers and insurance companies, and even society as a whole. There is no difference between autonomous vehicles and traditional cars in assuming responsibility. You simply shouldn't blame a driver for faults not committed by them."

Unlike Volvo, Carlos Ghosn, then CEO of Renault-Nissan, has stated that Renault-Nissan could be liable circumstantially and conditionally if an accident were to occur with an autonomous vehicle. According to Ghosn, car companies should be responsible for the products they develop, but drivers should be responsible for driving behavior. At the same time, car companies must take steps to educate customers about how much control they should give to autonomous driving. "Autonomous driving is not a substitute for a driver. Switching between the two modes is the key." An executive at Mercedes-Benz skirted the issue of responsibility and instead stressed that when compared with human drivers, autonomous vehicles have a stronger ability to avoid accidents. "In some traffic situations, the current drivers can not properly handle it in time. From the natural law, we can't forecast the traffic conditions at present, neither autonomous vehicles. Moreover, the handling capacity of self-driving cars will be far better than ordinary drivers."

The legal problems caused by autonomous driving have aroused the interest of Chinese and foreign legal workers. The revision or formulation of relevant laws and regulations has been launched. The UN approved amendments to the Vienna Convention and UN Regulation No. 79 in 2016, which means that most of the 72 signatory countries, including Europe and the US, allow cars with associated functions to be driven automatically during certain periods. The ratification of this amendment is the first time that autonomous driving has been permitted at the legal level. The one responsible for driving is no longer necessarily a person, but perhaps a car itself, opening the door to the legalization of autonomous driving.

The Federal Automated Vehicle Policy issued by the US Department of Transportation in September 2016 focuses on autonomous driving in terms of vehicle performance guidelines, state government regulatory models, NHTSA's existing and new regulatory methods, and provides a guiding pre-regulatory framework to traditional automakers and other institutions for the production, design, supply, testing, sales, operation and application of autonomous vehicles. In addition, 14 states in the US are trying to legislate for autonomous driving, yet there is no clear law to define liability.

Germany is now revising its domestic road traffic regulations so that networked autonomous vehicles are also incorporated into the legal framework. This means that if an automobile adopts the autonomous and interconnected driving mode, the driver is not responsible if an accident occurs.

Japanese National Police Agency issued the first guidelines for the self-driving car highway test conducted by a company in April 2016. The Agency has begun discussions at the legal level to complete the legislation by 2020, in response to issues such as the responsibility of accidents related to an autonomous vehicle without a driver and without steering wheel control.

A well-known lawyer in China has written a paper discussing the issue of responsibility determination in the context of autonomous driving. He said that in the current traditional driving situation dominated by human beings, the main legal relationships are as follows: first, an automakers takes legal responsibility for the buyer as the aggregator of automobile hardware and software; second, a driver takes legal responsibility for the consequences of automobile as a means of transportation; the third is the management of traffic order, which is centered on the above two, delineating the rights and obligations of drivers, pedestrians and other traffic participants, transportation facilities and order managers. Manufacturers' management addresses the problems of qualified automobiles; driver management addresses solves the problems of qualified drivers; traffic facilities and order management mainly addresses the issue of qualified hardware conditions and soft environment problems required for driving.

Under traditional driving conditions, an automobile is a machine operated by a person, and it is generally accepted that the person bears the legal responsibility for the consequences caused by the vehicle. In the case of autonomous driving, the situation changes. The first problem that may be encountered is that automakers may have a separation of software and hardware. An automaker is not a participant in the traffic behavior and driving on the road is a driver's behavior, but in the case of autonomous driving, whether or not the automaker is a participant in will become a topic of debate. If a traffic accident occurs, whether it is a vehicle hardware problem, a software problem, a transportation facility problem or something else, it needs to be determined according to the new traffic regulations. The resulting problem would be an increase in the number of traffic accidents, and the absolute future number will not be small even after the popularity of autonomous driving with accidents substantially decreased, contributing to the problem of liability and compensation that needs to be solved. In the past, traffic accidents were undertaken by the driver through insurance, or by himself. It seems unreasonable that in the case of autonomous driving, the driver is still responsible for the accident through inaction. Then, the risk of accidents is either borne by the insurance insured by the occupants, or by the hardware and software manufacturers that produce and sell the automobile products. The risk will be concentrated from the dispersed tens of thousands of drivers to a much smaller number of automakers and software manufacturers. Secondly, does the social role of a driver still exist? Autonomous driving and unmanned driving are not the same concept. Autonomous driving can have human intervention; whether it is autonomous driving or unmanned driving, there is still someone in control. The key is who controls and how to control it.

The above discussion only considers the manufacturers and drivers of autonomous vehicles and does not take the vehicle itself out as a responsible entity. Every taxi driver is an independent legal person and has his own insurance. Today, when robots are doing the same or even more difficult work as natural persons, should we consider whether these robots should also gain the legal status, let them buy insurance and participate in legal cases? Dehumanized legal liability has long existed, and a limited liability company is a good example. Of course, intelligent robots are different from natural persons and companies. Defining them as legal liability subjects will certainly challenge the existing legal system. Therefore, autonomous driving is not only a technical problem, but also faces greater challenges from ethics and social law.

In any case, autonomous driving cannot be barred from exploration for fear of accident. However, on the road of technology research and development and commercialization, more attention should be paid to safety and legal follow-up and formulation. These potential dangers require legal protection, and we can no longer allow the popular autonomous driving in a gray area.

8 Autonomous Driving Reshapes Auto Insurance Sector

Auto insurance is one of the pillars of the insurance industry. According to a study by Morgan Stanley and the Boston Consulting Group (BCG), auto insurance brings about $260 billion in premium income and $17 billion in annual profits to major insurance companies around the world. They estimate that the market value of the auto insurance industry is about $200 billion.

The development of autonomous driving technology will have a disruptive impact on the insurance industry. When asked if autonomous driving technology would become a problem in the auto insurance sector in an interview with CNBC in May 2016, Warren Buffett, Chairman of Berkshire Hathaway Inc., (also the world's top billionaire who has made a lot of money from his insurance company,) answered, The answer is yes. I think it will be a long process, but the result is beyond doubt. What makes cars safer is good for the society, but bad for the auto insurance industry.

Nowadays, more and more companies are beginning to get involved in autonomous vehicles. The promotion and entry of autonomous vehicles is now inseparable from the support of the insurance industry. For autonomous vehicles, insurance

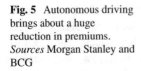

Fig. 5 Autonomous driving brings about a huge reduction in premiums. *Sources* Morgan Stanley and BCG

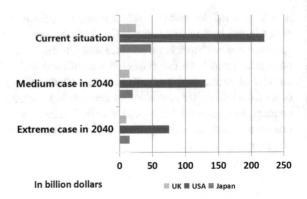

companies must set some standards to who should be held liable after a traffic accident. Some insurance companies believe that the assessment of liability is out of question. Conversely, some insurers are concerned that autonomous vehicles will make the car too safe after replacing humans. If this happens, it will directly affect their business (Fig. 5).

With the development of autonomous driving technology, changes in driving modes will lead to a decline or even disappearance of demand for some auto insurance products, and new types of insurance customized for autonomous vehicles will be born. For example, after the realization of autonomous driving, smart cars based on IoT applications will be able to be summoned anytime, anywhere, greatly reducing the probability of car theft, robbery and other accidents, and the demand for piracy and rescue will gradually diminish. The UK is negotiating amendments for autonomous vehicles in terms of insurance regulations and vehicle regulations. Adrian Flux of the UK has pioneered the "driverless driving insurance" business for autonomous vehicles, the first attempt of this kind by the insurance industry. The business includes a number of insurance terms that are exclusive to autonomous vehicles.

First, some system updates or security patches, such as driverless systems, firewalls, digital maps, etc., are not successfully installed within 24 h of the user being notified.

Second, when there is no satellite signal/signal failure, the navigation system fails or the vehicle operating system software is abnormal.

Third, when the collision prevention system, the operating system or the navigation system fails, human intervention is required but not executed timely.

Fourth, when the vehicle is attacked by hackers.

The British insurer's additional "driverless" clause is tailored to Tesla's autopilot, which has been the subject of recent concern.

Japan's Tokio Marine and Nichido Fire Insurance has included auto insurance coverage for traffic accidents during self-driving since April 2017. This is also the first insurance company in Japan to include autonomous driving in the scope of claims. After auto insurance signing and renewal, self-driving insurance will be given to users free of charge. This insurance is believed to support autonomous driving. In the case identifying defects in the autonomous driving system are defective, the insurance company will pay the insurance premium even if the cause of the accident is unknown and the driver is not liable. These situations are not uncommon, because different accidents such as hacking, braking failure, and sudden flameout are possible during autonomous driving.

In addition, California enacted the law on automated vehicles, which requires insurance coverage of more than \$5 million to apply for an autonomous driving license in order to deal with possible personal casualties and property losses caused by autonomous vehicles.

According to McKinsey's analysis, the full popularity of autonomous vehicles can reduce traffic accidents by 90%, reduce medical expenses and damages by \$190 billion annually, and save thousands of lives. Moody's analysis of self-driving cars released in March 2016 also supported McKinsey's conclusion of fewer accidents caused by cars as autonomous driving technology becomes more commonly used, which means that the demand for auto insurance would be reduced. By 2040, in a mature economy, the size of the auto insurance market may shrink by more than 80%. For insurance companies, the incidence of car accidents is reduced, and the car insurance claims rate is also reduced, thereby increasing their profit margin. However, on the other hand, the improvement of automation is increasing the cost of auto insurance, because the maintenance cost of a damaged autonomous vehicle, with its sophisticated equipment, is more expensive than that of a basic type vehicle. Comprehensively, the reduction of the incidence of traffic accidents caused by autonomous driving will inevitably lead to a decline in insurance rates, and fierce competition among insurance companies will even accelerate this process.

In addition, the insured will gradually change from owner to automaker, supplier of software and hardware and car rental company. At present, the policyholder of auto insurance is mainly an individual (owner). Auto insurance basically provides protection for accidents caused by human error. The sales model and profit model are based on the operation of individuals (car owners) as policyholders. With the application of autonomous driving technology, the role of the insured will gradually change: first, the responsibility of the individual owner will be transferred to the automaker, software and hardware supplier that provides autonomous driving technology, and the insured will be transformed into an automobile manufacturing enterprise, hardware and software suppliers; second, the rise of the car rental industry, a large number of car rental companies will provide self-driving cars, the insured will be converted into a car rental company. And as automakers, technology companies, and car rental companies improve their data collection and application, they may be in a better position to sell insurance than the insurance industry itself.

Moreover, the business model of auto insurance will also change. The UBI (Usage Based Insurance) insurance model is now receiving much attention and is considered an important direction for the future development of auto insurance. According to the analysis of the Insurance Society of China, UBI is very likely to subvert the traditional auto insurance. First, the "subversion" of insurance pricing, with the help of UBI, insurance companies can price based on more dimensions besides the number of risks; second, "subversion" in product form, first of all, there are auto insurance products charged according to the use of "quantity". The unit mileage rate varies according to the risk characteristics of the insured and the vehicle; the third, the "subversion" of the customer experience, with the leap of intelligent level, a car can be monitored and used without dead angle monitoring. For privately owned self-driving cars, shared autonomous vehicles and even driverless cars, UBI mode can be used for more scientific and accurate actuarial calculations to determine premiums individually and settle claims quickly.

What's more, insurance companies can participate intimately the auto industry in a broader context and greater extent. In addition to investigating the insurance related to autonomous vehicles, insurance companies can cooperate with automakers, software and hardware suppliers and car rental companies that develop autonomous driving technologies to establish strategic alliances to jointly understand and discover new insurance needs and provide insurance solutions and to advise self-driving cars entering the market on how to be accepted by consumers as soon as possible, and to help solve the pain points of manufacturers and consumers. In the gradually formed stage of the autonomous driving car market, insurance companies can cooperate with automakers and software and hardware suppliers to make recommendations to the government, formulate laws and regulations concerning the driving of self-driving cars, clarify the rules for defining insurance liability, and promote the introduction of laws and regulations. This would not only help drive the development of autonomous driving by eliminating the lack of laws and regulations or outdated laws and regulations, but also allow it to play a unique role in the insurance industry. In the process of adapting to the current self-salvation and finding new growth points, the auto insurance industry can be born again.

9   The Road to Commercialization

At the end of October 2016, on Interstate 25 in Colorado, the sharp-eyed drivers seemed to feel the breath of Halloween in advance because they found no one in the driver's position on an Uber Otto 18-wheeler that flashed past them. This was not a Halloween prank. This was Uber Otto's first unmanned commercial vehicle (Fig. 6), which was a big shock to everybody. The truck drove in the autonomous driving mode all the way from Fort Collins to Colorado Springs, 120 miles south, while the human driver monitored from the rear lounge area, and took over to drive manually in the urban area before the highway entrance and after the highway exit.

Anthony Levandowski, the technological genius of Google's driverless team, and Lior Ron, the head of Google Maps, left Google and founded the startup Otto in early 2016. Just six months later, Uber announced a $680 million takeover of Otto. Considering that the startup was only established for half a year, it can be concluded that they mainly relied on the technology developed at Google. Now, the commercial value of this technology has been recognized by investors, and the

Fig. 6 The first unmanned commercial transportation. *Sources* Uber

commercialization prospects are very clear—providing autonomous driving operation services for freight trucks. The test included 50,000 cans of Budweiser beer from Anheuser-Busch, which was installed on the freight truck just mentioned.

Switching from the human driver mode to the autonomous driving mode, the driver simply presses a button that says "Start". The modification of the Uber Otto truck costs nearly $30,000—two cameras for lane detection, a LiDAR to build a three-dimensional external environment, two front millimeter-wave radars to detect obstacles and other vehicles on the road, and a GPS to help locate the truck. The company carefully studied the high-speed route in advance to ensure nothing went wrong with the truck while on the road, and then carried out the transportation plan under good traffic and weather conditions. In the US market, a large truck is now priced at more than $150,000. Taking about the potential to cut cost, Otto's current autopilot kit is very likely to reduce from $30,000 to $10,000 in five years.

Autonomous cars are more popular than autonomous trucks, but the latter are more necessary now because they can have a direct impact on the current economy. Anheuser-Busch estimates that by deploying self-driving trucks in its distribution network, it can save $50 million annually in the US alone. Assuming that autonomous trucks eventually become popular throughout the freight transport sector, the total cost savings are estimated to be billions of dollars or more.

The US truck transportation sector is worth $700 billion. Highway transportation is well developed across the US, while railway transportation is relatively weak; 70% of the goods are transported by trucks. The total freight tonnage is expected to increase by 35% in the next decade. There were 1.6 million truck drivers in the US in 2015, accounting for 1% of the US workforce. The average age of truck drivers is 55 years old, indicating that the sector is not attractive enough for young people and it is difficult to recruit. According to the statistics of the American Trucking Association, there is a gap of 47,500 large truck drivers, which has become a difficult problem for logistics companies. According to this trend, this number will reach 175,000 by 2024. If logistics companies adopt autonomous driving, the labor cost savings would be significant, the service areas, motels, catering and other expenses basically not needed, and auto insurance costs greatly reduced; in terms of production efficiency, the autonomous vehicles can travel 24 h a day, 7 days a week. Including vehicle maintenance and cargo loading and unloading time are taken into consideration; the total driving time of 140 h per week is more than three times that of a human driver. This means a synchronous increase in capital turnover. Although accurate cost calculations still require more data, the above information already suggests that the cost of logistics based on autonomous driving will fall below a quarter, and the return on capital will probably increase by more than 10 times. The rise of self-driving trucks will also profoundly change the logistics industry, and places like standardized loading and unloading stations will emerge. In this location, human drivers will be responsible for the delivery of the goods to the self-driving trucks that have just been exited from highways, completing the final door-to-door transportation.

At this stage, the manned automatic driving truck technology can help release some of the driver's burden, rather than replace their work. Of course, the truck driver can take a break during the vehicle's travel on the highway, avoid fatigue driving, eliminate unnecessary hidden dangers, thus save valuable time and reach the destination faster. In addition, self-driving trucks may also make roads safer. According to the Insurance Institute for Highway Safety (IIHS), about one-tenth of the road deaths are related to large trucks. According to US Department of Transportation, the annual mileage of US trucks accounts for 5.6% of all vehicles, but the accident rate is as high as 9.5%. There are about 4 million truck traffic accidents each year, killing about 4000 people, and almost every accident is caused by human factors. Data from the Traffic Administration Bureau of the Ministry of Public Security also show that trucks account for 7.8% of the national motor vehicle ownership, but they produce 27.7% of fatal traffic accidents in China.

The entry of trucks into the field of autonomous driving has its unique advantages: because trucks are mostly seen on night and highway, road conditions are much simpler than urban roads; there are no parking signs, and very few pedestrians. If transportation time is not tight, there is no need to change lanes; the sensor can be mounted at a higher position from the ground and can detect farther. As a result, autonomous vehicles in this field can achieve the maturity required for commercialization in a short time, and the technical resistance in the early stage is smaller. Uber has been providing self-driving freight services since 2018, and more companies have started to follow up. Six major European truck manufacturers (Volvo, Daimler, Duff, Iveco, Mann, Scania) have already formed a driverless truck fleet of more than 12 vehicles for road testing.

China's truck market has great potential. As the world's second largest economy that is still under a rapid development, China has numerous transportation tasks to be handled every day. At present, there are about 7.2 million trucks in China, and the number of truck drivers is as high as 16 million. Many freight missions are extremely long and require two or three drivers to drive at once. Because of the country's size, the yearly mileage of these trucks is staggering. According to statistics from the

Ministry of Transport, the total length of highways in China has reached 117,000 km as of the end of 2015, ranking first in the world. The above factors suggest the market demand for self-driving trucks in China is very urgent. In the second half of 2016, Baidu and Foton launched an 18-wheel self-driving giant super truck, which is based on level-3 autonomous driving technology and Baidu high-precision map as well as the autonomous driving technology and commercial vehicle big data developed by Foton. This self-driving super truck is very practical, with a short-term target to solve the problem of logistics and transportation in the country.

Lior Ron, co-founder of Otto, said in October 2016, "The existing autonomous driving technology is fully qualified for commercial transportation services. In the future, we will continue to polish this technology in order to adapt it to various complex road conditions." Gong Yueqiong, Executive Deputy General Manager of Futon commented, "Futon Automobile has been developing driverless technology. This technology will be first applied to the Foton super truck. It is expected that the unmanned Foton super truck will be seen driving on the road by the end of 2016, as we are currently conducting the final field test."

Looking back at Otto's success and Futon's ambition, they are beautiful cases of technology and market integration. The development of a technology may initially face a very daunting goal, but after the technology reaches a certain level, it can produce commercial value in some scenarios. Therefore, by limiting the use scenarios, the difficulty of technology implementation can be reduced. In the case of highway freight, we can further reduce the technical difficulty by using formation driving with a leader, in this mode, a human driver drives the first car, followed by 5–10 driverless cars running at dense intervals, which means that labor efficiency is increased by 5–10 times, and its commercial value is evident.

In addition to the commercialization of autonomous driving for highway wagons, the commercial trial based on sharing economic concept is also being developed in the area of self-driving cars. Technology companies such as Google and Apple, despite their long R&D time and top-notch technology, have no business model in shape. Uber has first created a business model and quickly mastered the technology through cooperation and acquisition. In order to build the Pittsburgh unmanned fleet, Uber has invested $300 million in Volvo in addition to buying Volvo cars, and the two companies are cooperating in autonomous driving technology and car supply. Uber's daily operations generate approximately 100 million miles of driving data, which can help develop autonomous driving technology by means of high-precision maps and navigation. Accurate map navigation will have a crucial impact on the operation of the taxi. With this data, driverless cars can complement human drivers and sweep away their blind spots.

The world's first driverless taxi was unveiled in Singapore in September 2016, and members of the public will be able to call for rides on their smartphones. Running these driverless taxis is the startup nuTonomy, whose experimental run in Singapore is even a few weeks ahead of Uber's driverless taxi service in Pittsburgh, US. nuTonomy executives said the company's goal is to have a fully unmanned taxi fleet in Singapore by 2018; this would significantly reduce car ownership in Singapore and ease road congestion. If the commercialization direction of autonomous vehicles is recognized by the public, the auto service industry would find a new flash point—the above-mentioned driverless taxi is going to be the third space in addition to home and office, and could be used as mobile commercial real estate, mobile cinema, mobile office space, mobile cafes, and more.

On November 13, 2016, a passenger-carrying test of an unmanned bus was carried out on the highway in Senpoku, Akita Prefecture. Senpoku is one of Japan's "national strategic special economic zones" and has therefore become the first test site in the country to adopt steering wheel-free autonomous vehicles. The test used a "robot shuttle" developed by mobile game company DeNA, which can take 12 people without a driver's seat, steering wheel and accelerator pedal. The project team deliberately recruited more than 60 citizens to participate in the test ride. The bus determines the driving position through GPS and uses the sensor to sense the surrounding environment. During the test, the road along Lake Tazawa in Senpoku was blocked, and the unmanned bus passengers traveled four times along the one-way 400 m route and they felt good.

The autonomous driving sector involves three "trillion-dollar" markets: the global auto market, the travel market and the extra benefits that autonomous driving brings to the social economy, apart from auto industry. Morgan Stanley's research report points out that autonomous driving will bring \$1.3 trillion in revenue to the US each year, from fuel savings, congestion mitigation, accident reduction and productivity gains.

According to IHS, McKinsey and other forecasts, autonomous driving technology will be applied first to highways and fleets around 2020, and relevant laws and regulations and industry standards will be introduced. A completely driverless car will take longer to realize. BMI Research, a market research company, held a seminar in 2016, of which the most important conclusion was that it might take 15–20 years for a completely driverless car to come out.

Automakers have made it clear that high-performance cars will be the last to adopt driverless technology. BMI Research said in a report, "We expect that driverless technology will be mainly used in mainstream cars and commercial vehicles, and there are already freight vehicles that are basically unmanned. We expect that driverless technology is more suitable for shared travel services in the early stage, it is more convenient for fleet managers—sometimes automakers themselves, to deal with insurance, maintenance and other issues, the fleet can benefit from economies of scale."

Although experts predict that completely unmanned vehicles will not be able to truly become marketable products in the short term, this does not hinder the development of smart car research. The ADAS functions such as adaptive cruise control, automatic parking and risk warning have been implemented on some car models. Developed initially from traditional cars, modern smart cars rely on environmental perception technology and artificial intelligence to improve people's quality of life. They follow the rule of development from low to high, from few to many, from single point to multi-faceted, continuous integration and improvement. Its horizontal development is inseparable from the actual needs of various users, and its vertical vitality lies in continuous technological innovation.

In an article posted on leiphone.com in October 2016, Li Xingyu, Director of Autonomous Driving Business of Horizon Robotics, argued that there are many sectors with the potential for commercialization in a short time, such as:

Warehousing and logistics industry: Leading e-commerce companies such as Amazon and JD.com have deployed Automated Guided Vehicles (AGVs).

Autonomous vehicles for agriculture: including agricultural machinery that can be used for farming and harvesting. It is not difficult to move at low speeds on unpaved areas, and they can be transported by other vehicles to that area.

Partially closed places: such as resorts, tourist attractions, airports, mining areas, docks, construction sites, etc. Most of the vehicles in this application scenario are special vehicles, such as excavators, cranes and small electric vehicles. As technology advances, more autonomous driving scenarios will be possible.

Urban public transport system: There are fixed driving routes, such as bus lanes, which can selectively implement autonomous driving.

Commercial operating vehicles: such as taxis, company shuttles, etc.

For private vehicles, the popularity of autonomous driving applications can be gradually expanded according to the different scope of the scene, first on the highway, then on the parking lot, and finally on the open road. A survey from General Motors shows that in mega-cities, 30% of gasoline is wasted in the search for parking spaces, while the parking time in the central city usually exceeds 15 min. Autonomous driving within a parking lot has great potential. The driving environment in parking lots is relatively friendly as weather factor doesn't matter and the speed is low. It is a closed place, and its implementation difficulty is lower than that of the open road.

Once autonomous driving becomes popular among operating passenger cars and commercial vehicles, their success will invite private passenger cars to follow the trend. This allows end consumers to recognize the autonomous driving technology, and makes them more willing to purchase their own autonomous cars.

Another possibility is the rise of the repacking market. As long as self-driving software is on the market, people can use this technology to refit old cars and make them autonomous. Although NHTSA in the US has questioned the safety of modified cars, some companies have indicated that they will use autonomous driving technology to retrofit old cars.

The commercialization path of autonomous driving is bound to be different in different countries. Because the cost structure of the same application scenario is different, (such as in the taxi industry,) the labor cost of taxis in the US is much higher than that of China, which is one of the reasons why Uber's investment in autonomous driving technology is so radical.

The actual utilization rate of a private car is usually less than 10%, and the rest 90% of the time is in a state of suspension, which brings intractable urban diseases: parking difficulties, congestion, and air pollution. If driverless cars are accepted by most residents in the future and the sharing economy concept is deeply rooted in the hearts of the people, then the number of cars on the road will be greatly reduced. Secondly, due to the comprehensive coverage of V2X, the traffic flow can smoothed, congestion eased, and travel efficiency improved. It is also possible to cut fuel consumption and pollution by reducing the idle state and making cars run at economic speeds. The energy consumption and pollution could even be lessened, based on the widespread use of electric vehicles. Thanks to the formation of new traffic formats, traffic facilities including traffic lights, parking lots, etc. will undergo some adaptive optimization and upgrading, which will release more resources that benefit society, such as land resources. For mega-cities such as Beijing, this may be a good remedy for their metropolitan traffic problems.

In addition to the continuous maturity of technology and the continuous reduction of costs, the government's constant improvement of relevant policies and regulations are also indispensable to the commercialization of autonomous vehicles, especially for the standardization of V2X and the laying of infrastructure.

10 Excited and Confused Users

The era of autonomous vehicles is coming quickly. The World Economic Forum (WEF) and the Boston Consulting Group (BCG) collaborated on a survey in 2015, which is a large-scale survey of the world's autonomous vehicles, covering 5500 consumers in 10 countries around the world. The results of the survey might help to gain a deeper understanding of consumers' confidence in the development of the auto industry, to grasp the degree of customer acceptance of autonomous vehicles, and to provide unique insights into the support of policymakers in different cities around the world for autonomous vehicles. The survey shows that emerging market countries including China, India and the United Arab Emirates have the highest acceptance of autonomous vehicles, with as high as 75% of Chinese consumers in support; the acceptance rates in Britain and the US are about 50% while those in Japan and Germany are the lowest.

As early as the spring of 2012, when Google announced that its self-driving cars had been running safely for more than 200 thousand miles, a video was released showing a blind user called Steve invited to experience the self-driving car. He was sitting in the driver's position for the first time, and started the engine. Google's self-driving car first took Steve to a drive-thru fast food restaurant to buy the taco, and then went to the dry cleaner to pick up the clothes. When parking, the car automatically identified and parked in the disabled parking space accurately. Throughout the process, the self-driving car performed extremely well. The netizens who watched this video were amazed, and talked about it passionately, praising the car for its function.

It seems that people are immersed in the wonderful experience that autonomous driving brings to them, and the future of human travel will be full of pleasure and ease. From the realization of technical solutions, autonomous driving brings a new experience of human riding by liberating human hands, feet and even the brain, improving the comfort of travel.

Most people are anticipating such a revolution in the automotive field. However, they still hope to retain steering wheels in driverless cars out of safety considerations and nostalgia for the driving experience. In Volvo's survey on driverless cars, 72% of consumers believe that the option of human driven cars should be reserved, while more than half of consumers think that driverless cars should be equipped with steering wheels. According to BCG research data, among respondents who are unwilling to have a fully autonomous vehicle, 51% of them say that self-driving cars make them feel insecure and 45% think that the lack of control sense is the biggest obstacle. Consumer concerns include the inability to intervene in autonomous vehicles while driving, and the risk of cyberattacks.

Some people worry: if the technology content of a vehicle is too high and the electronic equipment is too much, will it be more prone to problems? Like smartphones, the more functions, the more problems. Some functions are useless, but come pre-installed. The experience is not satisfactory. Some also worry that cars will stumble on the road—if a small accident leads to sensor malfunction, would it bring a new safety hazard? This shows that people are not only excited about new things, but also a little confused.

There are also some people who instinctively resist the unknown. Just like when it comes to Google's driverless cars, what they care about is not how much the technology has improved, but what accidents have happened. Media networks have criticized them, stating "a driverless car would have serious judgment errors in the face of complex traffic conditions." The use of "serious" could be misleading, and make people insecure about driverless cars feel more anxious.

The public has no consensus on autonomous vehicles, and even public figures use different wording to describe this phenomenon. Former US President Barack Obama used the word "self-driving" in a Pittsburgh newspaper in September 2016, and his Transportation Minister Anthony Foxx used "automated vehicle" when he announced the Federal Automated Vehicle Policy in Detroit.

Some unreliable media and automakers are deliberately misleading consumers, blurring the different concepts of "autonomous driving," "driverless driving" and "assisted driving" on purpose, so that their products seem more magnificent. What consumers don't see is the small print list of restrictions telling you that you can't do one thing or another. Manufacturers use this method to evade responsibility to the maximum extent and attract potential customers with so-called new concepts and functions. In reality, it puts users in a dangerous situation.

Tesla's autopilot accident may not be entirely Tesla's responsibility. It may be because the user has not carefully read the small font notes and misunderstood the scope of Tesla's autopilot, thus failed to realize where the boundaries of Tesla's autopilot are. Faced with new technology or a new application, users, producers and technicians are often not at the same level of understanding. This level of understanding varies amongst users, an issue that should be solved slowly by slowly coming to a mutual level with the industry.

In an interview with reporters in the middle of 2016, Transportation Minister Anthony Foxx said that the acceptance of autonomous driving technology is an issue that should be paid attention to by the industry and the government. One of the most concerned topics for ordinary consumers is "whether the technology is safe enough." Though people will feel scared when they first try to ride a self-driving car, they will soon become accustomed to it. Foxx believes that people will have confidence in autonomous driving technology in the future. Moreover, this technical concept will soon appear in all aspects of people's lives, such as self-driving trucks, ships, and trains.

If a collision happens, drivers are usually responsible. But, once an autonomous vehicle-related accident happens, it is difficult to define the responsible party and pay compensation. Nearly 80% of consumers say they want automakers to come forward to help them settle claims or take responsibility in that case.

The governments of the US and Germany asked for autonomous vehicles to install recorders and share data under certain conditions. This practice has touched a nerve with modern people, with a high sense of personal privacy.

In addition to concerns about safety and privacy, some professional drivers have a mixed feeling of autonomous driving. For example, truck drivers in the US are very resistant to autonomous driving. A 32-year-old Texan Wade Dowden, who has been working as a truck driver for over ten years, bluntly said: "Let's buy a self-driving truck, and then we go to sleep? Damn it! It's impossible for the truck to drive on the road by itself. I can't do that." Dowden is not afraid of technology. He said that convenient digital maps and online entertainment have made truck drivers less stressed than before. Now Dowden listens to audio books and podcasts every day. When he is not driving, he watches Netflix videos and studies traffic management. However, he is not ready to hand over his truck to a computer, "it would make the job much less valuable. Once a driver only needs to drive a few miles in the city, we won't get any chance to raise our salaries."

The underlying problem is obviously not limited to the ability to raise salary. In the future, the popularity of driverless trucks will result in at least 1% of the working population being unemployed, although the arrival of this day should be very long. Some analysts believe that when the driverless car revolution gains a foothold in real life, it will not only have an impact on the auto industry, but also impact employment. According to the US Department of Labor, 884,000 people are engaged in the vehicle and parts manufacturing industry, 3.02 million are engaged in car dealers and automotive maintenance industry, and 6 million are professional automobile drivers. All this adds up to about 10 million people in the country whose jobs will be obsolete in the next 10–15 years. Extrapolated across the world, this number would be much larger.

However, there's no need to be too pessimistic. We will usher in an unprecedented era of high efficiency, and will generate new business forms, production methods, work patterns and employment opportunities. We can re-create employment, and society will surely reach a new balance in the new form. We should also see that high efficiency can not only require less labor, but also create the wealth that human beings need more quickly, thus improving the well-being of humankind as a whole.

New changes are taking place. The only thing to do is to embrace them.

Chapter 7
The Battle to Embrace the Trend

For the traditional automakers, the disadvantage of autonomous driving outweighs the advantage. Firstly, because the most critical software, algorithm and artificial intelligence technology are not their strengths, as autonomous driving matures, automakers will face sever competition from technology companies such as Google and Apple. Secondly, autonomous driving will promote the use of car-sharing, which will definitely affect the sales of new cars. Therefore, autonomous driving is almost a revolution for the traditional automakers. Without the pressure from encroaching companies as Google and Tesla, traditional automakers may lack the motivation to actively develop autonomous driving technology. Yet when the tide is irreversible, traditional automakers have no choice to follow this revolution.

The current development of autonomous driving technology is dominated by two camps: traditional automakers and technology companies. At present, technology companies are represented by Google, Apple, etc., while the traditional automakers are mainly Mercedes-Benz, BMW, Volvo and so on. The two camps are more competitive than cooperative due to obvious differences in respective technical solutions and implementation paths (Fig. 1), but limited cooperation has also begun to emerge, such as the case of Google partnering with Fiat, Chrysler and Honda.

Traditional automakers have adopted an incumbent approach towards driverless vehicles. They have upgraded and improved the assistant driving technology based on years of experience in the manufacturing and technology of the auto industry, that will eventually accomplish autonomous driving. By virtue of their own scientific and technological strength, technology companies try to achieve leapfrog development in unmanned driving technology by means of software technology, artificial intelligence, big data processing, high-precision digital mapping, diversification of research fields and global talents.

Chairman of Mobileye Amnon Shashua gave a speech at the International Consumer Electronics Show (CES) in 2016 outlining the different technological solutions adopted by traditional automakers and technology companies in the development of driverless technology. While traditional automakers begin with partial

© China Machine Press, Beijing and Springer Nature Singapore Pte Ltd. 2021
Z. Chai et al., *Autonomous Driving Changes the Future*,
https://doi.org/10.1007/978-981-15-6728-5_7

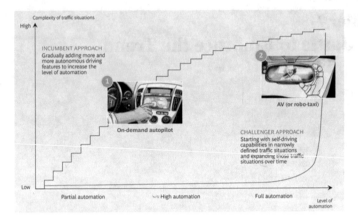

Fig. 1 Two vastly different approaches for developing autonomous vehicles. *Source* World Economic Forum and BCG analysis

automation throughout the vehicle, through low-precision maps and high-precision sensors and cameras, tech companies prioritize completely unmanned testing, aided high-precision 3D map and low-precision sensors (LiDAR) in fixed areas. Although these camps diverge in routing, the ultimate goal is the same. No matter which plan, the future goal is to achieve fully unmanned driving under all road conditions. Technology companies rely on the scalability and real-time updates of high-precision maps, while traditional automakers hope to improve map accuracy and artificial intelligence. However, a series of accidents in autonomous vehicles have also made clear that safety needs to be considered first.

Put simply, traditional automakers collect road data in real time through the high-precision in-vehicle cameras and feed them back to the vehicle which analyzes the data through algorithms and makes real-time decisions. Technology companies scan the road surface in real time through LiDAR, draws high-precision 3D topographic maps and uploads them to the in-vehicle computer. The computer then makes decisions by analyzing the data, so that the moving vehicle can act accordingly in real time.

Shashua believes that there are several practical obstacles in the current driverless solutions presented by technology companies. Firstly, he argues that the amount of data is too large. The roughly 3–4 GB/km of data generated by the moving vehicle needs to be compared with high-precision map data in real time. Yet, real-time transmission of such a large amount of data still has many obstacles in the existing wireless network environment. Intel recently predicted that by 2020, every smart car will generate about 4000 GB of data per day. Secondly, the real-time processing of data must be completed in-vehicle rather than through the cloud. Therefore, the unmanned cars currently being tested by Google and Baidu are basically high-performance mobile computers. Additionally, the cost is too high. The total cost of a set of Google's driverless systems was $300,000–400,000 at the beginning of

2016. Although the total cost has been brought down significantly due to the cost reduction of LiDAR, there is still a long way to go to achieve mass production.

In contrast, traditional automakers' technical solutions currently hold obvious advantages. Firstly, the traditional automakers have many years of auto industry manufacturing foundation; secondly, their high-precision real-time sensor acquires about 10 KB data per kilometer, which is very small and can be transmitted in real-time; thirdly, this solution can be used in almost any area, and does not require high-precision map data. The biggest advantage of traditional automakers is that most of the cars currently on the roads are made by those companies. They simply need to upgrade existing cars by adding unmanned driving technology to make self-driving cars in the future. In this sense, the set of solutions advanced by technology companies seems like starting from scratch.

From this strength comparison, it seems that the traditional automakers occupy an advantage at present. Although Google, Baidu and other technology companies have a prominent position in the field of science and technology, they are still weak compared with traditional automakers in the field of industrial production. Events like BMW's cooperation with Intel and Mobileye, GM's acquisition of Cruise Automation and Volvo's establishment of a joint venture with Autoliv have further consolidated the power of traditional automakers.

Nonetheless, as the industry develops, future infrastructure improvement might tip the balance in favor of technology companies. For example, future large-scale popularization of 5G or higher-speed network will solve the problem of large-scale data. In addition, the introduction of "solid-state" or "solid-state hybrid" LiDARs and a reduction in the number of lasers (from 64 to 16, or even lower), the cost of LiDAR alone can be significantly reduced. The total system cost is likewise expected to decrease progressively.

Whether traditional car company or technology company, future automakers will face a common problem in pushing their autonomous automotive: regulation. Industry experts fear that institutional barriers may even be slower and more difficult to overcome than technical barriers.

Extended Reading

Road Test Results in California

Many self-driving car R&D companies are in Silicon Valley, and companies conduct driverless road tests in California. The California Department of Motor Vehicles (DMV) has posted the 2016 autonomous vehicle disengagement report (Table 1), comparing autonomous vehicles of most enterprises.

The report has summarized the disengagements of the technology caused by the failure of the technology or when the test driver needed to take immediate manual control of the vehicle during testing submitted by each test permit holder. The reports include the total number of disengagements, the circumstances or testing conditions, the location, total miles traveled in autonomous mode on public roads, and the period between when the test driver was alerted of the failure to when the driver took manual control of the vehicle. In addition, it should be noted that the types of roads taken

Table 1 Autonomous vehicle disengagement annual report in California in 2016

	Autonomous miles	Number of disengagements	Rate	Road type
BMW	638	1	638.0	Mostly highway
Bosch	983	1442	0.7	Interstate road/highway
GM	9776	181	54.0	Urban street
Delphi	3125	178	17.6	Highway/urban street/suburban road
Ford	590	3	196.7	Highway
Google/Waymo	635,868	124	5128.0	Mostly suburban road
Mercedes	673	336	2.0	Urban street
Nissan	4099	28	146.4	Highway/urban street/suburban road
Tesla	550	182	3.0	Highway/suburban road/unknown road

Source California Department of Motor Vehicles, USA

by test permit holders are not the same. Generally, the ranking of road environments from difficult to easy are: urban street, suburban road, interstate road, highway, racing road. In order to be precise, there is no ranking in the report.

If state of the field five years ago could be compared to the Ancient China's Spring and Autumn period, Google's then-new self-driving car testing initiative could be compared to a lonely preacher with few followers. Now, we have come to the Warring States period, with numerous rising powers and almost all traditional automakers marching into the (autonomous driving) field. Five years ago, autonomous vehicles seemed like a far-off dream. Yet, it now appears that autonomous vehicles are no longer the things of science fiction, but a potential reality in the near future. Auto- and tech companies are moving towards autonomous driving, filling the whole industry with fierce competition, just like feudal lords vying for the throne. This chapter will describe important, active and representative, overseas enterprises that depict various aspects of global auto industry at the dawning of an era of autonomous driving, and use representative examples to provide readers with a panoramic view of the current situation of the industry.

1 Internet Car Making: To Stir or to Subvert?

1.1 *Google: An Example with Boundless Power*

Google Inc. was founded in 1998 in Silicon Valley, USA, and is known for developing the search engine Google. In 2008, Google launched the Android operating system, which remains the most used operating system in the world. In 2012, Google released a wearable smart device—the Google Glass. In 2016, Google reorganized, and its parent company Alphabet established itself as an artificial intelligence company. The founder of Google saw driverless cars as the transportation method of the future and secretly launched development into the technology in 2007, making it the world's first tech company to do so. To date, Google's research projects and technical reserves on driverless cars are far beyond that of other companies.

In 2010, the New York Times' report on Google's smart cars awakened established automakers in Detroit to focus on and invest in autonomous vehicles. As of the beginning of 2017, the total test mileage of the Google's driverless car project was 2.5 million miles, mostly on urban streets. The test mileage was expected to exceed 3 million miles in eight months. Testing mileage is critical to automation technology, not only in training systems, but also in demonstrating the reliability and safety of new technologies to regulators and the industry. Google's driverless cars were significantly improved in 2016, according to the record of the number of "disengagements" in the autonomous driving mode, the vehicle's performance has increased fourfold from 2015 in which the average number of disengagements per thousand miles in the test was 0.8, and in 2016 this figure fell to 0.2. There have been about 20 related traffic accidents since Google started its field tests, but Google's driverless car system was not at fault, except for a minor accident with a bus. Google has shown considerable confidence in its progress in unmanned driving research and development.

Google is committed to all-at-once full automation. Its first driverless car is based on a hybrid Prius, with a 64-line LiDAR mounted on the roof to create a high-resolution 3D environment model, or a high-precision map. The second generation of driverless cars came from a startup called 510 Systems, whose core technical team (led by Anthony Levandowski) developed a platform based on Lexus. This has remained the mainstream model of the Google fleet until today, with more than 20 vehicles on the road. Google co-founder Sergey Brin unveiled the third-generation Google's driverless car in May 2014 (Fig. 2). The design of the new unmanned car is bold and avant-garde, doing away with the steering wheel, rear view mirror, accelerator pedal and brake pedal of a traditional car. The occupants only control the car through two buttons: "start" and "stop". In order to avoid possible failures, Google set up two control systems in the third generation of driverless cars. Even if one set fails, there is a backup system that can control the car and prevent the car from running out of control. At present, there are more than 30 such cars on the road.

Extended Reading

510 Systems—The Unsung Hero of Google's Driverless Cars

According to the IEEE Spectrum magazine, almost all of Google's driverless cars and street view camera technologies were developed by 510 Systems, an unknown start-up company.

In 2005, 25-year-old Levandowski and several of his friends developed the Ghostrider driverless motorcycle, which has a high-precision GPS and stereo

Fig. 2 Google's third-generation driverless car. *Source* Google

camera. Despite having only two wheels, it has a first-class balancing ability and can correct itself when its body tilts. As the only two-wheeled driverless vehicle, the Ghostrider participated in the DARPA Challenge, coming in second. Levandowski later founded 510 Systems with several friends, together developing a machine-controlled smart camera that eventually evolved into the original Google street view system. They designed a processing system that combines data from the camera, GPS, and inertial sensors to encode the camera image with the positioning data. 510 Systems also pioneered the use of LiDAR scanners in the field of mobile mapping to perform 3D scanning of the surrounding space. This technology has also been adopted by major digital mapping companies.

They also developed an unmanned delivery vehicle for the famous Discovery channel. Using their accumulated experience in mapping, 510 systems used LiDAR to complete mapping for a 25-min journey in advance, stored the 3D map in the system, and then controlled the vehicle with centimeter-level positioning technology and control system. Under police escort, the vehicle crossed the Bay Bridge and completed a pizza delivery. Throughout the journey, the vehicle only slightly rubbed against the wall on a ramp.

A few months later, Google executives decided to develop their own driverless cars. Since then, 510 Systems has been working with Google, with the former responsible for hardware integration and the latter for software development. They designed five driverless cars within a few years, before Google offered to acquire 510 Systems. In the fall of 2011, 510 Systems quietly joined Google, becoming a key part of Google's mysterious X Lab. Google's X Lab's main research projects include autonomous vehicles, UAVs and smart glasses.

How do autonomous vehicles predict the motives and actions of pedestrians and other vehicles? How should they merge, or approach right of way? Both Google and Mobileye are experimenting with new algorithms, such as reinforcement learning and recursive neural network. They expect to solve all the perception-planning-control problems as a whole, with end-to-end deep learning. Designers believe that machine learning and expert knowledge can complement each other. The 2016 accident that led to Google's self-driving car crashing into a bus might even have been circumvented, if the bus driver's judgment was incorporated into the plan.

Google driverless cars use a combination of LiDAR and image acquisition technology. Google says that the combination of these sensors can make up for the deficiency of a single sensor. The LiDAR can detect the distances of objects within 200 m and create an environmental model. The front camera helps to identify objects in front of the car, including pedestrians, other vehicles, road conditions and traffic signals. Radars on the front and rear bumpers can ensure the car is at a safe following distance from other cars, and ultrasonic radars on the rear wheel ensure that the car does not go out of its lane, and measures the distance with the rear object or wall when reversing. Google's car is also equipped with high-precision equipment such as altimeters, gyroscopes and tachymeters to help measure various position data of the car accurately. In order to achieve accurate positioning, Google integrates the GPS positioning data with real-time data collected through aforementioned sensors. The real-time map onboard will be updated as the car moves forward, so as to display a more accurate map.

Google's latest strategy is to develop software and hardware simultaneously. This will allow it to better integrate sensor fusion software, image recognition and other aspects of autonomous driving system with sensor hardware. Google has an advantage in building all its sensor suites in-house, because it employs experts in every field to work closely with the AI engineering resources and ensure that the collected data is fully utilized. Additionally, Google's product development cycle can be shortened by its ability to control both software and hardware.

In March 2016, Chris Urmson, then Director of Google's Driverless Car Project, said at a media event, "the real emergence of autonomous vehicles may be much later than previous prediction, and it could take up to 30 years at the longest time." Larry Page, CEO of Google, apparently did not have the same patience. He recruited John Krafcik, the former CEO of Hyundai America, to lead Google's Driverless Car Project. It was apparent that Google had separated the project from the X Lab, paving the way to its commercialization.

Will Google build its own cars, authorize other companies to use its software products, or run business directly, like Uber? External speculation abounds. Google has stated several times that it does not intend to become a car manufacturer, but to create cars through cooperation. Yet, working with automakers is not an easy task. In May 2016, Google finally signed a cooperation agreement with Fiat Chrysler (FCA) to jointly develop autonomous vans.

In early December 2016, Google announced the divestiture of the driverless car project, and the establishment of an independent new company called Waymo. Waymo would shift its focus to the hardware and software needed to build driverless cars, and ensure that it can be scaled up. Waymo means "A new way forward in mobility".

"A few years ago, a LiDAR on the roof was worth $75,000." Krafcik, who is already the CEO of Waymo, said at the 2017 North American Auto Show, "Today, we've cut its cost by more than 90%. With scale, we can do a better job of making this technology available to thousands of people." In the early prototypes of driverless cars, Google used Velodyne LiDAR at gradually lowered price, ranging from $8000 to $30,000 (depending on the number of laser lines it emits). After Google broke with Velodyne in December 2015, it began recruiting mechanical engineers to develop new LiDAR systems. Waymo said it has newly developed two types of LiDARs, (short- and long-range), to allow cars to see obstacles that are very close, and far away.

Waymo is actively promoting its driverless cars on the market, working with FCA to retrofit 100 Pacifica vans, then conducting road tests in Mountain View, California, and Phoenix, Arizona, in late January 2017. The highlight of the van is that its autonomous driving technology was fully developed by Waymo itself. This was the first time that Waymo developed and manufactured its own cameras, sensors and navigation technology, instead of sourcing it from elsewhere. Independent R&D and manufacturing will enable Waymo to control more over the hardware of autonomous driving technology, while lowering costs.

Waymo, or Google, has recently adjusted its driverless car program, seemingly intending to speed up its move to the market. Waymo had previously pursued driverless cars without steering wheel, brake and accelerator pedals. However, Waymo stated in December 2016 that there was still a need to retain the driver's control due to regulations. The company made this decision partly due to its frustration in developing a driverless car without a steering wheel, and partly due to the growing pressure from Uber.

Waymo also said it would not be involved in car manufacturing but would prefer to offer driverless technology to other automakers, such as Chrysler and Honda, by buying cars or refitting driverless cars for better performance. The company is instead focusing on a range of products and services, including ride-hailing, logistics, personal transport and last-mile delivery solutions. Krafcik had already made it clear before that Waymo is not a car company. Its business is not to make better cars, but to make drivers better.

1.2   Tesla: A Web Celebrity with a Cute Face and Talents

On February 1, 2017, Tesla Motors Inc. officially renamed itself as Tesla Inc. By just taking off a word, Tesla has shown its high ambitions, unsatisfied with being a fish in the pond.

When it comes to autonomous driving, Tesla is an "influencer" that cannot be ignored. Based in Silicon Valley, Tesla was founded in 2003 by Elon Musk, a graduate school drop-out at Stanford University, and JB Straubel, a successful graduate from the same university. Tesla is also unusual because it is backed by numerous super investors, including Google's founders Larry Page and Sergey Brin, subsidiaries of traditional car giants such as Toyota and Daimler-Benz as well as Panasonic. Panasonic supplies the electric core of Tesla's lithium battery cells, and some of Tesla's car designs have been inspired by Benz.

Tesla's original entrepreneurial team came mainly from Silicon Valley, and thus used IT ideas to build cars rather than the ideas of traditional automakers, represented by Detroit. As a result, Tesla's endeavors are somewhat seen in a "David versus Goliath" light. Most of the electric car companies that started earlier have gone bankrupt, while only Tesla has survived in the market with its unique competitive advantage. Tesla has mastered the core technology of driving motor, motor control and battery, and can design vehicles completely according to user needs. Furthermore, it has established an excellent after-sales service network and charging facilities.

Tesla is indeed the most active manufacturer in the auto industry to promote autonomous driving. On the one hand, it needs autonomous driving—which also conforms to its brand image positioning—to act as a selling point; on the other hand, as a new car company with technology and Internet roots, Tesla is more radical in the application of new technologies than most of the traditional automakers. Buying a Tesla car will consumers cool and unique features, but not without potential risks. What technology can be put into mass-produced cars? What risks can be tolerated? And how risks can be managed? Only based on one-sided viewpoints, those questions cannot completely encapsulate these risks, while consumers act as the guinea pigs.

In contrast to Google's one-step approach, another autonomous driving path has gradually entered the mainstream. This approach begins with assistant driving, focusing on low-cost vision-based solutions in an attempt to achieve rapid commercialization. Among the leaders are Tesla and Mobileye. The two have partnered to create a driving assistance system that has been installed on more than 10 million cars. In October 2014, Tesla first introduced its Autopilot hardware for its Model S (Fig. 3). At the end of 2015, Tesla introduced the now-famous Autopilot system as part of the ADAS kit, with radars and cameras as the main technical means, including automatic lane changing, traffic-based cruise control, collision warnings,

Fig. 3 Tesla model S

and an automatic parking system. Strictly speaking, Tesla's autopilot function at that time was level 2. However, the Model S could handle most normal driving operations autonomously, and is fully capable of replacing the owner on the road, with its owner's hands and feet off the steering wheel and the pedals. It is worth noting that the Tesla autopilot function was still in the public test phase at that time.

Visual maps adopted by Mobileye are extracted from its visual data, suitable for real-time upload, and can be updated via crowdsourcing. Vision-based positioning is closer to the way people work. The approximate position is evaluated based on the signs on the road, and a real-time decision is made according to the change of the road lines (i.e. which lane is selected, to take the ramp or not, etc.). Those signs and lines are extracted from the visual and uploaded to the package to digital maps (only 10 KB level data for a kilometer) by crowdsourcing. The visual matching is then used to obtain positioning while driving.

While visual technology is booming, other technologies such as multi-sensor fusion of vision and radar might be more advantageous in certain situations. The advantages of vision are high resolution and rich semantics, yet it is susceptible to weather and illumination. Millimeter wave radar can only track objects but can't tell its size or shape. However, it is less affected by the environment. The fusion of the two has become the standard for assistant driving, as is the case for Tesla's Autopilot (which additionally includes short-range ultrasonic radar).

Director of Tesla Autopilot Project Sterling Anderson said at the EmTech Digital Conference in San Francisco in May 2016 that about 70,000 T cars already on the road supported Autopilot. These cars drive a total of about 4.18 million kilometers a day, well ahead of the 2.41 million kilometers of Google's driverless car project. Tesla will use the data collected from these tests to develop, improve and launch more features in the future.

Some Tesla owners have tried to drive their cars through morning rush hour, turning on Autopilot, and nap for half an hour, before switching to manual mode to leave the road. Drivers all around have posted videos of these conspicuously hands-off test drives, which have flooded the Internet. However, accidents have happened one after another. The world's first fatal autopilot accident occurred in January 2016, when a Tesla driver was killed following a collision with a road sweeper. The accident investigation showed that the car was in a "constant speed" state when the accident happened. This meant that the Tesla did not evade or decelerate before crashing into the other vehicle. On May 7, 2016, a Tesla Model S crashed into the bottom of a truck, totaling the itself and killing its driver. The driver had been watching a Harry Potter movie while his car was in autopilot. On July 1 of the same year, a Tesla Model X was overturned while in autopilot mode in Detroit, causing two passengers injured and hospitalized. Then on July 11, a Tesla driver drove from Seattle to Yellowstone National Park. On a two-lane rural road, the Tesla in autopilot state veered off the road and hit a roadside post. Fortunately, no one was injured—though the car was destroyed.

On August 6, a Model S retrofitted with a twenty thousand-yuan autopilot kit was involved in an accident in North Fifth Ring Road, Beijing. The autopilot system failed to identify a Santana parking in the inner lane, even as preceding vehicles went around. Instead, it accelerated, causing a serious accident. Fortunately, there were no casualties.

On September 29, a Tesla Model S collided with a Danish tourist bus on German Expressway A24. When the bus overtook and returned to the inner lane, the Tesla rear-ended it and injured the driver.

Some analysts working in the industry believe that the current Tesla autonomous driving technology is still not perfect, and far from the level necessary to handle complex traffic environments. The front-mounted sensors may have blind spots, which cause the vehicle to overlook obstacles in a certain range of angles. In the face of accidents, Tesla argued that each of the drivers did not follow its rules of use, which require drivers to maintain a hold of the steering wheel in Autopilot. However, from a human perspective, Autopilot is like an anesthetic that desensitizes drivers to potential risks. Tesla clearly wants its consumers to use Autopilot. However, it is also aware that as more consumers use this feature, they will face the danger of over-relying on Autopilot.

On August 30, 2016, Tesla improved Autopilot for users. Previously, if the driver's hand was away from the steering wheel for more than a few seconds, an alarm would sound. After the changes were implemented, Autopilot would be disabled if the driver does not heed the alarm. The driver would also be prevented from enabling Autopilot and other certain features until the vehicle was fully parked.

Tesla says that the Autopilots function is like an aircraft's automatic cruise and can be used under suitable conditions. However, the driver must maintain responsibility for the, car as its ultimate controller. In appropriate applications, Autopilot can reduce the driver's work and provide additional safety mechanisms compared with pure manual driving. It should be noted that "Autopilot" was one of the key points of sales promotion on Tesla's Chinese website at the beginning, but it has since been changed to "Autopilot assisted driving" due to frequent accidents.

Extended Reading

Tesla's War of Words with Mobileye

The dispute between Mobileye and Tesla became public in September 2016. Mobileye said in a document submitted to the Securities and Exchange Commission on September 16 that Musk had assured the company that a driver would not be allowed to take both hands off the steering wheel when Autopilot was in use. In the end, however, this was not the case.

Mobileye said that Autopilot was developed in collaboration with Tesla, and the company expressed concerns about the safety of Tesla's Autopilot during product planning discussions between the two companies in May 2015. "Our position has always been that Tesla's Autopilot should not allow the driver to leave the steering wheel with both hands without proper technical constraints," Mobileye said in the document. Tesla, however, has not kept the promise that Musk made in person. Autopilot, launched later in 2015, allowed users to activate it while their ahnds were away from the steering wheel.

Mobileye said that it terminated its partnership with Tesla after the company breached its safety bottom-line. Tesla retorted that Mobileye was a total loser and accused the company of insatiable greed. After discovering that Tesla was developing sensor technology internally, Mobileye had tried to force Tesla into abandoning internal R&D projects and continue to buy its products. After being rejected, Mobileye was forced to cancel the contract and sought an excuse to accuse Tesla.

Only the two parties know what caused the split. However, Mobileye's break with Tesla came after the fatal accident of the Tesla Model S in May 2016.

Prior to the release of Drive Pilot by Mercedes-Benz, Tesla's Autopilot was the only partial automation system in the market that allowed for autonomous lane changes. Some consumers have been calling for Tesla's Autopilot to be disabled and renamed before it becomes safer. Germany's Transport Minister, Alexander Dobrindt, demanded that Tesla stop promoting its driver assistant system using the word "Autopilot," as it might convey a message to drivers that they do not need to concentrate while driving.

Despite frequent traffic accidents, criticism in media and splits with its main suppliers, Musk has insisted on promoting autonomous vehicles, citing the famous quote, "Don't let perfect be the enemy of better." Based my knowledge, Confucius, the great Chinese thinker, Voltaire, the French enlightenment thinker, and Shakespeare, the outstanding British dramatist, have all said similar words. Perhaps the promotion of autonomous driving requires both caution like traditional automakers and radical strategies like Tesla, which could have a "Catfish Effect" on the development course of autonomous driving technology.

Despite the controversy, Tesla's business seems to have been a blessing in disguise—sales in the third quarter of 2016 rose 111% from a year earlier, the most profitable quarter ever. During this quarter, Tesla shipped 24,500 vehicles. Another 5500 vehicles have been ordered but not delivered, so they are classified as sales in the fourth quarter.

On October 19, 2016, Tesla released Autopilot 2.0, and announced that it could now be ordered by customers. All Tesla models (including Model 3) would have the hardware basis for full automation. Tesla believes that full automation will make Tesla a more reliable "driver" than human beings.

Although the hardware is ready, Tesla doesn't plan to put it to use right away. Instead, it will conduct field tests for at least a few million kilometers until it meets its safety standards. At present, Autopilot 2.0 will work in "shadow mode", running in the background and not interfering with daily driving. Autopilot 2.0 has the ability to self-learn and constantly improve itself. The data accumulated by each Tesla car is automatically collected and stored in Tesla's "fleet learning network", so that all Tesla cars can learn the corresponding "skills" and learn to react in different road conditions. When an accident occurs, Tesla compares the response of human drivers with that of program settings to see if the autopilot function can avoid the accident to some extent.

This also means that Tesla cars with new hardware can use fewer functions than the first generation of Autopilot at this stage. Temporarily disabled functions include automatic emergency braking, collision warning, lane keeping and automatic cruising. Tesla says that when these functions are tested and performed steadily in the future, they will be pushed to Tesla owners OTA (Over the Air). As Tesla cars are designed as pure EVs and can be controlled by electronic control signals, thy naturally synergize with OTA. Upgrades can be downloaded in the background through 3G or Wi-Fi, rather than visiting the service centers for software upgrade. If Autopilot 2.0 is a "weaponry" upgrade, then the OTA replaces the car with a new "brain," to improve the driving experience while adding in-vehicle functions.

1.3 Apple: Not Much Said, Not Little Done

Steve Jobs, the late Apple co-founder and former CEO, had been hoping to build a car. In an interview with the New York Times, he said that if he had more energy, he would be happy to challenge Detroit with an Apple car. To that end, Apple is rumored to have launched a car project, code-named "Project Titan".

"Project Titan" is quite mysterious. Apple is not the only technology company pursuing autonomous driving technology, but it has kept its lips shut while automakers and other technology companies report their progress. Although seemingly confirmed the existence of auto projects at Apple's shareholders' meeting in early 2016, the project has never been explicitly about car building. According to Apple Insider, a research and development base in another secret location outside Cupertino, Apple's main campus includes an "automatic operation area" and a "repair workshop". Apple intends to put some of its car manufacturing work in Ireland by increasing its investment in Cork, where a large-scale factory is planned. Sources also said that Apple had several driverless cars under test on restricted, closed roads. Apple's product launch date for "Project Titan" is set in 2020.

In the past two years, Apple has rapidly expanded the scale of this project, increasing the number of employees to more than 1000 in 18 months. Earlier in 2015, Apple poached Paul Furgale, an expert in autonomous driving and machine vision, to run "Project Titan", which had not yet been publicly named at the time. Furgale had worked as a consultant at the Autonomous Systems Lab of ETH Zürichand had also contributed to the development of automatic parking technology commonly used in the European Union. To fish for talent, Apple offered a $250,000 family resettlement allowance and a large salary for any engineer who has worked at Tesla. Apple even incited a lawsuit. A123 Systems Inc., a battery company, sued Apple over employee poaching to develop a "large scale battery division." A123 Systems went bankrupt in 2012, and its main business was subsequently bought by Wanxiang Group of China. Apple and A123 Systems reached a settlement in May 2015.

Apple has also negotiated with Daimler and BMW to establish a partnership for electric vehicle R&D but failed. According to reports, a similar disagreement in the negotiations has emerged over which company should dominate in joint R&D. To integrate Apple's services closely in a car, automakers would need to abandon their own services, which they are hesitant to do. Apple wants its electric cars to be tightly integrated with its Internet services and software, but Daimler and BMW argue that the safety of car users' data, as an important principle for future development, should come first. Apple's next target for cooperation could be Magna.

Extended Reading

Debate on Autonomy of In-Vehicle Systems

Although automakers have generally accepted Apple's CarPlay and Google's Android Auto, they have not given up their own in-vehicle systems. Ford and Toyota have teamed up to form a third-party service company called "Smart Device Link," whose in-vehicle system can be integrated into any car. It even allows iOS and Android Apps to scale to in-vehicle entertainment systems. While it may seem like an extension of mobile apps to onboard screens, many of its apps are separately overseen by automakers. In fact, Smart Device Link has already been installed on 5 million Ford cars. Toyota plans to integrate it into its models, and Mazda, Suzuki, Subaru, PSA and other automakers also agree to do so. Panasonic, Blackberry and other QNX-based in-vehicle service providers also said they would join this alliance.

Why so? Because automakers want to have more options, including the data generated by the Apps that Apple and Google cannot access. "The lifetime value of a customer is typically $1 million," said Roger Lanctot, an analyst at Strategy Analytics. This figure includes new car sales and services. "If they lose the control over data, the automakers lose half of the value, which is $500,000."

Data access rights ensure that automakers can push things like maintenance suggestions, insurance recommendations to users. This is just the beginning of value-added services. Therefore, automakers will not easily give up in-vehicle system platforms, or simply hand it over to Apple and Google. Of course, they are not fighting Apple and Google, but promoting the experience of their own in-vehicle platforms while offering compatible options. The two things are not mutually exclusive.

Apple has pumped many resources into "Project Titan". However, as the project grew, problems ensued. According to the New York Times, the team was unable to come up with clear product objectives, at the same time considered the goals set by the company "unrealistic", triggering a strong conflict between "reality and ideality". In the fall of 2016, Bob Mansfield, Apple's Senior Vice President of Macintosh Hardware Engineering, took over as head of "Project Titan" and was placed charge of R&D on autonomous driving, which he strategically scaled back. Based on Apple Insider, Apple Automotive Laboratory has experienced turbulence in the past few months such as strategic divergences, leadership changes and supply chain reengineering. Nearly 100 employees were laid off, for the purpose of "restarting the car project." This coincides with previous rumors of Apple plans to develop autonomous driving system before launching any car products.

Apple's earlier R&D plan had included not only electric cars like Tesla's, but also autonomous driving technology like Google's. Under Mansfield's leadership, Apple's plan has been adjusted. Its focus will be on the development of autonomous driving in the future, while "Project Titan" for developing electric vehicles still exists in a secondary position. The team's new focus is to give Apple the flexibility to work with existing automakers or return to car design in the future. There are also rumors that Apple has been recruiting new employees for the new mission, to keep the size of its auto team steady. Apple's executives have set a deadline for the team to demonstrate the feasibility of the autonomous driving system by the end of 2017, after which it will determine its final direction.

The news that Apple has quietly disbanded its auto team and turned its focus on in-vehicle autonomous driving systems sets off a ripple in the automotive community. Even the powerful Apple has abandoned car building! This news undoubtedly poured cold water on the "car-building fever" of Internet enterprises. Traditional automakers that have been faced the threat of Internet enterprises were likely finally relieved. Headlines such as "Apple retreats in the face of difficulties" and "domestic Internet car-makers are acting recklessly" also circulated publicly.

Was Apple's withdrawal the wrong general direction? The answer is no. Mobile Internet, Internet of Things, Big Data, Cloud Computing and commercialization of 5G network with transmission rate 100 times higher than 4G will make autonomous driving possible in the near future. At that time, smart cars—huge smart terminals—will become the third living space outside the home and office. This is almost universally acknowledged by the industry, which a major prerequisite for Apple to lay out its automotive projects ahead of time. From current trends in technology and market development, this is not wrong.

There are possible two reasons why Apple has shifted its R&D focus to in-vehicle and autonomous driving systems. First, cars are not easy to make. Building a car from scratch requires not only building a complete industrial chain from R&D to production, sales and after-sales, but also finding a balance among advanced technology, high quality, high efficiency and low cost. The trouble is, with such a complex business chain, Apple may not be able to find such a large-scale, highly efficient and low-cost automakers, (like Foxconn in its consumer electronics business) to achieve asset-light operation via outsourcing. Apple is smart to see that making good phones does not mean it can do well in manufacturing cars. The second possibility is that cars don't make much money. Auto giants like Volkswagen and General Motors, which sell tens of millions of vehicles annually, have an average profit margin of only about 6%. Toyota, which is recognized as the most efficient in the industry, has a profit margin of only 10%, only half of Apple's consumer electronics business. From the perspective of ROI, Apple's investment in automobile manufacturing is clearly not worth it. At the same time, competition in the smartphone sector, Apple's traditional business, is becoming fiercer, and Apple needs to spend more money on its R&D. In 2015, Apple's R&D investment was $8.6 billion, and its R&D expenditure increased by nearly 30% over the previous year. This ranked third among the top 10 most valuable companies in the S&P 500 index, while in 2016, its R&D investment was no less than $10 billion.

Although Apple's withdrawal from car manufacturing cast a shadow over the prospects of tech company car-making, the general directions of auto industry—electrification, intelligentilization, connectivity and car sharing—remain unchanged. This also does not mean that traditional automakers can simply rest without worries, protected by the "threshold" that they have built up so elaborately during the past

130 years. Without the direct competition with Apple, the auto industry still cannot avoid their fate of industrial transformation.

After all, Apple's withdrawal is not a complete departure in the automotive business but simply a decision to stop making its own cars. Apple has switched to autonomous driving software and supporting hardware. To this end, they set up a R&D center in Ottawa, Canada, and lured away QNX employees to participate in its software development. Dan Dodge, co-founder and CEO of QNX Software Systems, joined Apple's automobile project even earlier. Interestingly, Apple may integrate its autonomous driving products with augmented reality technology.

Because the Apple car project is so mysterious, it has piqued everyone's curiosity, with the media holding close to any potential clues. The Wall Street Journal finally found evidence in a letter from Steve Kenner, Apple's Director of Product Integrity, to the NHTSA of the USA on Nov. 22, 2016. The tech giant has finally seemed to publicly acknowledge that there is indeed an Apple car in the works, as Apple notes in the letter that it "uses machine learning to make its products and services smarter, more intuitive, and more personal," and further, that it is "investing heavily in the study of machine learning and automation, and is excited about the potential of automated systems in many areas, including transportation." Furthermore, Apple suggested that NHTSA further open its policies to attract more companies to the field and promote the rapid development of the industry. Apple went even further, calling for changes in policies that would give newcomers the same treatment as established companies and the chance to test autonomous driving technology on public roads.

On January 13, 2017, the US Department of Transportation announced the formation of a new Automation Committee to make recommendations on automated transportation. Lisa Jackson, Apple's Vice President of Environment, Policy and Social Initiatives and a former official in the Obama administration, joined the panel, a sign that the company still ambitious for future transportation and autonomous driving projects. Johnson was joined by Mary Barra, GM's CEO, and Krafcik, Waymo's CEO, among other executives on the 25-member committee.

1.4   Uber: Fighting for Tomorrow's Unmanned Ride-Hailing Services

Uber invented the ride-hailing business structure, and has become the largest ride-hailing company in the world at present. The company is also moving into autonomous driving field, with research collaborations with Carnegie Mellon University and the University of Arizona. Uber plansto launch ride-hailing services based on autonomous vehicles in the future, which will also solve driver shortages.

According to the Daily Mail, Uber began testing autonomous cars on the streets of San Francisco at the end of 2016, which passengers could book (Fig. 4). On their debut day December 15, one of the cars, a refitted Volvo, was photographed running a red light, almost hitting a pedestrian in the zebra crossing. Uber blamed the error on the engineer seated at the driver's seat to monitor the entire driving process and to

Fig. 4 Uber's autonomous carfleet

resume control whenever necessary. The California Department of Motor Vehicles (DMV) suspended the road test the same day and ordered Uber not to conduct any more tests until it was approved by authorities. According to the DMV, nearly 20 companies including Google, Tesla and Ford, have obtained special permission before conducting autonomous vehicle road tests and have promised to report any accidents. Uber adamantly said that California authority should not classify its cars as "fully autonomous" vehicles because they were in the early stages of testing as in-vehicle engineers were to monitor the whole driving operation. Uber then used its official Twitter account to urge the California authority to be more open to corporate innovation.

After this incident, Uber found that the navigation system on these cars did not identify bike lanes well. Uber confirmed that this was a bug in the program, saying it was being fixed to allow the vehicle to autonomously identify bike lanes and selectively switch lanes, rather than turning right regardless of whether there were bikes on the right side of the bike lane.

The Arizona government extended a helping hand to Uber after test qualifications for its 16 autonomous cars were rejected in California. Doug Ducey, Governor of Arizona, publicly stated that the state is providing various conveniences for new technology and new companies and was willing to embrace Uber's autonomous cars with wide roads and welcoming arms. Uber immediately asked Otto, its subsidiary autonomous truck company, to ship all its autonomous cars from California to Arizona, where it prepared to carry out relevant tests. American media have been quick to criticize the government of California. Despite having the world's high-tech industrial hub of Silicon Valley, autonomous vehicles field tests have faced challenges within California.

These are just episodes of Uber's story of autonomous cars development, which started in late 2014 when it poached dozens of researchers from the robotics department of Carnegie Mellon University in Pittsburgh and set up its own research center in the city. Travis Kalanick, founder of Uber, flew there to do recruiting personally. A forward-looking company must have a leader conscious of hardship (Kalanick left before the book was printed, but a successor has yet to be named). As the pioneer of ride-hailing, Uber sets a goal to do more than simply pushing the practice to the world. Kalanick made it clear that "Uber's mission—to provide transportation as reliable as running water, everywhere for everyone—is not possible without moving into this kind of technology." After raising more than $15 billion by selling shares and issuing bonds to become Silicon Valley's best-capitalized start-up, Uber is ready to spend big to make that happen.

Uber's bold investment in autonomous driving R&D has also provoked controversy. By the beginning of 2016, a total of hundreds of engineers, robotics experts and car mechanics were recruited by the company. Among them, 40 people were from the Robotics Institute of Carnegie Mellon University, accounting for one third of the staff. This massive departure was of great impact to the university. Although Uber later tried to make amends by funding a professorship for Carnegie Mellon University, their relationship with the university is still fragile.

In February of that year, Uber announced the establishment of the autonomous driving test base and set up the Advanced Technologies Center in Pittsburgh. The center has become the core focus of the company, which is trying to develop driverless cars. At that time, their goal was clear: to replace Uber's one million drivers with robots as quickly as possible. For this seemingly crazy plan, Kalanick said, "It's all about business. It's not just about science."

In July 2016, Uber bought Otto. Founded in January 2016, Otto had a total of 91 employees from Apple, Google, Tesla and other companies, aiming to change the existing mode of freight transport so that autonomous driving was to replace the hardworking truck drivers—The company has been testing an "interstate autopilot" system that allows drivers to snooze in their cabs. According to the insiders, Uber has spent a generous $680 million in order to acquire Otto. Otto's investors would receive an equity stake in Uber of slightly less than 1%, as well as 20% of profits from Uber's future trucking business. The founders of Otto said in a blog post that they were combining with Uber to "build the backbone of the rapidly-approaching self-driving freight system".

One component of Uber's driverless car research is developing its own maps and mapping software, and the company has spent half a billion dollars on mapping efforts around the world. In order to do this, Uber acquired the mapping software company deCarta, bought Microsoft's mapping technology, and signed an agreement with Digital Globe to use HD maps.

In middle of August 2016, Uber and Volvo announced a partnership. Like Uber, Volvo is also a keen to develop autonomous driving technology. The two companies will partner and integrate all their resources to accelerate the implementation of their autonomous driving technology. The world's first autonomous taxi fleet is part of a partnership between Uber and Volvo that would see the companies jointly invest

$300 m to develop a new self-driving vehicle. By the beginning of 2017, Uber's traditional automaker partners also included Mercedes-Benz.

In September 2016, Uber started testing the world's first autonomous taxi fleet in Pittsburgh, putting it ahead of Google and Ford in the race to bring self-driving car services to consumers. Uber has released a few details of its sensors and software used to guide autonomous cars. The new model Ford Fusion is equipped with 20 cameras, 7 LIDATRs, GPS and radars. Each vehicle is guided by three-dimensional HD maps independently developed by Uber, and the vehicle locates itself by matching landmarks and environment. Uber autonomous could drive autonomously and steadily and avoid obstacles automatically, but not perfectly. A *Business Insider* reporter experienced four times the amount of "disengagements" during the 8 km trial route. In one case, the taxi stopped itself and needed to turn to manual driving when crossing a bridge, and in another case, the taxi stopped for no reasons with no obstacles on the road. In the early days of opening up to the public, Uber's self-driving car could only carry two passengers, as a driver was required to drive manually in emergencies and an engineer was responsible for observing the driving conditions on board, and the route could not be changed in the middle of the journey. At first, all of Uber's autonomous taxis were Ford hybrids and modified Volvos, but now only the former is used in the market. Volvos are only used in experimental center. In the months to come, Uber and Volvo will gradually expand their fleets to 100 cars by the end of 2017, including the Volvo X90 plug-in hybrid SUV.

At the same time, the company is also building a new R&D center in Detroit to take advantage of the rich automotive talent pool there. Uber's takeover operation continues, and an artificial intelligence start-up in New York was acquired in late 2016. The acquisition led to the establishment of Uber AI Labs based on a team of 15 experts. The acquired AI startup has studied everything from improving traffic forecasting to developing flying cars. In the near future, Uber AI Labs will improve Uber's ability to predict car demand and collect data from multiple sources to help the company accelerate the launch of the autonomous taxi fleet.

From what has been disclosed so far, Uber is not the first company to enter the autonomous driving field, but it is probably the most active one.

1.5 Faraday Future: Fresh But Vital

We are experiencing unprecedented changes. Just a few years ago, perhaps no one could have expected such a disruptive change in our inherent lifestyle in the fields of travel, residence, shopping and so on. Most people cannot but learn to adapt, not even realizing that what they once imagined has become an unshakable reality. However, there are those who are unwilling to be carried along by the tide but want to lead the future, which is exactly Faraday Future's (FF for short) philosophy of survival.

Before 2017, FF was an unfamiliar name for many. But when its first production car, FF 91—an industry-leading product in intelligence, connectivity, performance and design—went global in January 2017, the young company stood under the public spotlights.

"Coming from nowhere," this is a phrase used by many media to describe FF's rapid development. Headquartered in Silicon Valley and Southern California, the EV maker has two home turfs—China and the USA—and its root in the Internet ecology. Now the company has more than 1400 employees from 36 countries and regions. With the Internet ecological economic theory, the company has made visionary plans and integrated diversified resources in fields such as strategic positioning, product development, productivity planning, business model, marketing model, UP2U user operation and car sharing economy to meet people's demand for travel in the future and redefine what the travel experience might be in the future.

However, as Albert Camus, the French philosopher, said, "All great deeds and all great thoughts have a ridiculous beginning." It's hard to believe that FF, now in its glory days, is less than three years old. When FF was first established in May 2014, it was a tiny company with two employees. By September 2014, when FF moved into the current vehicle R&D building (formerly Nissan North America headquarters), the whole team only had 10 people.

FF took only two and a half years to grow from 2 to 1400 people, the company's growth rate has shocked the public. The Faraday Future brand made its first world debut when they launched a concept supercar named FF ZERO1 at CES2016. Just 1 year later, FF launched their first mass production model FF91 (Fig. 5) in Las Vegas, with the first deliveries planned in 2018. That means Faraday Future, which started from scratch, can launch its production cars in the market in four years. In contrast, a typical traditional OEM with ready facilities often takes five years or more to develop a brand-new model and put it into market.

To achieve this radical goal, FF has planned a comprehensive layout in the automotive industry chain. As an old Chinese saying goes, "Provisions should be arranged before an army is mobilized", the company has already worked with many European and American tier-1 auto suppliers since 2015, making adequate preparations in the global supply chain. In the field of research and development of innovative technologies, including autonomous driving, FF obtained the California autonomous driving test license in June 2016, and accomplished the global R&D layout based on FF & Le Future AI Research Institute. FF successfully demonstrated its FF 91's unmanned valet parking function at the press conference in Las Vegas.

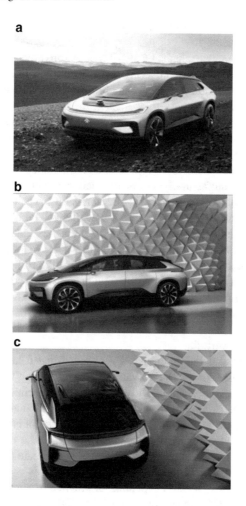

Fig. 5 FF91

As for the most concerned aspect, the production base, Faraday Future's Nevada plant was founded in April 2016.

However, "success is not winning the lottery", said Peter Thiel, a prominent Silicon Valley entrepreneur and investor. It is no accident that FF has achieved today's achievements. One of the important factors is talent. In addition to the well-known technologist who has contributed to the book, Jan Becker, senior director of Faraday Future Autonomous Driving, its executive team also includes many top talents in the industry such as Nick Sampson, former engineering director of motors and chassis of Tesla, Richard Kim, former main designer in BMW team, Peter Savagian, creator of the first mass-produced electric car EV1, and former director of electric drive systems and new products of GM. Most recently, Stefan Krause, the former CFO of

BMW Global and CFO of Deutsche Bank, became the new CFO of this emerging EV company.

In addition to automotive talents, FF also has numerous elites from the technology, energy and aerospace industries, covering the fields such as Internet, artificial intelligence, automotive technology research and development, manufacturing, design, marketing, sales service, finance, etc.

If talents are the most important cornerstone of FF's development, then organizational culture may be the origin of its huge difference from traditional carmakers. Bringing revolutionary changes to the industry also requires an unusual development path. As Hong Bae, FF's Director of ADAS and Autonomous Driving Technologies, once said, "Other manufacturers always try their best not to fail, but in Faraday Future, we are encouraged to keep trying and innovating to gain early experience. If you don't make mistakes, you can't explore new things." Perhaps it is this spirit of exploration that has enabled them to achieve what they are today.

Let's take technological achievements for example. In March 2016, Faraday Future announced that it obtained its first patent (#9241428 B1)—electric powertrain inverter (FF Echelon Inverter) in the USA. The patent was filed in just five months, quite exceptional in America, where it usually takes 18 months. More importantly, the patent breakthrough of electric powertrain inverter, the "brain" of motor control, meant that Faraday Future reached the top level in the world in terms of the core technology of EV powertrain.

At the FF91 global debut ceremony, FF revealed that it had filed 1940 patent applications so far. In addition, VPA (Variable Platform Architecture), one of their core technologies, has been applied to the FF91. This technology pioneered the use of universal powertrain. Universal modular framework and flexible powertrain design system can provide platform designers with a series of battery and wheelbase configurations, achieve the goal of building cars, SUVs and even high-performance racing cars on the same platform, which will greatly accelerate the launching of new vehicles.

At the same time, FF is also deepening its practice in automotive technology-related fields. In July 2016, the company announced that it became a core technical partner of the FIA Formula E Dragon Racing and co-founded the Faraday Future Dragon Racing team. Through the platform Formula E, FF's technology research and development efficiency has been greatly improved.

Compared with the development path of traditional car companies, FF is steadfastly advancing along a vertical evolutionary path "from zero to one". It is believed that before long, the transport ecology of the future that they depict will really appear in front of us.

2 Traditional Automakers: My Game, My Turf, My Rules

2.1 European Brigade

1. Mercedes-Benz: The "CASE" for a Forerunner

At the 2016 Paris Motor Show, Dieter Zetsche, Chairman and CEO of Daimler AG as well as President of Mercedes-Benz, put forward a comprehensive and informative e-mobility strategy CASE (Connected, Autonomous, Shared and Electric), what he called the four pillars for the transformation of the car product into an ultimate platform. "You have to be prepared for future investments in all of these fields." Zetsche said, "but it's exciting. All of these developments are not just technological changes, but they allow you to provide customers with better and more services."

Indeed, Mercedes-Benz is a bona fide pioneer in the field of autonomous driving. Mercedes-Benz was keen on the study of self-driving vehicles in the 1980s. It began to cooperate with the NavLab of Carnegie Mellon University in 1984. Jointly with the University of Bundeswehr Munich, it spent €750 million to kick off the Eureka-Prometheus Project in 1987. Mercedes-Benz's ambition can be seen from the name of the project. More than 2000 years ago, when the ancient Greek scholar Archimedes discovered the law of buoyancy, he blurted out "Eureka! (I found it!)" And in Greek mythology, Prometheus, one of the most intelligent gods and known as a "seer", who brought fire to humanity and was punished by Zeus. Prometheus faced his bitter fate firmly and never lost courage before Zeus, king of the gods.

The Eureka-Prometheus Project lasted for eight years and the lasting investment has been paid off. Mercedes-Benz autonomous driving technology reached its climax on the redesigned S-Class (W140). In 1995, an S-Class sedan achieved almost full autonomous driving from Munich, Germany, to Copenhagen, Denmark, and could overtake even at a speed of 185 km/h without or very little human intervention. The car's built-in equipment was developed by Ernst Dickmanns, a professor at the University of Munich, who pioneered dynamic computer vision and autonomous vehicles. It uses computer scanning technology to reflect real-time road conditions through a series of microprocessors and the probabilistic method. The computer controls the steering, acceleration and braking of the car by analyzing real-time images from four cameras with different focal lengths, and it can automatically track other vehicles and identify road signs. In fact, the car's VITA ("Visual Information Technology Application") reached the highest level of intelligent driving at that time. Some even feared that the use of the technology could be as dangerous as the revival of Skynet in the movie *The Terminator.*

Since 1995, Mercedes-Benz has independently studied driverless technology. It showcased the S500 Intelligent Drive concept car in 2013 as the result of its internal research and development project. The fully autonomous S-Class prototype traveled the roughly 100 km between Mannheim and Pforzheim in Germany, retracing a route taken by motoring pioneer Bertha Benz exactly 125 years ago when she set off on the very first long-distance drive.

Mercedes-Benz is also the leader in the research of passive and active safety systems, which are closely related to autonomous driving technology. The concept of Pre-Safe, formerly known as Integral Safety Concept, was first introduced in 1989 and it started to apply to the S-Class in 2002. Other Monumental progresses include: Brake Assist system in 1996, Brake Assist Plus in 2005, Pre-safe Brake Stage 1 in 2006 and Pre-safe Brake Stage 2 in 2009.

For its ADAS technology configuration, Mercedes-Benz takes a variety of routes. Take Adaptive Cruise Control (ACC) and Automatic Emergency Braking (AEB) as examples: the entry-level offering adopts a scheme with a single 77G millimeter-wave radar, while the medium and advanced versions also add two schemes from the base set-up, say, fusion with 24G millimeter-wave radar or fusion with monocular camera. For Lane Departure Warning (LDW) system, some vehicles use monocular camera, and some adopt binocular camera to identify lane lines.

Mercedes-Benz released its first Level-2 self-driving system with steering assistance, Distronic Plus, in 2013. It then launched a more powerful Level-2 system, Drive Pilot, which was applied to its E-Class for the first time in 2017, with radar and cameras as its basic technical configuration. It can accelerate or brake at a speed higher than normal highway cruise. It can also speed up to 81 km/h in 60 s without human interference. Drive Pilot can automatically change lanes if the driver instructs to do so, but it will refuse to execute the lane change order if the system detects a car in another lane. The 2017 E-Class is the only type of vehicle that has obtained an autonomous driving license in Nevada, which means Drive Pilot's capabilities are far superior than its competitors. However, Mercedes engineers have weakened the ability of E-Class's Drive Pilot function, to ensure that drivers understand that "people are ultimately responsible for the car safety". In addition, the 2017 E-Class is the first vehicle to adopt V2V technology, which means that all new E-Class cars can communicate with each other on road conditions and driving information. With the improvement of autonomous driving technology of other vehicles, new E-class cars will be able to communicate with cars of different brands in the future.

In addition to cars, Mercedes-Benz is testing semi-autonomous commercial vehicles in the USA and Europe. Earlier in 2016, Mercedes-Benz developed an autonomous concept heavy-duty truck—Future Truck 2025. Mercedes-Benz believes the conceptual design will eventually help reduce the number of highway accidents. The concept truck does not have the functions to navigate or shuttle autonomously on urban road like Google's self-driving cars, but can drive freely on highways. Mercedes-Benz argues that highways are precisely where trucks are most at risk, because the drivers are often sleep-deprived, and fatigue driving can pose a serious threat to other small vehicles on the highway. Mercedes-Benz connects existing assistance systems with enhanced sensors to the "Highway Pilot" system. Autonomous driving is already possible at realistic speeds and in realistic motorway traffic situations. The Mercedes-Benz Future Truck 2025 provides a glimpse of the future shape of trucks. The system includes an advanced set of cameras and radars. Blind Spot Assist warns the truck driver about other road users not only when turning; it also warns about imminent collisions with stationary obstacles and serves as an assistance system when changing lane.

On July 18, 2016, Daimler showed the outside world the result of its latest research—an autonomous bus with CityPilot system, named "Future Bus", and made its debut in Amsterdam, the Netherlands. The technology of the CityPilot is based on that of the autonomously driving Mercedes-Benz Actros truck with Highway Pilot presented two years ago. It has however undergone substantial further development specifically for use in a city bus, with numerous added functions. The CityPilot is able to recognize traffic lights, communicate with them and safely negotiate junctions controlled by them. It has a top speed of 70 km/h on the open road. It can also recognize obstacles, especially pedestrians on the road, and brake autonomously. It approaches bus stops automatically, where it opens and closes its doors. And not least, it is able to drive through tunnels. What's more, the "Future Bus" with this system can also automatically connect to the Wi-Fi in the surrounding infrastructure, receiving traffic information and traffic light status.

Extended Reading

Logistics Concept with Super Technology

In 2016, Mercedes-Benz released Vision Van, to demonstrate its highly efficient logistics concept (Fig. 6).The van features an extremely wide windshield, which curves around to the sidewalls like a high-tech visor, and its smooth surfaces make the silhouette more streamlined. The designers have done without a steering wheel,

Fig. 6 Vision Van. *Source* Mercedes-Benz

pedals and centre console in favor of drive-by-wire control by means of a joystick, thereby creating new design options.

The Vision Van merges numerous innovative technologies and serves as the central, intelligent element in a fully connected delivery chain. Innovative algorithms control order picking, the loading of packages, the fully automated cargo space management, route planning for the vehicle and the delivery drones. They also calculate ideal delivery routes for the package deliverer. Automatic order picking takes place at the logistics center, for example, and consignments are loaded into special racking systems. Driverless handling vehicles load the racks by way of an automated one-shot loading process. The intelligent cargo space management system automatically transfers packages for manual delivery to the deliverer at the delivery destination by means of a package dispenser on board the vehicle. At the same time, the system supplies two drones, each with a payload capacity of two kilograms, with consignments for autonomous delivery within a radius of 10 km.

The world debut of Mercedes-Benz concept autonomous car—F015 Luxury in Motion was the major highlight in CES 2015. The long-standing automaker fully demonstrated its new strategy for "Future Mobility" at CES 2017. Among the concept cars on display, Mercedes-Benz Generation EQ with a maximum running range of 500 km is equipped with autonomous driving features and OLED control screen.

Mercedes-Benz will concentrate its forces on artificial intelligence, connectivity, car sharing and service technologies, logistics industry and autonomous driving for the days to come.

2. BMW: In Another Journey with "Strategy Number ONE > NEXT"

On March 7, 2016, BMW released the new "Strategy Number ONE > NEXT" at its global centenary celebration, comprehensively promoting digital strategy by focusing on the development of science and technology reserves and expanding the range of digital connections among people, vehicles and services. It is dedicated to transforming from pure automaker to a high-end personal travel service provider by through an ACES (Autonomous, Connected, Electrified and Services) strategy that are more technically innovative and perceptually improved. Therefore, BMW spares no effort in developing autonomous vehicles. The concept car BMW VISION NEXT 100 (Fig. 7), which was launched on the same day, not only shows the its potential designs, but also its advanced technologies. "At the BMW Group we always strive for technological leadership. This partnership underscores our Strategy Number ONE > NEXT to shape the individual mobility of the future," stated Harald Krüger, Chairman of the Board of Management of BMW AG. "Following our investment in high definition live map technology at HERE, the combined expertise of Intel, Mobileye and the BMW Group will deliver the next core building block to bring fully automated driving technology to the street. We have already showcased such groundbreaking solutions in our VISION NEXT 100 vehicle concepts. With this technological leap forward, we are offering our customers a whole new level of sheer driving pleasure whilst pioneering new concepts for premium mobility."

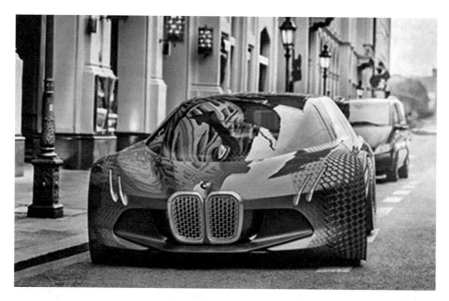

Fig. 7 VISION NEXT 100 concept car. *Source* BMW

Compared with Mercedes-Benz and Audi, BMW is not the most outstanding company in terms of technology, but it is indeed the most robust one. From carbon fiber frame technology to the promotion of electric cars, BMW has worked steadily and made every move count. Its future goals are clearly defined by its NEXT 100 concept car. The most futuristic part of the vehicle is BMW's innovations in driving mode. NEXT 100 has two driving modes named "Boost" and "Ease" respectively. In fully autonomous Ease Mode, the Companion takes over all driving duties. It ensures you and your passengers arrive safely at your destination and provides constant route, road condition and performance updates along the way. In Boost Mode, everything revolves around driving performance. That's why the Head-up Display shows ideal driving lines and braking, steering and acceleration points directly in your field of vision.

BMW's autonomous driving technology is defined as Highly Autonomous driving (HAD). BMW believes that future mobility solutions will not only "free the driver's hands", but also "free the driver's eyes (level 3), and brain (level 4)", and even make "driverless driving (level 5)" a reality. This will give drivers more freedom, when they can choose either to enjoy driving the car or work or just have some fun in the car. In 2013, BMW signed a cooperation agreement with Continental AG to jointly develop HAD system and electronic autonomous driving system, which can support HAD operations. BMW's earlier partners also include Cisco and IBM, which oversee analyzing the data and solving data compression and security problems.

BMW's production cars have equipped with rich ADAS functions like ACC and LDW since 2006, LCW since 2008, Collision Warning with Brake Support since 2012 and TJA since 2013. The launch of these ADAS systems has secured BMW's position among those powerful enterprises and brought many surprises to its fans. In order to implement those functions, BMW has explored various possibilities. For example, ACC can be realized through either a single 77G millimeter-wave radar or a monocular camera (like I3). For medium and high-end models, the technology scheme of the fusion of millimeter-wave radar and monocular camera is adopted. In AEB system, BMW only uses two configurations of monocular camera and fusion of millimeter-wave radar and camera. At the same time, BMW is developing HAD or even fully autonomous driving. In Germany, BMW's test cars have been able to drive on the freeway, from Munich to Nuremberg, at 130 km/h.

BMW is betting big on high-precision maps. It has acquired shares of HERE jointly with Mercedes-Benz and Audi to support future development of autonomous vehicles. In order to make self-driving possible in China, BMW has done a lot of localization work, including working with Baidu to develop high-precision maps. In September 2014, BMW and Baidu signed a cooperation agreement to jointly overcome the technical challenges faced by HAD in China's road environment in the next three years. According to the cooperation agreement, the two parties' R&D cooperation covers car use, driving strategies, maps, supporting infrastructure, relevant laws and industrial standards. On December 10, 2015, the first autonomous car jointly developed by BMW and Baidu completed its first road test under mixed road conditions in Beijing.

The test car was integrated with internal sensors and vehicle control interface provided by BMW. BMW provided technical support while Baidu was responsible for providing HD maps, vehicle positioning, environmental awareness, control system and other autonomous driving decision-making and control modules. It is reported that the HD maps collected and created independently by Baidu recorded complete 3D road information and could pinpoint the car in centimeter-level accuracy. Through object recognition and environment perception technology, combined with HD maps, the test vehicle can realize vehicle detection, recognition, tracking, distance and speed judgement of external objects, road segmentation, lane detection and other functions, thus realizing autonomous driving. In the end, the BMW 3-Series GT test car, without the driver's intervention, automatically completed a series of actions on the road in Beijing, including following the car to slow down, turn, overtake, and get off and on the expressway ramp.

At the same time, AutoNavi, another leading Chinese map company, started to woo BMW. On November 17, 2016, a German journal *Wirtschaftswoche* (Business Weekly) reported that BMW and Baidu had suspended cooperation due to "irreconcilable differences in opinion". However, according to a report of another German journal *Automobilwoche* (Automobile Weekly) on the same day, BMW denied that by saying that BMW and Baidu would continue to cooperate on map services and autonomous driving. With contrasting new sources, anyone with a discerning eye can see that BMW and Baidu's closeness from previous days seems to have lost in a cloud. Sadly, their friendship easily sunk. Part of the reason behind this is likely the conflict between Baidu Map and AutoNavi for title of "king of Chinese mapping".

BMW has provided customers with rich ADAS configurations in its latest production models, which not only ensure the safety of drivers at critical moments, but also provide driving support in relatively simply driving situations, such as in traffic congestion. Taking the brand-new BMW 5-Series as an example, through the upgraded ACC system, steering assist system and lane tracking assist system, the luxury model has also taken a new step towards autonomous driving. The driver can easily check the current maximum speed and adjust the cruise speed in the range of ±15 km/h. When the speed is below 210 km/h, the vehicle can automatically follow, accelerate, brake and maintain the lane, relieving drivers of the burden of driving at low speeds and in heavy traffic. In addition, the brand-new BMW 5-Series has made many achievements in human, vehicle and environmental networking. Drivers can operate vehicles through iDrive controllers, gestures, voice and direct touch of virtual buttons on the screen. In 2016, BMW demonstrated its one-stop digital concept through the launch of BMW ConnectedDrive. Based on the open mobile cloud, BMW ConnectedDrive seamlessly connected vehicles to users' digital lives through multiple access points, such as smartphones and smart watches.

After Mobileye announced the termination of the partnership with Tesla, BMW, Intel and Mobileye reached a strategic partnership in July 2016, cooperating in the next few years to jointly develop autonomous driving technology. In addition to the previous investment in HERE, BMW's partnership with Intel and Mobileye this time has taken the premium carmaker one step closer to the core technologies needed to put fully autonomous vehicles on the road.

On January 4, 2017, BMW AG, Intel and Mobileye announced that the road test of about 40 fully autonomous BMW cars would start in the second half of 2017. At a joint press conference in Las Vegas, the three companies further disclosed that the BMW 7-Series test cars powered with technology from Intel and Mobileye would begin a global road test tour starting from the USA and Europe. This release was the first progress announcement since the three companies announced their cooperation. Since then, they have been working on developing of an extendable architecture on that other automobile developers and manufacturers can implement their own designs and build differentiated brands. These products range from single key integration modules to complete end-to-end solutions, providing a variety of differentiated consumer experiences.

In order to further promote the development of fully autonomous driving platform, the three companies plan to release hardware samples and software updates in the next few years. Their ultimate goal is to release the BMW iNEXT model in 2021 as the cornerstone of BMW's fully autonomous driving strategy. At present, about 600 people are currently working on the iNEXT project, subsequently, HAD car models from various other brands under BMW AG umbrella will be launched too. At this point, the real intention of BMW to develop autonomous vehicles is finally revealed.

3. Volkswagen: Making Autonomous Driving a Mainstay of "TOGETHER— Strategy 2025"

As of 2015, The Volkswagen Group, based in Wolfsburg, is one of the leading automobile manufacturers in the world. The Group has 121 manufacturing facilities and 600,000 employees in 31 countries worldwide, with annual sales of more than

10 million vehicles and a turnover of €213.3 billion. The aim of the Group is to offer attractive, safe, environmentally friendly vehicles which are competitive on increasingly tough markets and represent the global benchmark in their respective classes.

On September 25, 2015, Matthias Müller took the throne of Volkswagen CEO after a fierce internal struggle. Saying goodbye to the era of Martin Winterkorn, this historic turnaround was particularly embarrassing due to Volkswagen's diesel emissions cheating scandal. The Group allocated €7.8 billion in emergency funds to deal with the crisis, for which it later paid a total of €18 billion.

At the same time, the external environment of the century-old auto industry is undergoing dramatic changes. The rapid development of the energy revolution, autonomous driving, ride-sharing and connectivity is reshaping the industry. Since the outbreak of the emission cheating scandal, the Group has also made a series of changes actively and passively. "We don't want to degrade Volkswagen to Foxconn of the IT world," Müller voiced his opinion in a forum on November 13, 2016, "if we don't change our thinking, we will become a supplier to Silicon Valley."

Volkswagen Group's "TOGETHER–Strategy 2025" plan released in June 2016 illustrates how it aims to be a global leader in providing sustainable mobility over the coming years. The guideline clarifies challenges and objectives in four key areas: transforming core business as an automaker, building new mobility solutions businesses including on-demand mobility and car sharing, strengthening innovation power in the field of digitalization and AI technology, and securing funding. The Group therefore intends to independently provide the resources necessary to address the future topic of autonomous driving and artificial intelligence. The aim is to license a competitive self-driving system (SDS) developed in-house by the end of the decade, on which Volkswagen plans to spend billions of euros. In November 2016, Volkswagen announced to sharpen the positioning of the Group brands and optimize the vehicle and drivetrain portfolio to focus on the most attractive and fastest-growing market segments. Correspondingly, Volkswagen will cut nearly 30,000 jobs worldwide, but at the same time, it has promised to create 9000 jobs in electric cars and mobile services in Germany.

The Volkswagen Group is preparing to make a rapid move towards autonomous vehicles by transforming its most important design center in Potsdam into the Volkswagen Future Center Europe. In the long run, the focus of Design Center Potsdam will be relieved from the production plan, which will give designers more time to explore possibilities for the Group brands' new UX (User Experience) and HMI (Human–Machine Interface) concepts. This move can help the Group build up their technical advantages in order to have a better understanding of the future products. Designers at the Design Center Potsdam will no longer be assisting other colleagues in the group's 12-brand network with their product projects, and the Volkswagen brand is expected to be the first to be affected. The move means that the internal teams of Audi, Seat, Volkswagen and other brands, as well as the senior design teams in California and Beijing, will have more work to do.

On the eve of the Beijing Auto Show 2016, Jochem Heizmann, President and CEO of Volkswagen Group (China) and member of Board of Management of Volkswagen

Group, announced that Volkswagen Future Center Asia would be located in Beijing. With this center in Beijing, Volkswagen is committed to building the products and services of the future, hoping to make a contribution to the research and development of digitization and autonomous driving technology.

In fact, Volkswagen established a digitization R&D department at a group level in early 2015. Later, Volkswagen Future Centers were successively established in Potsdam, Silicon Valley and Beijing, mainly developing future-oriented automotive technologies such as self-driving, HMI and digitization. In Silicon Valley, Volkswagen also set up the Electronics Research Laboratory (ERL), with the autonomous driving project being one of the biggest. In collaboration with Stanford University, the modified Tuareg and the Passat Variant won awards separately in the DARPA Grand Challenges. In addition, Volkswagen established an artificial intelligence center and formed an AI laboratory in Munich. Numerous professionals, including over 1000 software experts, have been recruited to boost software development so as to achieve the strategic goal of a competitive self-driving system (SDS) developed in-house by 2021.

Volkswagen has been thoroughly testing its self-driving cars in California's roads, albeit in relatively simple road sections. The car will send an alarm in time when the autonomous driving mode is shifted to manual operation, but there are still problems such as unqualified autonomous driving conditions, unsuccessful sensor identifications and software errors, which are also major problems faced by other companies such as Google and Mercedes-Benz. In addition to the USA, Volkswagen has also sought out a closed test site in Ehra-Lessien, Germany, to simulate more complex conditions of public transportation for its SDS's tests.

Volkswagen released its new self-driving concept car "Gira" on December 1, 2016, which is defined by the company as "a lounge on wheels". Once the road conditions permit, it will use the voice to remind the driver: "Master, you can give me the rest of driving work." When the driver presses the "AP" button to start the "automatic driving mode", the steering wheel and brake pedal will automatically shrink and hide in the body. Doesn't it feel like a sci-fi scene? At this point, the driver can relax. In AP mode, the seats can be adjusted to form a bed, and the front seats can be rotated back to form sofas in a living room. The console in the middle of the front and back seats can be moved to the center and 4 display screens can be pushed out to form a table. Passengers can surf the Internet with the screens or use the table to work or play. All these adjustments can be controlled by voice. According to reports, the above functions have been implemented. Self-driving technology has fundamentally broken the boundary of automobile interior design, making the spatial design of cars full of imagination.

Volkswagen revealed its second electric concept car, I.D. Buzz—essentially the VW Microbus reconstituted for the twenty-first century—at the 2017 North American International Auto Show in Detroit. Thanks to an overall height towering over that of the I.D. concept and fully autonomous operational capability, the Buzz can be a party on wheels with totally flexible seating arrangements, a center console that can be relocated, and practically all secondary controls live in a large touch screen. When the driver enters the Buzz, it will automatically set the seat position, music preference, commonly used navigation route, atmosphere lighting and commonly used contact information. The I.D. cruise is a new autonomous driving technology developed by Volkswagen. In the process of driving, by pressing the VW Logo button on the steering wheel, the squared-off steering wheel recedes into the dash and the autonomous driving mode starts to operate, and in-vehicle sensors, radars and cameras work at the same time, monitoring the road conditions in front of and around the Buzz, thus, the driver's hands will be completely free to do other things.

Shortly after that, "Sedric", which stands for "Self Driving Car", was unveiled by Volkswagen's parent company, Volkswagen Group, at the Geneva Motor Show in 2017. The concept car, hailed by Volkswagen as "the future of urban mobility", is capable of Level-5 autonomy, which means that the vehicle requires almost no human interaction. It has no pedals, steering wheel or brakes and Volkswagen noted that passengers don't need a drivers license in order to ride in a Sedric car. "Sedric will drive the children to school and then take their parents to the office, look independently for a parking space, collect shopping that has been ordered, pick up a visitor from the station and a son from sports training—all at the touch of a button, with voice control or a smartphone app—fully automatically, reliably and safely." Volkswagen said.

Sedric is able to 'recognize' its owners via facial recognition scanners, which also open the door for them. Sedric also has several built-in safety features, like LED signage on the vehicle's exterior that can notify people when to stop and wait for students to cross, as well as provide other useful information to other drivers.

In addition, Volkswagen invested $300 million in Gett, the largest on-demand mobility company in Europe, also invested in "Gofun", a business of Shouqi Car Rental in China, and signed a strategic cooperation framework agreement with DiDi Chuxing. Volkswagen hopes to transform itself into a travel service company, with annual revenues of billions of euros by 2025.

Volkswagen's unflattering "turnaround" will eventually sink into oblivion. What people see more now is its solid steps towards a glorious future.

4. Audi: Committed to Becoming the Jackpot in the Field of Autonomous Driving

In June 2016, Audi invited Chinese media to experience its latest achievements of autonomous driving on the A9 Highway in Germany. After the trial, a journalist from Sohu Auto wrote: "This A7 self-driving concept car, code named Jack, has two buttons below the steering wheel, one on the left and the other on the right. These are the start buttons of autonomous driving. Press and hold both simultaneously, the light band beneath the front windscreen will turn green, which means the autonomous driving mode has been started. An alert sounded and the steering wheel automatically retracted several inches, literally moving out of my hands. The car quickly accelerated to 100 km/h automatically, and then kept a distance of approximately 200 m from the preceding vehicle. When the radar detected that there were no vehicles in the fast lane, it suddenly changed lanes to overtake another vehicle and accelerated to 160 km/h. After overtaking, the vehicle automatically returned to its original lane.

It can accurately identify the lane on a bend. The car went through a toll gate on the highway and automatically slowed down to make way when the system detected vehicles passing by. When going through a road section under maintenance, where the two lanes combined into one, the car slowed down to follow the preceding vehicle, and even recognized the auxiliary line and merged into the single lane on its own. At the freeway exit, the light band under the front windscreen turned red and the alert sounded to remind me that the road was no longer suitable for autonomous driving. Here's what I liked: it was very easy to take back control. Simply grab the steering wheel and apply light pressure, and the system happily gives all controls back to you."

At present, Audi's autonomous driving technology is at Level-3. Its essence is to integrate automatic parking, lane keeping, cruise control, active braking and other technologies, and introduce products with mature Level-3 autonomous driving technology to the market eventually. Audi has been the most outstanding company on the development of Level-3 automation.

Audi's long-term commitment to the development of the self-driving technology, Piloted Driving, has been paid off. As early as in 2009, Audi conducted a road test in Utah with the TTS concept car combined with Google's HD maps, which established a new approach for the technological innovation of autonomous driving. Audi pioneered the industry when it first demonstrated its experimental self-driving technology in CES2011. In 2013, Audi took the lead once again by testing self-driving prototypes on real roads in Tampa, Florida. In 2014, Audi announced that it used zFAS, a chip set that integrates the heavy and complicated computer mainframe, as the core of the autonomous concept car, making great progress in autonomous driving technology.

In 2015, a self-driving Audi A7, code named Jack, arrived at the CES venue in Las Vegas from the Volkswagen Development Base in San Francisco, covering the 900 km journey without human intervention (Fig. 8). Jack's name is taken from "Jackpot" in gambling terms, which represents the first prize, in the hope of a smooth trip to Las Vegas. Before the opening of the CES Asia 2015 in Shanghai, Audi took Jack to the streets of the oriental metropolis, and invited the media to join the test drives, showing that Audi's self-driving cars could drive steadily and safely on urban roads of densely populated cities other than the original experimental city. The test was a great success and covered by a lot of media, which even became the talk of the town.

The tests proved that Audi's self-driving technology could basically cope with closed urban roads with clear signs on the ground and no traffic lights. The test on A9 highway in Germany was to explore V2X communications. Audi allows Jack to automatically obtain the information of the closed road/lane in front of it through the system setting, rather than relying on its own sensors. At the same time, Jack will automatically slow down when it knows there's a traffic jam or an accident ahead of it through information transmitted from other vehicles, and it can understand if a detour is already established. But Audi isn't alone in using cars as test platforms. Road infrastructure vendors are also exploring new technologies, including signs and

Fig. 8 The road test of Audi's self-driving car

signage that reflect radar signals, to allow autonomous vehicles to better position them from afar.

In addition, Audi's latest navigation system allows Jack to prioritize the route that allows him to turn on Piloted Driving for the longest time, rather than using the shortest distance or the highest fuel efficiency as the main criterion. More interestingly, Audi is trying to make Jack better co-exist with drivers on the road, by programming it to behave more like a human. For example, when a large vehicle passes by, it will make more room for itself. When changing lanes, it will not only turn on the turning light, but also move to the edge of the current lane first. It can also switch between a range of driving styles. When another car wants to merge into the line, Jack can either accelerate or brake by adjusting suspension, steering, power train or driveline systems according to different settings.

Audi has become the first car company in the USA to connect vehicles to transport infrastructures (V2I). In December 2016 in Las Vegas, Audi launched its Traffic Light Information system, which enables a car to communicate with the infrastructure in select cities and metropolitan areas across the USA. The car receives real-time signal information from the advanced traffic management system that monitors traffic lights via the on-board 4G LTE data connection. When approaching a connected traffic light, Traffic Light Information displays the time remaining until the signal changes to green in the driver instrument cluster, as well as the head-up-display (HUD), if equipped.

Audi strives to become the first automaker to achieve Level-3 automation by 2018, and apply it to the next generation Audi A8 full-size sedan. As an important part of the initiative, Audi is developing a new self-driving technology with Traffic Jam Assist (TJA), which will be first used on the new A8, A7 and Q8. Audi's TJA promises to relieve drivers from the tedium of slow-moving roads by taking care of braking, acceleration and staying inside of the lane—all with no input from the human behind the wheel. Enriched with environmental data from sensors, the "brain" is ready to take the helm at speeds from 0 to 60 km/h, with just the touch of a steering wheel mounted button. When the car senses the conditions are sufficiently traffic jamy (correct speed, no pedestrians detected, etc.), it alerts the driver that the TJA is available via the central display. Under special driving conditions, the car still needs the help of the driver. It needs to detect a driver on the driver's seat. More importantly, two cameras face the driver and analyze eye movement; once eyes have been detected closed for approximately ten seconds, they'll alert the system to bring the car to a halt, turn the hazard lights on and alert local authorities. Actually, law enforcement groups it has spoken with like this feature, too, as it should improve response time for emergency in-car medical situations (having a heart attack while driving, for instance).

Audi plans to reach Level-3+ automation between 2020 and 2021. As the first application level in two or three years after Level-3, Audi will launch it and define it roughly as a "more advanced" Level-3. This means it will expand the TJA application to the full highway speed range. In addition, at Level-3, Audi will install an event data recorder (similar to the black box on an airplane) in every Audi car. In the event of an accident, it will directly record the driving data before collision. Of course, Audi will not be recording all driving data. If there is no collision, the event data recorder will continuously delete the data it collects.

In the late 2020s, Audi will enter a period of Level-4 automation, which means it is equipped with complete highway and urban autonomous driving capability. These functions will only be allowed in designated mapped areas. That is to say, the cars are not completely free to drive anywhere. For example, the diver cannot require an autonomous vehicle travel from Los Angeles to New York automatically unless the route is set.

Extended Reading

AI Helps Audi's Self-Driving

Audi and NVIDIA are long-time partners, merging the best of automotive engineering and visual computing technologies on Audi innovations such as the Audi virtual cockpit. To showcase their progress, Audi and NVIDIA demonstrated an Audi Q7 Piloted Driving concept vehicle in CES2017, which uses neural networks and end-to-end deep learning on NVIDIA's artificial intelligence platform to navigate a complex course. The vehicle relies on its trained AI neural networks to recognize and understand its environment. With no driver behind the wheel, it performs several laps on a closed course, where the configuration of the track will be modified in the middle of the demonstration. The course features a variety of road surfaces including areas with and without lane markings, dirt and grass, as well as a simulated construction zone with cones and dynamic detour indicators.

5. Volvo: Exploring Sustainable Solutions for the Future with Drive Me

Volvo is one of the traditional automakers who is at the forefront of autonomous vehicle development. Volvo has been deeply involved in the development of autonomous driving technology, including the establishment of a future traffic test site in Gothenburg, Sweden, and conducting related road tests in Sweden, China, UK and the USA. Volvo is the first automaker to express willingness to assume responsibility in the event of an accident in autonomous driving mode. Unlike other technology companies, Volvo is not only focusing on its own technologies and products, but on building an intelligent transportation ecosystem.

While gradually improving the level of automation, Volvo also prioritizes people's driving experience. Volvo firmly believes that the only solution for the development of autonomous vehicles is to take the driver-oriented technological innovation and establish an intelligent transportation ecosystem that all road participants adapt to and accept. "The concept of autonomous driving of IT companies and that of our automakers are very different." Volvo Chairman Li Shufu said in an interview. "The technologies are the same, but the market positionings are different, because there are still some people want to drive. Just like horses. Why do people keep horses when they don't have to go to war? Just because they like riding horse. It's the same with cars. There are still people who enjoy driving. They want to have such a pleasure."

Volvo also classifies the autonomous driving into four stages. The first stage is driver assistance, corresponding to SAE's Level-2, which can help drivers with necessary information collection when driving, and provide clear and definite warnings when necessary, such as Lane Departure Warning (LDW), Forward Collision Warning (FCW), Blind Spot Information System (BLIS) and so on. The second stage is semi-autonomous driving, corresponding to SAE's Level-3, which can allow the vehicle to do things such as Automatic Emergency Braking (AEB), Lane Change Assistance (LCA) and other actions where drivers fail to respond to warnings. The third stage is highly autonomous driving, roughly corresponding to SAE's Level-4, including high-speed automatic cruising and Traffic Jam Assist (TJA) systems, which can allow the car to drive automatically for a period of time under the surveillance of the driver. Finally the highest stage—the fourth stage, corresponding to SAE's Level-5, allows the car to be fully autonomous without human surveillance. This means that the driver can be engaged in other activities, such as work, Internet surfing, rest and sleep or entertainment.

Drive Me is also an autonomous driving related project that has recently come into the spotlight. Drive Me is the world's first large-scale autonomous driving test project, which has been approved by the Swedish Government and has been implemented since the end of 2013. The project is led by Volvo, with the full support from the Swedish Transport Administration, the Transportstyrelsen, Lindholmen Science Parkand the Municipality of Gothenburg. The Drive Me project is not just about testing and validating Volvo's self-driving technology, but also about exploring solutions to create a sustainable transportation in the future.

In 2015, Volvo moved the test site of Drive Me to Beijing, China, and chose a 15-km highway section of the West Sixth Ring Road. Ranking amongst the most complex and challenging traffic roads in the world, Beijing is a path that Volvo must pass. The road conditions were relatively good and there was a short tunnel. Outside the tunnel, the vehicle could slow down in time when it overtook occasionally or the vehicles in front slowed down, and automatically accelerate to the set speed of 70 km/h when there was no abnormal situation. Comparing to the test in Stockholm, there were no lane changes during testing in Beijing. Technicians from Volvo explained that they didn't have very precise map of this area. But once in the tunnel, since the lane markings could not be identified, the vehicle went out of autonomous driving mode and the testing staff had to take over.

Volvo officially launched Auto Pilot in 2015, integrating those ADAS systems into one platform. In terms of technology, Volvo cars have adopted sensors (including radar, LiDAR and high-definition camera), HD map and GPS technology to perceive the outside world and integrate them into the body of the vehicle. Through the integration of various information, the car can identify and control the road situation. In addition, in this car with Auto Pilot system, there are also four intelligent and safety technologies including Park Assist Pilot (PAP), In-Car Delivery (ICD), Pedestrian Protection System (PPS), Slippery Road Alert (SRA), as well as other active and passive safety technologies such as Lane Departure Warning (LDW) and automatic braking at intersections. The system allows the driver to temporarily hand over control and conveniently switch to autonomous driving mode. In case of emergency, the cars ends an alarm to remind the driver to take back control. If the driver's hands leave the steering wheel for over 15 s, the system will automatically take over to ensure that the vehicle does not lose control. The dashboard displays the actual time to destination, and the system can notify the driver in advance whether overtaking is required. When the car has one minute left to reach its destination, the Auto Pilot system starts to remind the driver to take over the car again. If the driver does not respond, the system will automatically drive the vehicle to a safe area.

In December 2015, Volvo released its latest flagship sedan, S90, which is equipped with Pilot Assist II, a driver assistance system that expands the speed range for adaptive cruising to 130 km/h at maximum. The S90 also updates the City Pilot system, which can not only identify pedestrians during the day and at night, but also large animals such as deer and horses, providing drivers with assistance to slow down or stop if a collision is about to happen—the performance in emergency steering and braking is improved. On the S90, Volvo's Car Navigation system, Sensus Navigation has been iterated to the third generation, and it is one of the few brands that can realize real-time update of data and services through OTA. Volvo plans to equip the new XC90 with Pilot Assist II in 2017 and make it a standard feature. In the future, all the new cars under Volvo's Scalable Platform Architecture (SPA) will come with autonomous driving technology.

In 2015, Volvo also participated in a fully autonomous vehicle project called the Safe Road Trains for the Environment (SARTRE) led by the Europe Union, aiming to develop a wireless system that will allow cars on a public highway or motorway to join in a platoon, or semi-autonomous "road train" of vehicles. In this research project, the TJA system developed by Volvo successfully helped the vehicles on traditional highway to form a platoon and travel orderly. This system allows a vehicle to automatically follow the preceding vehicle when traveling at a speed lower than 50 km/h. The system is also known as the "road train" system. It resembles a train in a way, except this train is made up of single vehicles, and all you need is let a driver to drive the head vehicle. A long line of vehicles keeps close formation and consistent pace and is strictly under the leadership of the first vehicle to form one integrated "train". Just think about it! The scenario envisaged here is the autonomous driving relying on the development of V2V technology. The V2V technology can reduce the distance between vehicles, thus easing traffic congestion. The highly uniform driving behavior of a road train is also the ultimate form of intelligent transportation in the future.

In 2016, Volvo established a partnership with Uber to jointly develop the next generation of autonomous driving technology and start to work on basic models with full autonomy. In September of that year, Volvo and Autoliv Inc., a tech company in Sweden, announced their intention to form a joint venture to develop the next-generation autonomous driving software.

Volvo announced at the 2017 North American International Auto Show that it would set up Drive Me test sites in China, UK and the US, and planned to launch its first production car with full autonomy in 2021. Volvo launched 100 Drive Me XC90 cars to put into test, allowing motorists to use advanced self-driving features on public roads. This functionality was limited to specially designated autonomous driving zones in and around Gothenburg. It was the world's first public test involving real-life users, police, governments and universities, rather than private tests run by engineers, or pure driverless test by Internet companies.

Volvo's exploration for autonomous driving is praiseworthy. Compared to other automakers, Volvo's current functional advantages include a greater combination of active safety as well as the speed and scale with which new technologies are being applied to production cars. Although Volvo is expected to produce fully autonomous vehicles by 2020, its focus is still on driving safety. "Our vision is that by 2020 no one should be killed or seriously injured in a new Volvo car.", as claimed by Håkan Samuelsson, President and CEO, Volvo Cars.

6. PSA and Renault: The French Legion Will Not Be Outdone

Thanks to the launch of Fukang sedans early days in China, Citroen is no stranger to the country. The Groupe PSA behind the brand, namely PSA Peugeot Citron, has strong ties to China. Groupe PSA, Europe's second-largest automaker, suffered heavy losses a few years ago, with a net loss of once more than €5 billion, and was on the verge of bankruptcy.

PSA formally launched its revitalization plan soon after receiving strategic capital injection from China's Dongfeng Motor Corporation in March 2014. At that time, Carlos Tavares, newly appointed Chairman of the PSA Group Managing Board, put forward the group's medium-term strategic plan for 2014–2018—"Back in the Race", trying to make drastic adjustments in four aspects: brand positioning, product planning, global layout, enterprise upgrading and transformation, so as to push Groupe

PSA back to the track of making profit. The plan has been a phased success, with PSA turning a profit in 2015.

China Dongfeng Motor Corporation has become one of the three major shareholders of PSA after investing €800 million in PSA. The cooperation between the two sides has also risen from the Chinese market to the global strategic perspective. At that time, there were three main points in the bilateral agreement: first, the joint venture's sales target in 2020 would be 1.5 million units; second, establish R&D center in China; third, set up an overseas sales company to handle the Asian market outside China. In July 2016, the two sides pledged to deepen their cooperation in areas such as autonomous vehicles and car sharing.

Groupe PSA has kept on the straight and narrow in autonomous driving R&D. PSA Peugeot Citroen has divided autonomous driving into five stages: stage 1, auxiliary driving functions, such as adaptive cruise; stage 2, driving assistance system, which systematically integrates adaptive cruise, active braking, automatic steering and other functions; stage 3, partially autonomous driving, which is autonomous driving under certain road conditions and under the monitoring of the driver; stage 4, highly autonomous driving, which is autonomous driving in specific road conditions that does not require the driver to monitor; stage 5, the final stage is to realize autonomous driving during the whole trip. At present, PSA is working to bring together a large range of functions that will gradually lead to the partial, and later total, delegation of the driving to the car itself, if the driver wishes so. Scheduled to start in 2018, automated driving functions will include "The Connected Pilot" which offers assistance in speed and trajectory control, parking without the driver help, identifying obstacles and increasing night visibility.

Groupe PSA also announced that it would equip its vehicles with "Highway Chauffeur", an automatic assistant driving system, starting in 2018. The system allows the vehicle to keep lane driving at speeds below 130 km/h via a suit of radar, ultrasonic radar, cameras and GPS systems on the outer side the vehicle. The driver's driving behaviors could be constantly monitored by the sensors inside the car among which a capacitance sensor on the steering wheel can detect whether the driver's hands are on the steering wheel, and a camera can judge whether the driver is distracted by aiming at the his/her eyes. In addition, in terms of HMI, PSA has also introduced the Scoop@F system, redesigned the user interface after taking autonomous driving and V2X technologies into account, by highlighting the road condition warning and active safety measures, so as to give drivers early warning through images, sounds and tactile sense.

Groupe PSA was the first manufacturer to test its autonomous vehicles on public roads in France. In July 2015, four modified prototypes set off from Paris and travelled 580 km in autonomous mode to Bordeaux in southern France. During this process, the vehicles were able to make turns following lane curve, change lanes to overtake and so on. The vehicles were also equipped with V2X technology, which can realize pedestrian identification, forward emergency avoidance and other functions through vehicles ahead and roadside infrastructure. In April 2016, two Citroen test cars set off from Paris and drove 300 km to Amsterdam in autonomous mode without any control or operation from the drivers.

Extended Reading

French Self-Driving Electric Buses Drive into Las Vegas

According to The Verge, an American technology journal, Las Vegas, the US gambling capital, officially launched a self-driving shuttle bus service on a designated route between the Las Vegas Strip and the Eighth Street in January 2017. For safety reasons, the maximum speed of the self-driving electric bus is only 43 km/h and will be further limited to 20 km/h in the trial period. The car takes the "Internet business model" and wants to ride free but make money through advertising. The monthly cost of $10,000 can be offset by in-car advertising. "Hopefully this summer or fall we will have a fleet operating in the city, and not limited to the current route." said Las Vegas Mayor Carolyn Goodman.

Interestingly, the self-driving service provider is not Google, Uber or other companies from the USA, but Navya, a French leader in autonomous driving systems. The French company "designs, manufactures and commercializes autonomous, driverless, and electric vehicles that combine robotic, digital and driving technologies at the highest level", as indicated in its website. In addition to this, another French company EasyMile also has used its self-driving electric buses to carry passengers on short rides to try out the technology in Wageningen, the Netherlands and Paris, France. The French, it seems, have made a remarkable comeback in the autonomous vehicle field.

Renault is another prominent French automaker. Since it has a partnership with Nissan and Mitsubishi, they share some autonomous driving technology. Some of Nissan's autonomous driving technology will supposedly be ported to Renault models in the future. Renault is committed to introducing "hands-off/eyes-off" autonomous driving technology in its main stream affordable car models. This technology aims to make roads safer and more pleasant to drive on, and to give people more time to use on the road, ultimately allowing autonomous driving to serve everyone. After 2020, if new road traffic regulations around the world permit, Renault plans to equip its core series of vehicles with "eyes-off" technology, which will be used first in situations with heavy traffic that requires frequent stops and starts, and then in more complex conditions.

In December 2016, Renault Group established an autonomous driving demonstration zone in Wuhan with its Chinese partners. An autonomous prototype based on Renault's battery electric car ZOE could be tested and demonstrated for two years on a 2-km stretch of road in the zone located in the Sino-French Ecological Demonstration City in Caidian District, Wuhan.

7. Jaguar Land Rover: Going Off-Road with Autonomous Cars

Jaguar Land Rover (JLR) is a British car manufacturer that was formed by the merger of Jaguar and Land Rover. Jaguar has a glorious history as the world's leading manufacturer of luxury sports cars, while Land Rover is a global manufacturer of top luxury all terrain 4 × 4 cars. In history, Jaguar and Land Rover have been owned by different owners in different time, among which included Ford and BMW, and eventually merged into JLR in Ford's hands. In 2008, India's Tata Motors bought it from Ford for $3 billion. Despite its noble birth, the company has been reduced to a trading object.

In early 2016, JLR announced the launch of the Connected and Autonomous Vehicle (CAV) project with large-scale autonomous driving and Internet of vehicle technology. The three-year project, with a budget of £6 million, is part of the UK Connected Intelligent Transport Environment (UK-CITE) project, the first of the kind in the UK. The project, known as "the Living Laboratory", consists of a new 100-vehicle fleet of semi-autonomous and connected cars from JLR, which are tested in the real roads about 65 km around JLR's base in Coventry, UK. Five of the vehicles in the fleet are based on existing JLR models, while the others are test prototypes under development with V2X technology and a high automation.

In addition, Move-UK, another project that JLR took part in at the same time, was launched by Bosch, UK Transport Research Laboratory. The project also lasted for three years and was funded at around £5.5 million. This project aimed to make autonomous vehicles more humane, without the roughness and abruptness of computer-controlled cars. The testing vehicles used by JLR were driven by some of the company's employees in Greenwich, London in their daily life. After the vehicles went through bad weather or complex road conditions such as traffic jams, busy intersections, roads under construction, special vehicles showing up behind it, the in-vehicle sensors and computers would record the drivers' driving behaviors and choices, and then the system would gradually improve itself by using the machine learning technology, so that the vehicles can cope with more complicated situations in the autonomous driving mode, and achieve more natural driving behaviors.

In addition to the adaptive cruise and lane keeping functions, through V2X technology, JLR's self-driving cars can also change lanes and exit the highways smoothly if they're greeted with smooth, flat roads, and structured environments. In the next three years, more communication facilities will be installed on both sides of the test road, and the test vehicles have the ability to identify traffic lights, limit viaducts, signs and other infrastructure, so as to test simultaneously the transmission efficiency and reliability of V2V and V2I technology at high speed on the real road. JLR is committed to making cars communicate with each other in order to improve their safety. Vehicles can connect to each other, sending information about their position, wheel slip, suspension height, wheel hinges and so on, so that if one of them stops, the others will be alerted.

While today's self-driving cars can travel on paved roads, JLR wants to crack the incalculable surfaces with autonomous off-roaders. In this regard, JLR has demonstrated its own self-driving technology. In mid-2016, JLR announced a new technology of its own—a 4X4 SUV that can autonomously travel on muddy cross-country roads. JLR uses a system called the Off-Road Connected Convoy for its autonomous driving technology, which allows a fleet of vehicles to travel off-road by communicating with each other. The autonomous off-road vehicle is installed with built-in ultrasonic radars, which can detect the terrain five meters in front of the vehicle and send the information to the terrain-sensing device, to control the speed of the vehicle.

This terrain sensing device integrates ultrasonic, camera, radar and LiDAR detection technology, which can give the vehicle an ultra-wide field of vision and also help choose its own road to pass through the unknown road through appropriate skills and methods.

Land Rover also demonstrated a Range Rover sport that can be driven remotely from a mobile phone in 2016. Just like in the James Bond movies, all the core functions such as going forward, backward and steering can be controlled through the mobile App. The maximum speed of the vehicle is set at 6.4 km/h and the effective range is within 10 m. One of the advantages of remote control is that, for example, on a rugged mountain road, when the driver needs to get off the vehicle frequently to observe the ground clearance, or tire contact surface, the driver can realize the vehicle's drive-through remotely by the mobile App outside the vehicle. Or, when the parking space is too small or even insufficient for the driver to get off, the car can be parked remotely through the mobile phone. In the future, remote control will also include voice control.

2.2   The US Squadron

1.   GM: An Evangelist of Intelligent Connected Vehicles

General Motors aims to be the first company to launch an autonomous car, as Mary Barra, its CEO, revealed in a conference call with analysts in February 2016. Facing the surging tide of industry reform, General Motors is laying out its forces, gathering resources, and defining the future personal travel mode through its own scientific and technological strengths. It has made great efforts to build a new momentum in the four pillars of intelligent and connectivity, car sharing, autonomous driving and electrification,

In fact, in the past few decades, GM has been working hard on the development of autonomous vehicles and relevant technologies. Back in the 1950s, GM developed the concept car, Firebird II, which was described "A Jet Plane on Four Wheels" because it looked like it belonged to the sky rather than the earth. In theory, the Mission Control system of the car can realize autonomous driving. At the time, many of the car's technologies were thought to exist only in science fiction, but they have come true step by step after 60 years. In 1959, GM released the Cadillac Cyclone concept car, enriched by futuristic design elements and forward-looking technologies. The Cyclone concept car featured a collision avoidance system based on radar technology, and radar sensors were installed at the dual "nose cone" position of the car front.

In 1998, GM launched an automated highway research project with a Buick LeSabre and released the XP2000 concept car. GM installed high-strength magnets on a 12 km highway in southern California to show that the modified Buick LeSabre can turn, accelerate and brake automatically without driver's intervention. XP2000 concept car can liberate the driver completely in the more advanced automatic highway. The vehicle can not only drive to the destination automatically, but also pay tolls and fuel bills automatically through a "smart card" on the way.

In 2000, GM equipped sensors and other equipment for a range of Buick LeSabre models, enabling them to better sense the surrounding traffic environment and alerting drivers if potential hazards are detected, marking a big step toward the ultimate goal of developing automotive connectivity technology. In 2005, GM demonstrated the V2V system, an omni-directional object-positioning sensor based on GPS and wireless technology that would give a car "the sixth sense". The system provides automatic safety features such as Lane Change Warning (LCW), Blind Spot Detection (BSD), Automatic Emergency Braking (AEB), Forward Collision Warning (FCW) (automatic braking) and Cross Traffic Alert (CTA). In 2008, GM showcased a V2V transponder, making it possible to communicate between the car and any object through the handheld sensor. The 2017 Cadillac CTS would be the first to apply this technology.

At the 2010 Shanghai World Expo, GM demonstrated the first Electric Networked-Vehicle (EN-V) that combines battery-powered systems, short-range communications, sensors and GPS platform in a compact and flexible manner. It promises to solve personal traffic problems in the future city. Through the integration of advanced technology, it has not only improved the V2V networking function, but also reduced the demand for parking space. In the future, "zero fuel consumption, zero emissions, zero blockage and zero accidents" in urban traffic will be a reality. The EN-V 2.0 fleet was also put into operation at Shanghai Jiaotong University in the first half of 2015, becoming part of a comprehensive transportation system consisting of bicycles, cars and buses.

In 2012, GM first demonstrated the Super Cruise technology, which has the functions of automatic following, braking and speed control in the highway driving environment. This technology that can improve the comfort of drivers driving through congestion or long distances could be available for the first time in the 2017 Cadillac CT6.

In 2014, General Motors launched onboard 4G LTE services, making it the first automaker in the world to offer such kind of technology. The technology provides users with a continuous and powerful Wi-Fi hotspot, allowing users' personal mobile devices to maintain a fast connection to the Internet. The high-speed, stable and safe interconnected environment based on this technology can provide a foundation for building a networking ecosystem and intelligent transportation in the future. Currently, nearly 3.5 million vehicles with onboard 4G LTE services are on the road.

GM has sped up the development of autonomous vehicles since 2016. In January, GM set up a special R&D team for autonomous driving. It immediately bought Sidecar, a ride-sharing company, and launched Maven, a car sharing service, which offered hourly rental car services to many cities in North America. Almost at the same time, GM invested $500 million in Lyft. The two sides plan to jointly establish a comprehensive network of autonomous vehicles in the USA that provides users with car rental services and eventually deploy driverless electric fleets through Lyft. In March, General Motors spent another $580 million to acquire Cruise Automation, a company that develops and tests autonomous driving technology, to work

Fig. 9 The self-driving Bolt electric car for testing

together to build a self-driving BEV—Chevrolet Bolt. GM has fed various data of vehicle software and hardware as well as enhanced engineering cooperation to Cruise Automation, which greatly improves the R&D efficiency. At present, its development and testing work has been extended from San Francisco, California, to Arizona, and now to icy and snowy Michigan.

The test in Michigan takes place on GM's Technology Center in Warren and downtown Detroit. Bolt can be driven in autonomous driving mode in the campus of Warren Technology Center. GM employees can book the car through a car-sharing mobile app, which allows the car to travel to its destination and park itself with the help of autonomous driving technology. The project will provide data and practical experience to accelerate the development of its autonomous driving technology.

The self-driving Bolt electric car (Fig. 9), like the standard Bolt electric car and the Chevrolet Sonic, will be built at the same GM assembly plant in Michigan. The self-driving version additionally features LiDAR, cameras, sensors and other hardware to support autonomous driving that require special assembly.

In the USA, GM has been one of the important OEMs actively participating in the development of V2X technology for over ten years. GM, together with relevant enterprises and organizations, helps the country establish DSRC technical standards. GM believes that DSRC has some technical advantages, such as fast network access, low communication delay rate and high reliability. At the same time, it also prioritizes the installation of security applications with interoperability, security and privacy. At present, the USA and Europe have developed mature and royalty free DSRC technical standards. In collaboration with GPS, DSRC can detect the positions and activities of other vehicles within 300 m.

In recent years, GM has also accelerated its development in China. In October 2016, GM demonstrated eight safety applications based on ICV technology for the first time in the National Intelligent Connected Vehicle (Shanghai) Pilot Zone, including six V2V safety applications, such as Intersection Movement Assist (IMA), Emergency Electronic Brake Light (EEBL) and the Control Loss Warning (CLW), as well as two new V2I safety applications of Red Light Violation Warning (RLVW) and Workspace Slowdown Warning (WSW). GM also worked with Tsinghua University and Changan Automobile to take the lead in drafting the V2X Application Layer Standard of China, which was completed and submitted at the end of 2016.

2. Ford: Let Autonomous Driving Technology Benefit Global Consumers

A Ford Fusion hybrid self-driving car without opening headlights successfully completed an automatic navigation test in the dark night on the deserted desert highway, a very dangerous driving task for manual drivers. This nighttime road test carried out in Ford's automobile proving ground in Arizona is another important milestone in the development of the company's fully autonomous driving technology, and another step forward to realize the commitment of "bringing the convenience of fully autonomous driving technology to global consumers".

The self-driving car utilizes high-determination 3D maps, complete with data about the street, street markings, geology, geography and historic points like signs, structures and trees. The vehicle utilizes LiDAR heartbeats to pinpoint itself on the guide progressively. Extra information from radar gets melded with that of LiDAR to finish the full detecting ability of the independent vehicle.

For the desert test, Ford engineers, wearing night-vision goggles, checked the Fusion from inside and outside the vehicle. Night vision permitted them to see the LiDAR doing its employment as a network of infrared laser bars anticipated around the vehicle as it drove past. LiDAR sensors shoot out 2.8 million laser beats a second to precisely filter the encompassing environment. "There is no doubt that even on winding roads, it stays precisely in the lane," a Ford engineer said.

In fact, Ford has been working on the development of autonomous vehicles for more than a decade and is one of the pioneers of this field in the auto industry. "The next decade will be defined by automation of the automobile, and we see autonomous vehicles as having as significant an impact on society as Ford's moving assembly line did 100 years ago," said Mark Fields, then Ford president and CEO. "We're dedicated to putting on the road an autonomous vehicle that can improve safety and solve social and environmental challenges for millions of people–not just those who can afford luxury vehicles."

Since becoming the first automaker to begin testing fully autonomous vehicles inside Mcity, the University of Michigan's simulate urban environment, Ford has made enormous strides in researching how these vehicles operate in hazardous conditions, such as snow and complete darkness.

"The effort to build fully autonomous vehicles by 2021 is a main pillar of Ford Smart Mobility: our plan to be a leader in autonomy, connectivity, mobility, customer experience and analytics. "Ford's" official website says. In 2016, Ford tripled the size of its autonomous driving test fleet and conducted tests for 30 self-driving Fusion Hybrid cars on streets in California, Arizona and Michigan.

Extended Reading

"It Was A Really Cool Challenge."

In recent years, LiDAR is considered the most effective way to sense the autonomous driving environment. Ford's dream is to use drones to help guide self-driving vehicles, including on off-road adventures, by mapping surrounding areas beyond what the car's sensors can detect. "At some point, people are going to want to take their autonomous vehicle into the woods or off road where the drone could guide them," said Alan Hall, spokesman for Ford's in-house technology department. Hall argued that the drones also could prove useful in areas beyond the digital maps of urban and suburban areas and inter-city highways.

The idea for using drones came out of a 'brainstorming' session of researchers and engineers working on Ford's autonomous vehicle. Ford and Chinese drone maker drones DJI held a competition for programmers to see if they could teach a drone to fly from and return to a moving vehicle. The idea was to see if a drone could use its cameras to guide a vehicle into and out of a disaster area where communications and roads have been destroyed or disrupted. Only one of the 10 participants actually succeeded, with a drone launched from a moving Ford F-150 pick-up truck which returned after completing the assigned task. Ford has joined forces with a team of researchers in the Silicon Valley Research Center in Palo Alto, California, working with the idea of find way drones could help autonomous vehicles solve future navigation problems.

According to Ford's five-year plan for autonomous driving development, they will introduce two Level-2 ADAS systems by 2019—Traffic Jam Assist (TJA) and Full Assisted Parking Aid(FAPA)—with technologies that rely heavily on radars, cameras and ultrasonic sensors. Confusingly, Ford's Level-2 system will be equivalent to Audi's at Level-3, since both have adopted TJA that can automatically queue up, brake, accelerate and stay in lane in the case of traffic jams. Ford's main technical requirements for its Level-4 autonomy, which will be mass-produced in 2021, include LiDAR and radar, camera, 3D map, 4G accessible and possibly V2I communication technology. Instead of gradually leveling up autonomous driving technology, as Audi and Nissan have been doing from Level-2 to Level-3 and then to Level-4, Ford will go skip directly to Level-4, taking away the steering wheel, gas and brake pedal. Whether driven by the need for market positioning or the demand for profitability, it makes strategic sense for Ford to be the first to make a fully autonomous vehicle,

Foreign media has reported that Ford will first sell autonomous vehicles to ridehailing service companies such as Uber and Lyft to provide ride- or car-sharing services and realize the commercialization of autonomous driving as soon as possible. Similar to General Motors, Ford also said that due to the expensive prices of selfdriving vehicles, it may take a while to enter the daily lives of ordinary users. It is more feasible to use the ride-hailing platform for promotion.

To that end, Ford is working with four startups, either through investment or partnership, to promote the development of autonomous vehicles. Ford is also expanding its Silicon Valley presence by creating a dedicated campus in Palo Alto to ensure that these innovations will be made, and the current Palo Alto staff will be doubled by the end of 2017.

At present, in the field of autonomous driving, major global automakers are seeking to combine internal independent R&D with external M&A. Ford has committed to expanding its research in advanced algorithms, 3-D mapping, radar technology and camera sensors. To help accelerate the development of these new technologies, we have announced four key investments and collaborations with Velodyne, SAIPS, Nirenberg Neuroscience LLC and Civil Maps.

Ford, together with Baidu, has invested in Velodyne, the Silicon Valley-based leader in LiDAR sensors. The aim is to quickly mass-produce a more affordable automotive LiDAR sensor. Ford has a longstanding relationship with Velodyne and was among the first to use LiDAR for both high-resolution mapping and autonomous driving more than 10 years ago. Ford has invested in Berkeley-based Civil Maps to further develop high-resolution 3D mapping capabilities. Civil Maps has pioneered an innovative 3D mapping technique that is scalable and more efficient than existing processes. This provides Ford another way to develop high-resolution 3D maps of autonomous vehicle environments.

Ford has acquired the Israel-based computer vision and machine learning company to further strengthen its expertise in artificial intelligence and enhance computer vision. SAIPS has developed algorithmic solutions in image and video processing, deep learning, signal processing and classification. This expertise will help Ford autonomous vehicles learn and adapt to the surroundings of their environment.

Ford has an exclusive licensing agreement with Nirenberg Neuroscience, a machine vision company founded by neuroscientist Dr. Sheila Nirenberg, who cracked the neural code the eye uses to transmit visual information to the brain. This has led to a powerful machine vision platform for performing navigation, object recognition, facial recognition and other functions, with many potential applications. For example, it is already being applied by Dr. Nirenberg to develop a device for restoring sight to patients with degenerative diseases of the retina. Ford's partnership with Nirenberg Neuroscience will help bring humanlike intelligence to the machine learning modules of its autonomous vehicle virtual driver system.

On February 10, 2017, Ford announced a $1bn investment in Argo AI, an artificial intelligence company which will produce the software needed for a new generation of self-driving cars. Ford's investment over five years would see the new company develop the software needed to make self drive cars a reality, initially in cities and then across a wider area. It has expected to profit from not only having its own autonomous car on the road in 2021, but by licensing the technology to other companies. This billion-dollar acquisition has been the single largest investment in autonomous driving technology by a traditional automaker so far. The current team developing Ford's virtual driver system—the machine-learning software that acts as the brain of autonomous vehicles—would be combined with the robotics talent and expertise of Argo AI. This innovative partnership aims to work to deliver the virtual driver system for Ford's Level-4 self-driving vehicles.

All of this shows that Ford, a veteran auto giant, has realized the profound changes that the upcoming technological wave will bring to the auto industry. In this critical period the traditional automaker has the opportunity to compete on the same platform with emerging tech companies in Silicon Valley through aggressive acquisition and investment. Ford's frequent "assassinations" are an attempt to avoid the embarrassing situation in which hardware enterprises become "labors" of software companies, just like in the mobile phone industry.

2.3   Japanese Squadron

1. Toyota: Starting from Safety and Aiming towards the Blue Ocean of Autonomous Driving

Toyota has been developing autonomous driving functions since 2013 or even earlier. Its original intention is to ensure driving safety, so autonomous driving has been a part of "Toyota Integrated Safety" from the very beginning. In October 2015, Toyota developed its "Mobility Teammate Concept", an autonomous driving technique very close to the production standard. Prior to this, Toyota had launched the Level-2 Automatic Highway Driving Assist (AHDA), which consists of three parts: Dynamic Radar Cruise Control (DRCC), Lane Trace Control (LTC) and Human–Machine Interface (HMI). The HMI also includes path preview. On the basis of the original map, the calculated track is configured to monitor the vehicle control, and if necessary, detect and intervene the driver's status to prevent the driver from distracting and dozing off. Toyota has basically ensured that it is at the same level with its competitors in terms of autonomous driving, especially regarding the combination of horizontal and vertical directions on a single lane as a mass production node.

For Highway Teammate at Level-3, Toyota improved the perception by fusing more sensors and endowing the vehicle with a more powerful decision-making ability. At the same time, V2X systems (V2V and V2I) were also added to improve the level of intelligence when switching between manually driving and autonomous driving.

The onboard detection system of Toyota's autonomous driving includes front and rear LiDARs, front and rear millimeter-wave radars and a front-facing camera. Together with the vehicle's own features of Dynamic Radar Cruise Control (DRCC), and Lane Trace Control (LTC), the system is guaranteed to work properly. In addition, high-definition 3D maps, unobstructed satellite signals and a well-established transportation infrastructure are also important aspects of autonomous driving.

In order to commercialize its test autonomous vehicles, Toyota has conducted several real-road tests on highways. The test vehicle enters the highway entrance ramp and the driver can switch to the autonomous driving mode. Subsequently, the in-vehicle system accurately pinpoints the position of the vehicle based on the high-precision map, recognizes the situation of the surrounding obstacles or vehicles by a suit of in-vehicle sensors, and selects the corresponding route and the lane according to the destination. Therefore, the test vehicle can intelligently set the driving route and the target speed, and automatically operate the steering wheel, the accelerator pedal and the brake pedal. Thus, a series of operations can be realized, such as shunting and merging, lane changing at the exit, lane keeping and safety space keeping.

In terms of safety, Toyota has also proposed another auxiliary safety system called Toyota Safety Sense, or TSS, which is the collective name for Toyota's safety concept. It is a bundle of active safety features, including parking, active/passive safety, pre-collision system with pedestrian detection and the rescue in the late stage, in order to reduce the number of traffic accidents and minimize the damage to people and vehicles caused by accidents. The TSS system, to be first launched in Japan, the USA and Europe, is scheduled to be delivered to all Toyota and Lexus models by the end of 2017. Toyota says its ultimate goal is to achieve "Zero Casualties" from traffic accidents.

Toyota is also striving to make progress in the Internet of vehicles. Their ITS Connect system is a coordinated safety technology, a kind of V2V and V2I communication technology that can convey road traffic condition reminders, formations and cruising, etc. Said technology has been applied to the Japanese version of Crown models as an optional feature from 2015 an after. The intersection detection equipment used in ITS Connect system is funded by the Japanese Government and is currently installed only in Tokyo and Toyota City in Aichi Prefecture in limited quantities.

In November 2015, Toyota announced plans to build a R&D center on artificial intelligence in Silicon Valley, and invest one billion dollars into it within the next five years. Toyota had been cautious about driverless cars previously, and this new move has suggests that Toyota would officially enter the increasingly competitive field of autonomous driving. Akio Toyoda, President of Toyota, said that he changed his mind when he realized the need for cars for the disabled and elderly during his preparations for the 2020 Tokyo Olympics and Paralympics.

Toyota's autonomous driving technology is being developed by three main forces: Toyota Research Institute (TRI), the Collaborative Safety Research Center (CSRC), and Toyota Connected (TC), founded in 2016. Typically, an automaker works with different suppliers to get parts and sub-systems. Toyota, however, has integrated forces internally for autonomous driving technology. The concept set by TRI is independent, with its own procedures and processes, but its connection with Toyota is very close. The code written by the institute is used to produce cars.

TRI has two projects on autonomous driving: code-name "Guardian" to develop a new generation of ADAS—"a super advanced driver assistance," and code-name "Chauffeur" to develop truly autonomous driving. The "Guardian" was led by the TRI laboratory in Massachusetts and "Chauffeur" was led by the TRI office in Michigan. The goal of these two projects is to achieve Level-4 autonomous driving. Gill Pratt, TRI's CEO, said in June 2016 that the advanced security features being developed by the "Guardian" project could be the first to generate revenue. He added that the breakthrough point may be concentrated in the evasion technique, a technology in which a car perceives and avoids danger before collision.

Extended Reading

Toyota Autonomous Driving with a Two-Track Approach

Gill Pratt, TRI's CEO, explained Toyota's strategy of "walking with two legs" in a speech, China Automotive News reported. He thinks there are three types of unmanned vehicles currently. For the first type, the vehicle's autonomous driving system acts like a robot, executing all the instructions from the driver. The second type allows the autonomous driving system to "share" control of the vehicle with its owner. The system and the owner can play the roles of driver and co-driver alternately, and the system hands over the control to the owner in case of problems or failures. For the last type, the driver and the autonomous driving system will control the vehicle simultaneously during the driving process, allowing the system to learn from the driver's behavior for future reference.

Based on these three different types, Toyota has built two different product modes—"Chauffeur" and "Guardian". Pratt says mistakes are not allowed in the first mode, while in the second mode the system will try to avoid accidents, which tends to make cars safer.

  Pratt asserts that, given the immaturity of autonomous driving and the fact that many drivers still seek the pleasure of driving, the two modes will co-exist for a long time. As a result, Toyota has adopted a two-track approach in developing unmanned vehicles. Pratt thinks the two models are actually derived from the same technology; thus, they are complementary with artificial intelligence as the core, and vehicles adopting the "Guardian" mode will dominate the industry.

  Toyota unveiled the Concept-i concept car (Fig. 10) at CES 2017, which emphasizes the bond between car, driver and society, and building a mobility experience

Fig. 10 Concept-i concept car

suitable for users in the twenty-first century. Concept-i has a striking design, resembling that of the space shuttle, with three very eye-catching gull-wing doors. Yui is the nickname for Concept-i's artificial intelligence system, which acts as a liaison between occupants and the car. Yui, in tandem with AI, anticipates people's needs and informs the car so that Concept-i can consider and execute that next action accordingly. Even the exterior of the car elevates the relationship between the car and the world around it, communicating with others on the road while expressing motion and excitement. When Yui takes over the vehicle, a bold sign declaring "Automated" shows that the car is running automatically.

2. **Honda: Honda Sensing, Low-Keyed but Formidable**

In Honda's planning, autonomous driving technology has always played an important role. In October 2015, Honda officially announced it would commercialize its autonomous vehicles by 2020. Limited by the status of the technology, however, Honda Sensing is still an "assistant system", but it can be regarded an "embryonic form" of Honda's autonomous driving technology in the future. Maybe a few years later, it would be the technical code for Honda's autonomous driving.

At the Intelligent Transportation Systems (ITS) Conference in 2014, Honda has demonstrated its autonomous driving research progress and demonstrated some of Honda Sensing's features on a highway in Detroit, including convergence, separation, lane changing and other operations on the highway through autonomous driving. In terms of Internet of vehicles, Honda has also demonstrated their coordinated autonomous driving technology, including the latest technologies and concepts such as vehicle-to-vehicle interconnection, vehicle-to-pedestrian and bicycle interconnection, and vehicle-to-motorcycle interconnection. In this system, the test car can automatically turn, avoid, stop, start, and achieve high-level coordination between traffic participants through V2X technology, differential positioning and other technologies.

Honda Sensing adopts the scheme of a monocular camera + millimeter wave radar, which mainly features six functions: Collision Mitigation Braking System (CMBS), Road Departure Mitigation System (RDM), Adaptive Cruise Control (ACC) with Low Speed Follow (LSF), Lane Keeping Assist System (LKAS), Lane Watch (LWC), and Traffic Sign Recognition (TSR). When in use LKAS + ACC could give drivers a hint of an "autonomous driving" feeling at high speed. In terms of publicity, Honda has made Honda Sensing a big selling point, all while keeping the typical caution of Japanese manufacturers.

Honda Sensing's "Virtual Tow" feature is also unconventional in terms of driving safety. In our daily driving, we often encounter emergency braking of the preceding vehicle, which may be caused by sudden illness or personal injury during driving. If that driver is alone, they may not get help quickly, which will likely cause incalculable consequences. Honda has cleverly designed a "Virtual Tow" program to address this situation. The distress signal sent by the driver will be "bridged" through the V2X technology and spread through cars. Vehicles receiving the signal can provide help, or act as a "guide car" leading the endangered vehicle to emergency service places such as first aid centers, so that people in distress can get help promptly.

Honda has a unique advantage of being able to reduce the price of its autonomous driving sensor kit to a level that most people can afford. Take for example its Civic compact car, which sells for $20,000 in the US market. What's more, Honda's sensors can be activated at any speed. It can be used not only on the highway, but also on urban roads.

Honda unveiled its innovative interiors at the Los Angeles Auto Show in November 2016. The center control screen consists of a curved touch panel mounted between the front seats, similar in style to BMW's iDrive system. Through this screen, the driver can experience semi-autonomous driving techniques and can also use a variety of related functions. The patterns and colors displayed on the screen can vary depending on the driver's mood and driving mode. The screen can display the status of nearby vehicles and identify whether they are is in autonomous driving mode or manual driving mode. Vehicle systems can also use artificial intelligence to identify pedestrians and cyclists and predict their direction of travel. Honda said the technology could incite consumer confidence in autonomous driving technology, and demonstrate the reliability of their driving systems.

Honda announced in December 2016 that it has begun to explore the possibility of developing a fully autonomous driving car with Google that does not require a steering wheel, accelerator or brake pedals. If this kind of cooperation were possible, it would make Honda the first Japanese manufacturer to partner with Google. If Honda and Google could overcome the obstacles of different industries and promote cooperation, it would also encourage other companies to pick up the pace of cooperation, or intensify R&D competition. This was Honda's first concrete details of its fully autonomous driving technology. If all goes well, Honda and Google will finalize more details of their joint development in the future. Their cooperation is considered faster and more economical than other partnerships. Waymo's autonomous driving technology is much more advanced than Honda, and cooperation can speed up the development process. Likewise, Honda's technical reserves (such as the Honda Sensing system and hydrogen fuel vehicles) can also be used by Google. It is reported that Honda will integrate technology from Google's Waymo car-building project into its vehicles, including sensors, software and computing platforms. Honda, for its part, is expected to offer Waymo test cars adapted to its autonomous driving technology, starting with the upcoming Odyssey in America.

Honda is very hesitant to commit to a timetable, reluctant to give a specific deadline given the potential for unforeseen technical or legislative obstacles. However, Honda reveals that it will also roll out vehicles that integrate V2X and V2V communication technologies based on DSRC in 2020. Allowing cars and infrastructure to communicate with each other will advance automation. In the end, no matter at what level of autonomous driving technology has been developed, their goal has always been "zero accidents." However, Honda has not yet released a timetable for when this goal might be reached.

3. **Nissan: From ProPilot to "Zero Fatalities"**

For Nissan Motor, autonomous driving technology falls into its category of Nissan Intelligent Mobility, which is a suite of integrated technology that is designed to increase safety, comfort, and control while driving, connecting you with your vehicle and the world around you. The main reason for Nissan's development of autonomous driving technology is to achieve "zero emissions" and "zero fatalities". Nissan believes that electrification can better support vehicle intelligence and eventually lead to autonomous driving. Cars with a high degree of electrification are not only compatible with accessories like navigation systems, smart keys, and power seats, but also more suitable for installing electronic active safety devices.

Nissan has been developing electronic safety assistance systems for more than 17 years, dating back to the launch of Cima FY33 in 1999. The latest ADAS technology has just been introduced in China. As far as Nissan is concerned, ADAS is the basis for autonomous driving with "zero fatalities". As the technology develops, new applications are constantly updated, to achieve the ultimate goal of autonomous driving.

Nissan's current models are already using some ADAS technologies, such as Forward Collision Warning (FCW), Lane Departure Warning (LDW)/correction, and vehicle collision prevention systems. Nissan has also introduced ProPilot semi-autonomous driving technology, (or Level-2 by SAE) to the market, based on years of technical reserves and market experience in ADAS and automotive electrification. It is also one of the key factors for Nissan to realize intelligent mobile planning. ProPilot was first applied to the fifth-generation Serena minivan launched in Japan at the end of July 2016, then Qashiqai launched in Europe in 2017, and soon will be found in models sold in China. There are not many production cars on the market with similar technology, so Nissan is arguably ahead of most of its competitors.

Unlike most companies, which use fused camera and millimeter wave radar technology, the ProPilot is only implemented with a monocular camera mounted inside the front windscreen. The monocular camera monitors the preceding vehicle and reads lane markers to maintain appropriate distance and centering. The throttle opening, braking and steering are controlled based on this received information. Nissan's monocular camera has a horizontal viewing angle of 52° and a vertical viewing angle of 40°. The technology for identifying vehicles and pedestrians with a monocular camera is provided by Mobileye, while Nissan further develops its own image analysis technology. For example, when it recognizes a vehicle merging ahead, it will determine whether and when to change the vehicle it is following.

The ProPilot system on the latest Serena models is still in its infancy, and it can only be driven automatically on a single lane of the highway. ProPilot allows the vehicle to travel in the middle of the lane, corner smoothly, adjust speed to maintain a safe driving distance, and brake or stop the vehicle completely when necessary. ProPilot technology can be activated or deactivated by a button mounted on the steering wheel, and the vehicle's operating status can be seen on the on-board display. The driver can preset the speed between 30 and 100 km/h, which the system will follow. There's a lot that ProPilot can do for drivers, but there's even more that it won't. People living where the roads are mainly composed of sand and salt for months at a time may find that the camera has trouble sorting out lane markers and struggles finding its way. Similarly, the system doesn't work as well at night as it does when the roads are well-lit. It won't work in a rainstorm, or when the vehicle's wipers are actively being used, but can operate when they are set to auto/mist and activate for small splashes of water.

ProPilot will not drive the car without active management from the driver, nor can it change lanes automatically. If a lane change is needed, the driver will have to take control. The system will disengage until it locks onto the new lane. To make sure the person behind the wheel is ready to take control at any moment, the car will beep at the driver to keep at least a finger or two on the steering wheel. If no hands are detected, the system will sound a beep and flash a nastygram in the gauge cluster. Though both Nissan and Infiniti use the big bad word (autonomous) in their marketing materials, they go to great lengths to make it clear that the systems are put in place to help the driver maintain control and exist to support, not replace the driver.

Nissan has established an autonomous vehicle development center in Sunnyvale, California—the Nissan Research Center Silicon Valley (NRC-SV). According to *China Daily*, two Nissan intelligent driving cars based on the Leaf model gave autonomous rides around the NRC-SV to 25 reporters from China, Japan, France, Italy, and the USA in January 2016. The car carried three reporters on each of the roughly 25-min, 16-km urban road tests. The car accelerated, braked, stopped at signals, switched lanes, made turns and, finally, stopped in front of a police car invited for the purpose of the test.

The development of Nissan's autonomous driving technology can be divided into three stages, with plans to raise a level of autonomy every two years. Reaching Level-2 in 2016, the Nissan models at this level can cope with ordinary traffic jams and drive autonomously in one lane on the highways. In 2018, it will reach Level-3 with more advanced systems, enabling autonomous driving in multiple lanes. By 2020, it will reach Level-4 with completely matured technology. At this stage, autonomous driving will be possible even on urban streets and intersections. Nissan plans to bring the autonomous vehicles at Level-4 to downtown areas, though it has not said exactly where. However, it is likely to be somewhere cars like Audi and Ford cannot go. A Nissan engineer admitted that autonomous driving system at Level-4 requires a HD map, suggesting it would also face limitations in some operational areas.

The ProPilot, now on the market, is the first stage in Nissan's three-stage goal. Nissan has planned autonomous driving vehicle testing in Japan, and the upcoming release of a new Leaf electric car with the ProPilot system. Nissan is committed to developing autonomous vehicles promptly and will work with DeNA, a Japanese tech company, to test self-driving cars. In addition, Nissan has also announced the "Seamless Autonomous Mobility" (SAM) system, developed jointly with NASA. When a self-driving car encounters an unmanageable situation, the SAM system will issue a warning and provide a manual solution and feedback to the car, while the driver does not have to manually operate.

Microsoft is building an Internet car platform that integrates features of Skype and Office 365 to provide cloud services to vehicles. Nissan plans to be the first automaker to use the platform. With Microsoft's cloud technology, the platform allows automakers to access vehicle data to predict service time, collect driving data for research, and provide real-time voice navigation. Nissan vehicles using Microsoft technology are expected to be available as early as 2018.

2.4 South Korean Squad

1. Hyundai: A Low-Cost Solution by Ioniq

Hyundai Motor has been very active in adopting software like Apple's CarPlay, Google's Android Auto and Baidu's CarLife to integrate smartphones into vehicle dashboards. However, compared with other traditional automakers such as Mercedes-Benz, General Motors, and Ford, Hyundai has been relatively cautious in autonomous driving. It is now is striving to catch up. The goal of Hyundai is to realize semi-autonomous driving by 2020 and sell fully autonomous vehicles by 2030.

In recent years, Hyundai's attitude toward new types of cooperation is rather conservative, and it has avoided building joint ventures to produce high-cost, low-output electric and sports cars. Hyundai is now moving away from its go-it-alone strategy to find technical resources and catch up with its rivals. Earlier in 2016, Hyundai announced that it would invest approximately $1.7 billion in autonomous driving development. To make up for the lack of technical reserves, Hyundai is trying to partner with emerging tech companies and Silicon Valley giants. Inspired by Deep Mind's AlphaGo, Hyundai hopes to develop an AlphaGo driving system that automatically navigates on the road and provides intelligent services—for which Hyundai seems to call for Google's participation. Jeong Jin-Haeng, General Manager of Hyundai Motor, said publicly in August 2016 that Hyundai shared similar business directions with Google, calling for the two companies to cooperate. A partnership between Hyundai and Google is not impossible, for Mark Krafcik, Waymo's CEO, was also CEO of Hyundai Motor North America and a key player in Hyundai's growth in the USA.

Hyundai is currently working on the application of ADAS technology. G90, the flagship sedan of its high-end brand Genesis has applied the Genesis Smart Sense (Genesis Smart Sense), which is a combination of driving assist functions such as ACC, LKS and AEB. At the same time, Hyundai was also the first to show interests in Willow Run, Michigan as a testing site. The test site of Hyundai Motors in California is likely to shift its focus to in-depth assessments of technical prototypes, while its work in Michigan and Wisconsin will focus on developing autonomous driving electronics and testing vehicles for the market.

Hyundai has tested its self-driving car—an Ioniq sedan equipped with autonomous driving software and hardware—on the streets of Las Vegas. This Ioniq test car is very different from a typical self-driving vehicle in appearance, as we find no externally mounted cameras and sensors. What's more, relatively cheap radars and cameras have replaced expensive 360°-rotating LiDAR. The low-cost autonomous driving hardware system includes three Ibeo LiDARs in the front and side of the car, a four-camera array on the top edge of the windshield and forward-facing and back-facing millimeter wave radars that provides the car with a surround view. The most remarkable of these technologies is the Ibeo LiDAR. In contrast with the common 16, 32, or even 64-line 360°-rotating radars, Ibeo has only a 4-line set-up, and does not require the entire sensor to rotate at a high speed. Hyundai hopes that autonomous driving technology developed in this way will reduce its reliance on computing power, which in turn will make this type of cars affordable for more consumers.

By working with the onboard software, the Ioniq self-driving car can automatically detect and track the location, shape, speed and direction of the objects at its front and both sides, whether they are cars, pedestrians or other things. The radars can also detect surrounding obstacles, while the cameras are responsible for tracking pedestrian, traffic signs and signals. In the test drive, the sensors allowed the car to automatically change lanes and stop in case of red lights and pedestrians. In addition, Ioniq can sense the acceleration and braking of the proceeding vehicle and take corresponding measures, to avoid congestion and collision. Hyundai is optimizing the computing procedures and equipment to reduce the cost to collect data in the sensors and convert it into instructions in attempt to popularize their autonomous vehicles in a large scale.

Hyundai said on January 17, 2017 that it would invest $3.1 billion in its existing manufacturing facilities in the US. In addition to building new plants, the said funds will be used to develop new technologies (such as autonomous vehicles) and upgrade existing facilities over the next five years. The South Korean automaker rarely discloses plans to invest in specific countries.

2. Kia: Autonomous Driving Brand "Drive Wise"

In early 2016, Kia Motors, a subsidiary of Hyundai Motor Group, officially launched "Drive Wise", a sub-brand of its autonomous driving technology. Prior to this, Kia had developed and promoted ADAS systems that prevents accidents, such as the BSM and AEB. Drive Wise is a technical upgrade in driving assistance and autonomous driving. Building off the original ADAS functions, information technology and navigation system were added to give more precise sensing and control.

Although Kia is in its infancy in developing semi-autonomous and fully autonomous technologies, it already has a plan. According to reports, Kia plans to apply its series of semi-autonomous driving features to its model lineup by 2020, including Highway Driving Assist (HDA), Preceding Vehicle Following (PVF), Emergency Station System (ESS), Traffic Jams Assist (TJA), Smart Parking Assist System (SPAS), Remote Advanced Parking Assist System (RAPAS), etc. It plans to launch cars with Level-4 automation in 2030, with technologies such as Urban Autonomous Driving (UAD) and automatic parking. In terms of technical require-ments, HDA relies on radar and cameras, while UAD requires GPS and external sensors to position itself on the road, along with real-time traffic updates. Like Audi,

Kia has also equipped a driver status monitoring system that can pull over if it detects a distracted driver on its self-driving cars. As a key part of the development of "unmanned vehicles", fully autonomous vehicles can realize safe and efficient driving and parking by establishing direct communication links with vehicles and surroundings.

3 Tier-1 Suppliers: The "Gunmen" are not to be Trifled with

3.1 Bosch: All Count—From Perception to Decision Making and to Execution

Speaking automotive components, Bosch is undoubtedly a heavyweight competitor. As the inventor of Anti-Lock Braking System (ABS), Electronic Stability Program (ESP) and Adaptive Cruise Control (ACC), Bosch is one of the industry leaders in the research and development of autonomous driving technology. Bosch separates automated driving technology into three levels. The first is partial automation, in which the system can control the vehicle's horizontal and longitudinal motions and autopilot in these directions under specific working conditions. The so-called longitudinal automatic pilot can control the speed of the car through torque and braking, while the horizontal pilot means that the driver does not need to hold the steering wheel (However, they need to maintain vigilance at all times.). Bosch has already implemented traffic jam assist in 2015, with plans to implement integrated cruise assist in 2017 and highway assist and automated valet parking in 2018. After that step is high automation, where the system can control the vehicle's motion in defined operating conditions, though the driver still needs to intervene after a certain interval of time. At this level, Bosch plans to implement a traffic jam pilot in 2018 and highway pilot in 2021. Bosch hopes to reach full automation after 2025, in which the system can fully control the vehicle to deal with different conditions and no driver's supervision is needed.

Yudong Chen, the President of Bosch China, states that the company has good solutions in all problems related to autonomous driving. Indeed, Bosch has the R&D capacity and experience for everything from cameras, radars and other related components, to electronic control units (ECU), and artificial intelligence algorithms. At the beginning of 2015, Bosch acquired ZF steering systems, further strengthening the foundation of Bosch's autonomous driving technology.

Bosch provides technology to Google's prototype and provides ADAS-related products to Tesla, Mercedes-Benz and Volvo. In terms of localization, Bosch also keeps a close cooperation with the well-known map company TomTom. Several years ago, Bosch set up business unit for driver assistance, employing about 200 engineers to work on ADAS. Bosch has a series of in-depth partnerships with

China's automakers as well. Bosch's Highway Assist, equipped with five medium-range radars and a monocular camera, facilitated Changan's 2000-km sensational unmanned, partially autonomous highway journey (at a speed of 0–130 km/h). In addition, Bosch cooperates with automakers such as Great Wall for the parking sensors on Haval H7, and Geely for a complete set of ADAS systems, including ACC, LDW and PEBS, on Geely Boyue.

One of the biggest news at CES 2017 was NVIDIA's announcement of cooperation with Bosch. NVIDIA will provide deep learning chips, while Bosch would figure out how to better use these deep learning chips to run its software and algorithms. As a Tier-1 supplier, Bosch focuses on how to further integrate and optimize existing technologies such as sensors and artificial intelligence to deliver an overall solution to the OEMs. Providing a holistic solution is Bosch's ambition.

3.2   Delphi: Building Up Super Autonomous Driving Technology by M&A and Cooperation

Delphi has been highly active in autonomous driving, developing new technologies while acquiring, working with, or investing in emerging tech companies to consolidate its position in the industry. In July 2015, Delphi acquired Ottomatika, a self-driving car software company with Carnegie Mellon's genes for $32 million (which was a pretty good deal, considering General Motors and Ford had poured hundreds of millions of dollars in similar acquisitions.). In the same year, Delphi became a shareholder of Quanergy, a 3D LiDAR startup. The solid, jointly developed LiDAR "S3" could reduce the cost of a single sensor to $250, and exemplifies current trends in automotive LiDAR towards becoming "solid, small and cheap". In 2015, Delphi also acquired Control-Tec, the only supplier in the industry able to provide a complete set of automotive business intelligence systems. This system improves the efficiency of problem detection and troubleshooting, shortening the testing cycle of vehicle procedures.

At the beginning of 2015, Delphi used a modified Audi SQ5 to successfully conduct a long-range autonomous driving test across America from San Francisco to New York. Up to 99% of the trip was handled by autonomous driving. After Delphi and Mobileye announcing a collaboration in August 2016, they demonstrated the jointly developed the Centralized Sensing Localization and Planning (CSLP) autonomous driving system (arguably at Level-4 or -5 automation), on a public road. The CSLP system is very powerful, ensuring that the vehicle's positioning accuracy is within 10 cm, with or without GPS signals. It can guide the vehicle through complex lane forks or areas without lane markings and features 360° pedestrian sensing and 3D vehicle detection. This allows the vehicle to behave like a human and determine the best path in advance.

Delphi is full of confidence, and plans the SOP of its CSLP system in 2019. This will enable vehicle customers to adopt Level-4 or 5 automation without large investment, thus to improve the technological process and popularization speed of the whole industry and accelerate the pace of fully autonomous driving to the market. With the launch of CSLP, Delphi has become one of the few global suppliers that can provide integrated autonomous driving solutions to OEMs.

In addition, Delphi has carried out "mobility-on-demand", a pilot project of autonomous driving base on cloud computing in Singapore. To this end, they have conducted tests in point-to-point, low-speed, urban automated mobility-on-demand services in designated autonomous vehicle test areas. In the future, this technology could have numerous other innovative applications, such as Delphi's "mobility-on-demand" for off-peak vehicle operations. Delphi hopes to apply "mobility-on-demand" in reality within five years.

3.3 Continental: Focus on the Six Building Blocks of Automated Driving

Continental has earmarked "automated driving" (this is the term it uses rather than autonomous driving), as a major trend in the global automotive, and one of its long-term goals in its technology strategy. In 2012, Continental was the first automotive supplier to receive a license to test automated driving technology on public roads in Nevada. The company has completed approximately more than 100,000 miles of testing in highly automated driving mode, accompanied by a test engineer monitoring the vehicle's performance. Continental also started an Automated Driving Initiative in Japan and China, in addition to the US and Germany.

Continental is also developing an Automated Driving (AD) Development Vehicle for the Chinese market. The Chinese team is responsible for the development of components, algorithms, as well as validation of AD systems and functions in their region. The Cruising Chauffeur® is one important automated driving feature of the vehicle. It delegates all driving related tasks to the vehicle, such as automated lateral and longitudinal guidance to keep the car within the lane as it travels around bends and executes lane changes requested by the driver. A light band at the bottom of the windscreen will shine to notify the driver when Cruising Chauffeur® is enabled. The vehicle can then take over from the driver, cruising along the highway by adjusting its speed in response to traffic conditions and transport regulations. The driver does not have to take over again until exiting the highway after vehicle notification. The Development Vehicle is designed to handle the driving in a highly automated driving mode for speeds of up to 130 km/h on freeways at these settings.

Nowadays the safety function of ADAS system has been widely recognized. Automated driving represents further development of Continental's ever-improving ADAS system. In 2016, Continental completed the acquisition of the Hi-Res 3D Flash LiDAR business from Advanced Scientific Concepts, Inc. (ASC). This innovative technology will further enhance the company's Advanced Driver Assistance Systems product portfolio with a forward-looking solution, enhancing the surrounding sensors needed to achieve highly and fully automated driving.

Continental is working on the following six building blocks to drive the technology forward: sensor technology, swarm connectivity, human–machine dialogue, system architecture, reliability and the public acceptance of automated driving.

Continental has created an open and independent global dialogue platform on automated driving: 2025AD.com/2025AD.cn. On this platform, experts and consumers can discuss future mobility from the technical, legal and social perspectives to help consumers better accept and trust automated driving technology. The website offers manufacturer-independent information, easy-to-understand images and case studies of automated driving. After reading this book, readers may wish to visit the website.

3.4   Valeo: Open-Minded Innovation and All-Round Development

Valeo, a French high-tech company, is in the midst of autonomous driving research and development, and should not be underestimated.

Valeo is comprised of four Business Groups (Powertrain Systems, Thermal Systems, Comfort and Driving Assistance Systems and Visibility Systems) and "Valeo Service." The Comfort and Driving Assistance Systems Business Group focuses interfacing between humans, vehicles and the environment. It is dedicated to the research and development in three areas: Autonomous Driving, Human–Machine Interface (HMI) and Vehicle Connectivity. In 2005, Valeo established Shenzhen plant for Comfort and Driving Assistance Systems, bringing more innovative products and technologies to China to provide customers with fast and effective support.

Amidst growing discussion of autonomous driving in public life, Valeo officially unveiled its plans in December 2015. These plans were a self-driving prototype called Cruise4U, with perpetual Level 3 automation–whether it is on a highway, normal or congested traffic, or on a daily route. As long as conditions permit, the driver is free to choose manual driving mode or Cruise4U autopilot mode. In the Cruise4U autopilot mode, the driver can spend time on work or entertainment. The Cruise4U self-driving car has currently traveled a total of 125,000 km in automatic mode, on two continents (Europe and North America). The core technology of this Cruise4U system is Valeo's ScaLa "hybrid solid state" LiDAR with Ibeo, which can accurately scan any obstacles in front of the vehicle. All scanned and collected data from the Valeo in-vehicle camera are integrated to create a complete driving environment map which provides accurate analysis and judgment of the driving environment for the vehicle intelligent system.

As the world's largest supplier of active safety automotive parts, Valeo has a broad product portfolio of sensor products. Its development is divided into three major stages: in the first stage, a single sensor to achieve a single function, such as parking assistance, adaptive cruise, automatic and emergency braking; in the second stage, complex sensors to achieve more complex functions, such as lane-change assistance based on front cameras, and forward collision warning. The sensor would likely develop with more functions. In the third stage, high-intensity fusion of multiple sensors completes a complete function, such as fully automatic parking (parking in various scenarios) through a combination of 360° camera, ultrasonic radar, short-range radar and ECU. This stage would also enable autonomous driving on highways, and during traffic jams or in urban conditions.

In addition, Valeo focuses on intelligent HMI and vehicle connectivity. In terms of HMI, Valeo's technology helps drivers switch quickly, safely and smoothly between autonomous driving and semi-autonomous driving; towards vehicle connectivity, Valeo provides telematics connectivity solutions from millimeter-level to infinite distance.

3.5 Magna: A Canadian Giant that Should not be Overlooked

Magna, the world's most diversified auto supplier, ranks #1 in North America and #3 globally, but has kept a low profile. With a focus mainly composed of premium models, Magna is an all-round player in the industry—its product system covers from components as small as a mirror, latch, or camera, to large structures like door frames, hood, powertrain, transmission, seating, front/rear exteriors module, and complete vehicle engineering and manufacturing. Although Magna's capability is comprehensive, the company holds strong positions in each every of their product areas. The company takes advantage of acquisition and integration and keeps a good balance of manufacturing and R&D. In the area of autonomous driving, Magna has been steadily conducting projects after projects as always.

Frank O'Brien, Executive Vice President of Magna Asia, said Magna started with vision system in vehicle electronics, particularly in autonomous driving. He believes that a car, like a person, needs an "eye" to send signals to the brain to process and to respond. Mirrors and cameras are these "eyes". Magna has been working on this since the 1980s. For example, it tried to extend the function of mirrors, by embedding a compass function, lighting and even a camera screen in the vehicle's rearview mirror. Magna anticipated that cameras would one day replace mirrors to become the true "eyes" of a vehicle. One secret that most of the industry doesn't know is that Mobileye's first automotive customer was Magna. Magna's research on autonomous driving over the last two decades has focused on vision-based safety systems through cameras.

Magna, through its patented EYERIS system, has so far sold nearly six million cameras per year to automotive customers worldwide. It leads the ADAS industry with a commanding market share of about 21%. EYERIS is primarily intended to address vehicle safety; and Magna hopes to leverage the system like granting the vehicle with an additional intelligent eye. EYERIS helps drivers avoid accidents through both passive and active safety. Magna's ADAS system has made progress in both near field and far field, image vision and machine vision. These functions include object recognition, lane keeping assistance and lane departure warning, traffic sign recognition, light adjustment, self-parking and adaptive cruise control.

As the auto industry evolves rapidly, Magna is making strategic efforts to lead the future development of autonomous driving. Magna has been cooperating with IT companies for more than a decade. The corporation funds Israeli software companies on the safety of vehicle connectivity and driving data privacy. In recent years, Magna has been investing more heavily on collaboration in this area. At the end of 2016, Magna and Innoviz Technologies Ltd. announced a collaboration to deliver LiDAR remoting sensing solutions and implement autonomous driving features and full autonomy in future vehicles. Considering that LiDAR is imperative to achieve the desired levels of performance and safety, the cooperation helps Magna integrate Innoviz's technology into its autonomous driving systems to provide a complete sensor-fusion solution to automakers (Fig. 11). In April 2017, Magna invested 5 million CDN into the Vector Institute to boost Canada's research in artificial intelligence, to enhance competence in the area. As future mobility continues to evolve, artificial intelligence will play an important role in the dynamic decision-making of automatic driving. Object detection and classification and scene segmentation, including traffic volume, speeds and road conditions (weather, light and visibility), will be key attributes to self-driving capabilities.

At the same time, Magna understands that autonomous driving cannot be realized without a regulatory framework, infrastructure and ultimately the implementation of industry standards. Thus, Magna has completed numerous confidential projects with the US, Canadian and EU governments. The 60-year-old Canadian company, like the country where its headquarter is based, marches forward unnoticed, while beside the main artery.

3.6 Visteon: Consolidating Integrated Domain Controllers and Focusing on Automotive Neural Network

Visteon is known as the only pure-play supplier of automotive cockpit electronics, the fastest-growing segment in the auto industry. Relatively speaking, Visteon is less famous as Delphi. However, by comparing the relationship between Visteon and Ford to that between Delphi and GM, it's easy to understand Visteon's position in the auto industry as a parts and components subsidiary spun off from a major U.S. auto company. Visteon's strategy towards autonomous driving technology aims at the next generation vehicle structure: a single domain control solution integrating the sensors connected via Ethernet and the centralized computing approach featuring sensor fusion, object detection and tracking, situation analysis enabled by artificial intelligence to realize autonomous driving in application. Visteon will first develop the integrated domain control hardware based on the current SmartCore™ Cockpit

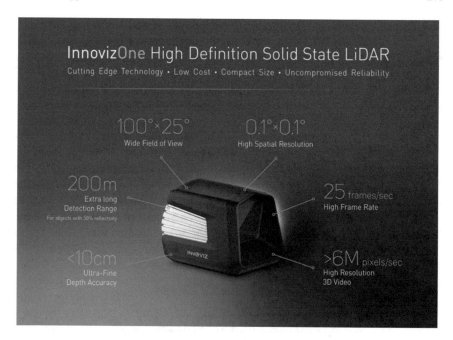

Fig. 11 High-definition solid LiDAR that Magna helped to develop

Domain Controller's default automatic protection function, then develop the automotive neural network technology, provide the open software development kit, and gradually move forward to the Level-3 and -4 autonomous driving function settings. One of the key differentiations of Visteon's autonomous driving technology is artificial intelligence, specifically the application of deep machine learning, object detection and classification. Other key driving and monitoring functions, also help it become a completely autonomous driving control system.

At CES 2017, Markus Schupfner, CTO of Visteon, shared some insights on artificial intelligence and ADAS technologies that Visteon is developing and testing. He believes that LTE is rapidly becoming the standard for wireless communications to vehicles, particularly in the US and Europe. The growth in automotive LTE is largely driven by connected services and features that can be offered through telematics and infotainment systems. As a result, infotainment systems are fast becoming the digital core of vehicles, and a stepping stone towards the future of autonomous driving. In order to break down the barriers between a vehicle's native apps and apps running on personal portable devices (smart phones or home devices). Visteon has developed a second-generation infotainment platform called Phoenix, which complies with open standards such as W3C and GENIVI. The Phoenix infotainment platform allows third-party software developers to easily build new applications using HTML5 and rich JavaScript-based application programming interfaces. This eliminates the need to rewrite the application when porting to other automakers' infotainment systems.

Visteon and cooperating automakers can provide value-added services through software over-the-air upgrades and updates, for example: firmware patches and data, or map updates. Vehicle diagnosis and live troubleshooting can the vehicle to the nearby service station before the component fault occurs, and so on.

In the safety of V2V and V2I, absolute interoperability and standardization between connected vehicles are needed to ensure the effectiveness of V2X technology. Markus Schupfner believes that the proposed V2V draft rules issued by NHTSA are a step in the right direction. Visteon has been researching and developing V2X technology for years to meet the unique needs of a global marketplace. Visteon is a supplier to the U.S. National Highway Traffic Safety Administration (NHTSA) Safety Pilot Model Deployment in Ann Arbor, MI, and recently joined the Car-2-Car Communication Consortium. In addition, Visteon actively cooperates with university research institutions and the autonomous driving alliance with the goal of introducing the latest technology into the automotive market.

3.7   HARMAN: More than an Audio Company

Many people consider HARMAN, a recently listed audio company with annual sales of $7 billion, to be deeply appreciated by both fans and musicians. The company has been a pioneer in the field of connected automobiles over the past 10 years and has achieved remarkable success.

HARMAN now has over 2000 basic patents in key domains of connected vehicles and created many key industry benchmarks.

In 2010, HARMAN launched the "Aha" car cloud service platform that incorporated Google Earth into the car navigation system. In 2014, HARMAN released the Aha Analytics car cloud service solution, which supports over-the-air (OTA) map updates based on the Navigation Data Standard (NDS). To enhance the connected car space ecosystem, HARMAN acquired Symphony Teleca and Red Bend in 2015 to bolster cloud services and OTA update solutions. In 2016, HARMAN acquired Tower Sec and released a 5 + 1 security architecture for connected cars. The security architecture is considered the strongest in-deep security architecture in the connected car sector. It delivers strong performance in terms of detecting and fending off attacks, protecting security content and privacy, as well as downloading and updating car ECUs over the air. In the same year, HARMAN unveiled the end-to-end Life-Enhancing Intelligent Vehicle Solution (LIVS), encompassing augmented reality navigation, connectivity security, connected ADAS, amongst others. LIVS is a holistic solution that helps users embrace the autonomous driving era. The smart cockpit, a flagship product on the LIVS platform, combines various meters and security features, and supports multi-screen interaction that boosts the integrated cockpit experience.

HARMAN connected ADAS is especially designed to test the infrequent bugs that are difficult to detect in the development stage due to limited time, resources and driving distances. This tool monitors the data of vehicles in service and assists carmakers in assessing their ADAS performance after running for billions of kilometers. HARMAN also integrates various camera horizon solutions, including lane departure warning, pedestrian detection, augmented reality navigation, and a 360° camera, to name a few.

For many years, HARMAN has been a leader in the connected car area. HARMAN's Telematics Gateway leverages a series of powerful and scalable technologies and options to continuously drive connected car development. Its updated gateway supports FOTA (Firmware Over-The-Air) and innovative vehicle ECU. HARMAN's V2X system can support both DSRC and LTE-V, and its design can be upgraded to support 4.5G and 5G.

Extended Reading

Differences between the Internet of vehicles and Telematics

Internet of Vehicles (V2X) refers to networks and applications that use advanced sensing, network, computing and control technologies to comprehensively perceive roads and traffic, enable large-scale and large-capacity data flow between multiple systems, control vehicles throughout the driving process, and control each road environment. Its main purpose is to improve traffic efficiency and safety. The Internet of Vehicles is an umbrella category, with an ultimate goal to achieve intelligent transportation.

Telematics is a part of V2X and came into being before V2X, though it is seldom mentioned nowadays. Telematics is an in-vehicle information system with unlimited communication functions. It is a typical network that "links" vehicles, and is used for remote control of vehicles and remote reading of information. At present, there are two ways of implementation, either through built-in ports or OBD ports. Some implemented functions include remote detection, remote control, and call center. Mature products in this category include Bluelink, On-star, Snap-on and Verizon Wireless. Telematics' main service targets are drivers and commercial vehicle managers. Its purpose is to improve safety, fuel economy, entertainment and fleet efficiency.

After more than two years of collecting and learning NDS-based data from production cars, HARMAN launched the Deep Learning Dynamic Map Layers solution in March 2017. It uses data collected from cameras and other car sensors to recognize road signs from the surrounding environment and compares them with the digital map information from the onboard navigation system. If a difference is detected, the information is anonymized and sent to the cloud, where HARMAN's scalable cloud platform analyzes the data collected from other similarly-equipped production vehicles. Using spatial machine learning techniques, the solution can then, in real time, deliver critical updates back to the road network.

The solution will keep ADAS and navigation systems up-to-date with speed limit changes, and warn drivers of upcoming construction zones and any other signs they may encounter on the road. HARMAN's solution is based on the NDS, meaning that dynamic map layer information can be shared among different vehicle makers and models. The solution has been used by a German car brand.

"As more cars on the road begin using this technology, our deep neural networks algorithms will develop protective driver skills that we will all grow accustomed to using in our everyday drive experience," said Mike Tzamaloukas, Vice President, Autonomous Drive Business Unit, a HARMAN Connected Car Division.

3.8 Denso: Cooperating in a Japanese Way

Denso was originally one of Toyota's parts factories. After December 1949, Denso was separated from Toyota Group and is now one of the world's top suppliers, ranking first in Japan. Denso has received special attention from Toyota for its ADAS. However, recently, its throne has been threatened by Continental—specifically, by Continental's windshield sensor unit.

In October 2015, Denso and Toshiba teamed up to develop machine-vision ADAS products. Denso uses Toshiba's TMPV7506XBG image recognition processor, which supports a variety of camera-based ADAS functions, including lane detection, vehicle detection, pedestrian detection and traffic sign recognition. Denso monitors the driver's face and eyes through its HMI system to ensure safe driving.

In February 2016, Denso and NTT Docomo, Japan's largest mobile communications operator, announced that the two companies would cooperate on control technology, using 5G technology and mainstream LTE-V to develop ADAS and autonomous driving technology. Denso has been developing technologies towards accident prevention through V2I and V2V. With the help of Docomo, Denso will be able to accelerate the development of highway convergence and collision prevention technology at junctions with poor visibility. In September of the same year, Denso acquired 51% of the stake in Fujitsu Ten, a Japanese company that produces radar systems and develops autonomous driving systems.

In early 2017, Denso cooperated with NEC to develop safe driving technologies and information communication countermeasures using artificial intelligence. The images collected by Denso's onboard cameras will be analyzed through NEC's artificial intelligence deep learning technology. The autonomous driving technology developed will be more capable of intelligent risk identification, to avoid people and obstacles. The two companies also collaborated to address the increased risk of hacking, as connected vehicles grow in popularity.

4 Autonomous Driving Concept Companies: Focused and Capable

4.1 NVIDIA and Its Giant Competitors

NVIDIA was founded in 1993 by Jensen Huang, a Chinese-American scientist. Before 2016, NVIDIA's share price never exceeded $40. In 2016, It suddenly gained ground in technological fields, especially in autonomous vehicles. The company's share price climbed to a high of $120 in February 2017.

NVIDIA has long thoroughly mastered parallel computing technology based on Graphics Processor Unit (GPU), and has also adapted it to use in some self-driving cars. In the past few years, Google's self-driving cars have been using NVIDIA's Tegra processors. In early 2016, NVIDIA introduced the DRIVE PX 2, which has a performance equivalent of 150 Mac Pros and processes at an unprecedented rate of 24 terabytes per second.

The DRIVE PX 2 combines NVIDIA's processor and sensor platforms, including cameras, LiDARs and situation-aware radars. The early judgment of road conditions and pedestrians is situational awareness, one of the most important features in an autonomous vehicle. This situational awareness will only increase, with the accumulation of data, driving experience, and computer analysis. Although many scenarios can be programmed in advance, deep learning capabilities are required to summarize and analyze the situations for use in future driving tasks. In addition, autonomous vehicles can share this experience with each other in order to improve their respective situational awareness.

NVIDIA already has a variety of hardware platforms such as DRIVE PX/CX/PX 2/Auto Crusise/Xavier. Currently, NVIDIA's platforms have been used or tested by more than 50 automakers and suppliers, including BMW, Mercedes-Benz and Ford. The general architecture of these platforms is similar, namely, self-retrofitted ARM processor + NVIDIA GPU. In addition, NVIDIA also provides the DRIVE WORKS, an autonomous driving platform development kit, which can collect the data of all vehicles through the cloud, re-learn and form rules before returning to each self-driving car. This can lead to holistic machine learning processing, and self-development of autonomous driving ability.

With such a huge market, Intel and Qualcomm will naturally refuse to stand by and merely watch NVIDIA's domination. As NVIDIA's only competitor in computational power, Intel plans to launch its Intel Xeon Phi processor (codenamed Knights Mill) in 2017, as the latest generation of product designed for deep learning. The processor claims to have double the computing power of its rivals, and spearheads Intel's attack on NVIDIA. Intel had previously spent $400 million to acquire a tech company called Nervana Systems, with the goal of owning its deep learning accelerator chip presumably available in 2017. In March 2017, it spent $15.3 billion to acquire well-known Israeli tech company Mobileye, further consolidating its power.

Qualcomm has the greatest discursive power in the connection and transmission capabilities of mobile network. Therefore, its strategy is to "make full use of data in the cloud." At medium to long distances, he local processor can calculate the direction and speed of the vehicle from its displacement from a certain point, at the same time pinpointing its GPS position on the high-definition map. It is certainly difficult to completely store HD map data locally; therefore, transmission through the network is ideal. Other core functions of autonomous driving, such as lane, vehicle or pedestrian detection are completed separately with other Qualcomm processors. However, Qualcomm is vague about its plans to complete data fusion and make decisions. In October 2016, Qualcomm acquired Dutch automotive and security chip manufacturer NXP Semiconductors for $37.3 billion, making it the largest acquisition in the semiconductor industry. This acquisition will help Qualcomm occupy a key position in the automotive chip field.

4.2   Mobileye: "Eyes of Automobiles" that are Sparkling

Co-founded by Amnon Shashua, a professor of computer science at the Hebrew University of Jerusalem, Israel, Mobileye is currently recognized as the most successful tech company in the field of visual algorithms and ADAS research. Mobileye's strength comes in part from its early realization that a single-lensed camera (mono-camera) would become the primary sensor to support ADAS and eventually autonomous vehicles. "The Mobileye mono-camera was inspired by human vision, which only uses both eyes to obtain depth perception for very short distances," says Amnon Shashua, "Therefore, the added benefit of a second camera lens is only relevant for short distances. Driving-scene interpretation is based on much longer distances. All depth-perception cues for farther distances—such as perspective, shading, texture, and motion cues, that the human visual system uses in order to understand the visual world—are interpreted by a single eye. Therefore, Mobileye understood that a single-lens camera could be the primary sensor to enable autonomous driving."

Years of accumulated experience has allowed Mobileye to offer far more content than competitors in terms of environmental models. While others are still trying to improve the accuracy of single-lane detection, Mobileye can already provide semantic level descriptions of road features, such as left and right lane markings of the current driving lane and adjacent lanes, and road bifurcation, identified by deep neural networks. In 2016, Mobileye launched the Road Experience Management (REM), which is to be implemented in a way similar to Baidu's Learning-Map, but does not use complex 3D LiDAR for date collection and mapping. Instead, it utilizes its camera and the visual sensing system's ability (algorithms and control software) to identify information such as lane markings or road indications, and generates maps based on rich texture and color features in the image. The advantage is that it requires a small data bandwidth (10 KB/1 km) at a low cost. For generating and maintaining maps, Mobileye takes a hive intelligence approach to solve problems. Distribution of its collections and data updates are done through a large number of Mobileye existing products that are mounted on production vehicles.

Mobileye is also making preliminary attempts to develop decision-making and planning, (the most challenging problems in autonomous driving technology) through deep learning. Unlike Deep Mind's DQN network, Mobileye has also accounted for time sequences during driving.

Mobileye may be known to the public for its break with Tesla in 2016. Before, the image processing chipEyeQ3 used by Tesla ADAS system was provided by Mobileye. Eyeq4/5, which Mobileye is working on together with Delphi, is an upgraded form of EyeQ3.

According to news media, Mobileye has signed contracts with many global companies for 273 Start-Of-Production (SOP) models totally by the end of 2016. Volkswagen, BMW and General Motors have purchased Mobileye's system that allows vehicles to travel automatically at high speed, to be produced in 2018. In May 2016, Mobileye reached agreements with two car companies to provide fully autonomous driving systems in 2019. In July 2016, Mobileye, BMW and Intel jointly developed autonomous driving technology. In January 2017, a full set of CSLPLevel-4 to 5 autonomous driving solutions co-developed by Mobileye and Delphi automation was launched. In early 2017, Mobileye partnered with Atlas Financial Holdings, an insurance company, to install an advanced collision avoidance system on 4500 rental cars in New York to prevent traffic accidents.

Intel, the world's largest chip manufacturer, acquired Mobileye in March 2017 at the price of $ 63.54 per share. Mobileye reportedly has a fully diluted equity value of $15.3 billion and an enterprise value of $14.7 billion. At this point, Mobileye, having waited for the highest bid, finally became "a respectful guest of a powerful family" through its own strength.

4.3 Three LiDAR "Masters": Smaller, Cheaper, More "Solid"

As one of the most important sensors for autonomous driving, LiDAR has long been the focus of the industry. Large companies such as Google, BMW, Mercedes-Benz, Audi and Volvo, automotive suppliers such as Bosch, Delphi, Continental and Valeo, and start-ups such as Cruise Automation, and nuTonomy, have all adopted LiDAR in their autonomous driving systems. Therefore, some insiders have seen LiDAR as the pillar of the self-driving sector. However, at CES2017, high-specification LiDAR sensors were largely replaced by 4-line and 16-line products, with the exception of Waymo and Uber (which are still using 64-line LiDAR).

Velodyne, Ibeo and Quanergy are the key representatives amongst the companies that are developing LiDAR. Velodyne undoubtedly plays a leading role. Velodyne LiDAR received a $150 million joint investment from Baidu and Ford in August 2016. In December 2016, Velodyne developed a monolithic gallium nitride (GaN) integrated circuit, developed in partnership with Efficient Power Conversion (EPC). The design consolidates components and results in significant advances in sensor miniaturization, reliability, and cost reduction. Each integrated circuit is less than 4 mm squared, the size of George Washington's nose on the U.S. quarter. Velodyne showcased the new LiDAR named Solid-State Hybrid Ultra Puck Auto at Ford"s booth during CES2017. The company is currently working closely with 10 high-tech companies and 9 OEMs for a total of 19 autonomous driving projects. It is expected that its product orders will jump from the current few thousand to millions in 2020. So, it is not inconceivable if a LiDAR could only cost $50 by then.

Quanergy's competitiveness and innovation come from three factors: laser phased array, optical integrated circuits and far-field radiation. Among them, laser phased array is an important technology to achieve "solid state". In 2015, Quanergy received a strategic investment from Delphi. Consequently, the latter soon released the world's first integrated solid-state LiDAR for autonomous vehicles at CES2016. The LiDAR is based on 8-line S3 Optical Phased Array solid-state LiDAR, covering a detection range of 10 cm—150 m. Quanergy plans to cut the LiDAR price to about $250 in the future. In July 2016, Quanergy received a B-series funding of $100 million, padding its pockets deeply.

In August 2016, ZF announced the acquisition of a 40% stake in Ibeo, a German LiDAR company. The two parties will cooperate to develop the next generation of compact and affordable LiDAR systems. The system is capable of 3D image reconstruction of the surrounding environment without the need for rotating mirrors that are included in current LiDAR systems. Hyundai Motor's Ioniq has gone with the economical choice: Ibeo's LiDAR.

At present, there are two technical solutions on information acquisition: LiDAR and visual capture. LiDAR technology and effects are relatively mature, yet expensive. Visual capture is inexpensive, and cameras and platforms are readily available, but continuous capability improvement is necessary. In the field of LiDAR, there are many companies beyond the above-mentioned three "masters", such as Innoviz from Israel and SLAMTEC from China, which have also received financing and committed to use innovative technology to reduce the cost of LiDAR. This field is an ocean of opportunity.

4.4   Car Concept Startups on "AI 100": Are They "One-Trick Ponies"?

CBInsights, a leading big data research firm, released a list of companies known as "AI 100" in the Innovation Summit 2017 in the US. Many of these companies are small in scale but unique in skills, and they might also have their shares of the autonomous vehicles in the future.

Zoox: Zoox is a startup that develops fully autonomous driving technology and aims to provide modern cities with the next-generation "Mobility-as-a-Service (MaaS)" technology. The company's vision is to preserve people's most natural and safe way of life through the transition to self-driving vehicles. Zoox plans to run fully autonomous taxis by 2020. With a valuation of $1.55 billion, Zoox is already a tech unicorn.

nuTonomy: nuTonomy is a start-up company separated from MIT. Its goal is to provide more convenient transportation service, improve traffic congestion and reduce carbon emission pollution. It is well known that its self-driving taxis were put into trial operation in Singapore in August 2016. nuTonomy received its A-series funding of US $16 million in May 2016.

Aimotive: Aimotive, a Budapest-based startup that uses artificial intelligence algorithms to realize Level-5 automation, claims to have the cheapest solution yet—the cost of its test system is only $2000. Aimotive mainly uses real-time simulator to simulate different road and weather conditions to test and improve its deep learning algorithm. All the test and driving data will be recorded as learning materials for the algorithm. The company received $10.5 million in financing from NVIDIA, Bosch and Draper.

Nauto: the system developed by Nauto uses camera and artificial intelligence technology to analyze driver's behavior and driving data. The system requires a device costing just $400 mounted on the windscreen, with small-sized cameras, computer vision system and machine learning technology to collect and process data. The system can detect behaviors like drunk driving or phone usage and alert the driver to stop. This will help auto insurance companies assess risk, prevent fraud, and reward high-quality customers. On October 7, 2016, Toyota, BMW and Germany's Allianz jointly invested in the company.

Chapter 8
Vital Opportunities in China

According to the "Auto Blue Book: Report on China's Auto Industry Development in 2016", China's economy is entering a new period of growth. China's industrial economy has maintained a rapid growth rate of above 9% in the past two decades, due to the continuous demands in infrastructure, real estate and automobiles. This has been further promoted by rapid urbanization and steady growth of household income. From an international perspective, when the average per capita income reaches approximately Intl. $12,000 (in 1990 Intl. $), the auto market will go into a period of low to medium growth. Although China's economic growth has slowed in recent years, its GDP per capita is close to Intl. $12,000, with some coastal cities clearly exceeding this value. It is also normal for China's auto market to enter a period of slow growth, which is also in line with the development path of automobile powers. The arrival of autonomous driving vehicles is redefining a new standard of normal for China's economy.

The capital-, technology-, and labor-intensive auto industry has become a pillar industry of the national economy, and its international competitiveness has a major impact on China's overall economic strength. At the same time, the auto industry is also in the process of globalization. With China's accession to the WTO and booms in economic globalization and trade liberalization, overseas auto giants entered the country's auto industry in droves, spurring its integration with the global auto industry. China's auto production and sales volume has been ranked first in the world for many years. The huge auto industry has become one of China's largest economic categories, second only to real estate. The Chinese market consists of 85% of the global market, larger than that of the US, Japan and Germany combined. Despite its size, China's auto industry is not that strong. The localization strategy of "trading market for technology" has proved unsuccessful. As autonomous driving becomes a new technology benchmark for auto companies around the world, it is likely another strategic opportunity for China. How would this opportunity best be used?

At present, the auto industry faces four major challenges—energy shortage, environmental pollution, traffic congestion and accidents—all over the world, but most seriously in China. Given the scale of China's auto industry, the global problem would

Z. Chai et al., *Autonomous Driving Changes the Future*,
https://doi.org/10.1007/978-981-15-6728-5_8

largely be resolved by addressing its role in China. In this regard, industry experts have conceded that these four issues need to be solved with a three-fold approach—reduced carbon output, informationization and intellectualization reduced carbon output is a technical issue, informationization is a support to the industrial foundation, and intellectualization is the strategic commanding height of industrial development. This "three-fold approach" is concentrated in two of the most important areas: energy-saving, and new energy vehicles and ICVs. The former links to reduced carbon output which mainly addresses energy and environmental problems, while the latter connects informationization and intellectualization, making cars smarter and providing more efficient and safe travel for humans, as well as saving energy and reducing emissions. It should be noted that ongoing industrial innovations are achieved not only through developing products and technologies, combining user experience and application scenarios, but also through utilizing business models and financial capital as powerful leverage, so as to collaboratively realize the great development of the auto industry.

ICVs will rewrite the "Smiling Curve" of China's automobiles as the country has obvious advantages in the field of autonomous vehicles. In terms of EVs and smart cars, the core technology has shifted from engine and transmission to artificial intelligence, a field in which China is competitive with the US. Car electrification has filled the technical gap between China and auto powers on the internal combustion engine, and patent barriers have also been completely removed. With a clear strategic target, a strategic layout of basic technical system made up of "Three Longitudinal Developments and Three Horizontal Developments" (Three Longitudinal Developments refer to fuel cell, hybrid and battery electric vehicles; and Three Horizontal Development refer to multi-energy powertrain system, motor drive system and control unit as well as battery and battery management system), top-down strategic actions, and systematic fiscal and tax incentive policies, China is in position to overtake the US in just three or four years as the world's largest and most diversified EV consumer market. However EVs alone are not enough, as the global auto industry is being changed by the era of ICVs, and revolutionary breakthroughs in a new generation information technologies such as mobile Internet, big data and cloud computing. However, EVs are the best carriers for automatic driving, and their development can complement and facilitate each other.

According to a forecast from IHS Automotive in July 2016, annual sales of autonomous vehicles will reach nearly 21 million units globally in 2035, influenced by recent research and development by automotive OEMs and supplier and technology companies that are investing in this area. Given China's dominance of the global light-vehicle market, it may not be a surprise that IHS Automotive forecasts more than 5.7 million vehicles sold in China in 2035 will be equipped with some level of autonomy, the single largest market for the technology. The development of self-driving cars and V2X technology is also benefiting China in more ways than just meeting the needs of consumers. The development of such a huge cross-border integrated industry is a significant driving force for China's innovation-driven development strategy. This will help China open new sources of revenue, create new investment opportunities and jobs in relevant businesses, including infrastructure construction and software and hardware developments. At the social and environ-mental level, ICV development effectively reduces traffic accidents and casualties, alleviate traffic jams, reduce energy consumption and environmental pollution. From a reform perspective, the development of new technologies and new business models, especially the impact of new enterprises, can open a window to promote systematic and institutional reforms in relevant fields. More importantly, it will help the China's auto industry complete the fundamental transformation from traditional automobile manufacturing to providing transportation services.

1 China's Opportunities and Challenges

1.1 Stepping into "No Man's Land"

People are still used to viewing automobiles as a means of transportation. Yet, thinking outside the box, cars have increasingly become large mobile devices with huge information processing capabilities. Current technology allows the onboard computer to access the cloud processing platform promptly, integrate data instantly, plan routes and travel to the intended destination automatically. By 2021, driver-less cars will enter the market, marking the beginning of a new phase. In this process, China will face historic opportunities and challenges. It is worthy of careful consideration by all people working in auto industry in China.

The concept of "no man's land" was put forward by Ren Zhengfei, Founder and President of Huawei, at the 2016 National Science and Technology Innovation Conference. Ren said that Huawei was gradually stepping into "no man's land" in its industry by setting forth with no guided navigation, no established rules and no one to follow. Huawei's historical technique behind its high growth—"opportunistically" following others—would be obsolete, as it forges new ground. When an enterprise or industry enters the "no man's land", it needs to change from a strategic follower to a strategic leader. Like Huawei, China's auto industry is entering "no man's land". China's market, the largest in the world, has unique and complex Chinese char-acteristics. Due to the rapid development of electric vehicles, autonomous driving technology and car sharing, the auto industry is on the verge of change. The original business models and competing patterns will change dramatically, and the indus-trial structure will be reconstructed. Previous experiences may fail either at home or abroad. There are no rules to follow. This technique of "opportunism" may not benefit us anymore, as the need arises to explore the development path of the industry.

In this "no man's land", China is seeking for solutions and accumulating expe-riences for the development of global auto industry, playing a unique role in the progress of the global auto industry. This is an obligation of the emerging auto power towards the global auto industry, and an opportunity offered to China by the world. There is no need for China to unduly humble itself, for chances are that China will generate the best practice. Some other countries have begun to refer to China's experience and study China's case when formulating their auto industry policies. For example, the German government has "benchmarked" China's subsidy policy when promoting the EV development. China's current traffic situation makes it long deeply for self-driving cars, and people around the world await the verdict on what prospect autonomous driving can bring to China, and the salutary lessons China will provide to the world.

The time for autonomous driving vehicles is coming soon. Chinese drivers appear more open to vehicular experimentation. A World Economic Forum survey found that "75% of Chinese say they are willing to ride in a self-driving car." This view was echoed in a separate survey undertaken by the Roland Berger consulting firm. It found that "96 percent of Chinese would consider an autonomous vehicle for almost all everyday driving, compared with 58 percent of Americans and Germans." PwC's survey in China in 2017 also reached a similar conclusion. Among the factors that affect consumers' purchase decisions, Internet connectivity (51%) has surpassed traditional major concerns such as the price (46%) and engine performance (46%). 30% of economy car buyers and 40% of high-end car buyers would change to other brands for better intelligent and connected features, even at a higher price point. Furthermore, over 85% of the respondents expressly expressed desire for a fully autonomous vehicle. The agitation of many tech-savvy consumers and the expansion of ICV market have given China the potential to stand at the forefront of innovation in the global auto industry.

1.2   Brookings Institution's Analysis on China's Pros and Cons

In September 2016, the Brookings Institution, an authoritative American think tank, wrote of the status quo of autonomous driving. In a report named "Moving forward: Self-driving vehicles in China, Europe, Japan, Korea, and the United States", Darrell M. West, Vice President and Director of Center for Technology Innovation at Brookings, pointed out that China is expected to surpass the US in the field of autonomous driving, but the government sector faces enormous challenges. The thought-provoking report discusses China's situation at considerable length and is worth a read.

According to the report, the biggest challenge in China is the need to develop a national framework for autonomous vehicles. Chinese policymakers must be careful regarding how they regulate autonomous vehicles. They need to balance innovation on the one hand with societal values designed to protect drivers, safeguard data, and protect security. What they decide sets the broader framework in which businesses operate. China has an advantage in that most of its regulatory processes related to autonomous vehicles operate at the national level. Its top-down approach has the benefit of simplifying the maze of regulatory rules and procedures that exist elsewhere in federal systems. But there remains fragmentation over who will oversee regulation and development. Formal jurisdiction is split between the General Administration of Quality Supervision, Inspection and Quarantine (which handles product recalls), the Ministry of Industry and Information Technology (which makes industry policy), the Ministry of Transport (which makes plans for the transportation development), the Ministry of Public Security (which is in charge of vehicle registration, license management, and traffic safety supervision), and the National Administration of Surveying, Mapping, and Geo-Information (which enforces rules on map data collection). Other agencies handle environmental protection, recycling, commerce, and finance. In all, nearly 10 ministries and departments have jurisdiction over some aspect of autonomous vehicles. Getting these agencies to coordinate and work together is the task of current planners.

In addition, the central government needs to invest in research and infrastructure development that aids autonomous vehicles. Having resources that make it clear this sector is a priority is important for the future of the industry. It is a way to signal to the world and the domestic sector that intelligent cars are important, and the country is committed to their development. As of 2014, China's expressway mileage reached 106,000 km, surpassing the US to rank first in the world. One year later, this figure increased again, totaling 117,000 km. This is China's foundation. As China moves towards an era of autonomous driving, expressways will be the first applicable road type. Yet improving roadways should be a high priority for autonomous vehicles. Both semi-autonomous and fully autonomous vehicles need roads that allow their cameras and sensors to operate effectively. If car cameras cannot read lane markings, 3D high definition maps are not as useful. Also it would be helpful to install smart traffic lights that emit electronic signals to autonomous cars on whether the light is green, yellow, or red. Putting money into ITS is another way the government can help the sector.

The report also calls on the Chinese government to allow road testing and accurate mapping. Accurate maps are vital to the future of autonomous vehicles. Existing technology can graph roads down to several centimeters in accuracy. But for reasons related to security, government regulations mandate that public maps cannot be more accurate than 50 m. Cars simply cannot operate safely with that degree of imprecision. Since most Western countries do not have these kinds of restrictions, it places Chinese businesses at a competitive disadvantage.

To develop fully, the autonomous vehicle sector must address issues of legal liability. Right now, insurance companies undertake elaborate risk assessment based on driver age, gender, experience, and the like. Since most accidents are the fault of humans, they focus on who is at fault and who should be held liable for accidents caused by speeding, drunk driving, going through spot signs, or hitting another vehicle. It is less clear how to evaluate liability when there is no driver or the driver is relying upon automated controls. In China, policymakers are considering rules that shift legal liability away from drivers. These decisions are important because Chinese roadways feature a range of people and vehicles.

The industry and governments should consider a public awareness campaign to inform people about the benefits of autonomous vehicles, and their contributions to long-term job creation and economic development. Such a campaign would

help people understand what autonomous vehicles are, how they differ from traditional cars and trucks, and the differences are between semi-autonomous and fully
autonomous vehicles. Public engagement is important so that everyone understands
the impact of intelligent cars and traffic systems on society, the economy, and life
itself.

Darrell M. West has taken an optimistic stance on China's autonomous driving. He
put forth pertinent suggestions for the development of China's self-driving cars indif-
ferent aspects, such as national framework, cross-department coordination, infrastructure, standard system, public road test and mapping, legal liability, or public
awareness campaigns. Some hit the nail on the head, though others might not suit
China's national conditions. After the report was disclosed, it caused heated discussion among industrial and economic circles. His opinion roused echoing sentiments
from the majority, but rebuttals towards his arguments are also worthy of attention. Some think that West is a bit too optimistic about autonomous driving for the
following reasons.

First, autonomous driving without intelligent roads is just a lame duck.

Second, autonomous driving without V2V communication assistance is unreliable.

Third, autonomous driving can only achieve reliable driving under a quarter of
meteorological conditions.

Fourth, the cost is not yet low enough to entice consumers to buy without
consideration.

Fifth, the basic laws of information security have not yet clarified the boundary
between the public and private.

Sixth, China should beware of being misled towards autonomous driving.

Last but not least, for China, where agriculture is predominant and the urbanization
rate is low, the government faces more pressing issues such as employment, pension,
medical insurance, anti-terrorism. It does not seem like a proper prioritization to
invest a large amount of money into autonomous driving in such a period. Given that
we have not even fixed the chip and operating systems of mature products such as
mobile phones, autonomous driving has a long way to go in China.

Although opinions are biased and the information used may not be completely
accurate, they are not all a puff of air. Some may be even enlightening.

1.3 Let's Check Our "Inventory"

In his report, West did not cover technology, which is of vital significance to the
development of autonomous driving in China. In all parts of the industrial chain of
autonomous driving, Chinese enterprises are actively carrying out R&D to narrow
the gap with the top players in the world.

The perception part of a self-driving car is composed of components such as
cameras, LiDAR, millimeter-wave radars, ultrasonic radars and so on. For the time
being, multi-line LiDAR is perhaps necessary for future autonomous vehicles. The

foundation of opto-electromechanical technology in China is as good as that in any other countries, and the industrial chain is relatively complete. Therefore, there are considerable opportunities for many companies carrying in-depth research in LiDAR area such as RoboSense, Hesaitech, SureStar, and Benewake. In addition to LiDAR, millimeter-wave radar and ultrasonic radar have gradually become the sensing devices for multi-sensor information fusion in autonomous vehicles in recent years. In mainland China, Autoroad is a strong player, though it is a little weak when competing with global leaders such as Bosch and Continental. Currently, one of the main perceptual approaches for autonomous vehicles to determine the environment in front of the vehicle is to film the situations, and then perform image and video recognition. There are a dozen tech startups in binocular and monocular camera environment recognition field, such as Smartereye, Exeye, Zongmutech, and Minieye, which are challenging established company Mobileye.

In terms of satellite navigation, GPS occupies an absolute market position at present, though China's BeiDou Navigation may be a promising challenger. The successful launch of the 22nd BeiDou satellite on March 30, 2016 indicates that China's BeiDou navigation system has entered a stable operation state. Of the BeiDou Satellite Navigation Network, the ground-based enhancement network, the indoor positioning network, the Internet and the mobile communications network have basically been completed. Considering future development, the intellectualization and networking of location information services is a cornerstone of smart city construction. In light of its unique advantages in China, we should wait and see whether BeiDou Navigation System can occupy a share in the application of autonomous vehicles in the future.

HD maps are the basis for route planning for autonomous vehicles. China's HD maps are dominated by Baidu Map, Amap and NavInfo. In December 2016, Navinfo and Tencent announced their strategic cooperation in making a joint stake in the mapping company HERE. Audi, BMW and Daimler, the original shareholders of HERE, reduced their shares accordingly. NavInfo and HERE will jointly develop location technology and provide services for the Chinese market, and work on HD maps for autonomous driving. This strategic cooperation has not only expanded NavInfo's global business, technical data resources, and global OEMs, but also enabled HERE to circumvent the hurdles facing foreign companies entering the Chinese market independently, allowing it to quickly penetrate the Chinese market. At the same time, this development has also intensified competition in China's HD map business.

As for V2X, with the birth of the concept of the Internet of Vehicles, automotive electronics has changed its focus from machinery and safety to system integration capabilities and the synergy of vehicle-to-vehicle as well as vehicle-to-environment. However, Chinese companies do not possess the core technology of the chips used in high-end sensors for information acquisition. At the same time, the communication network bandwidth has become a technical bottleneck in V2X. China has not yet developed its cloud computing and ultra-large data processing technologies, as the collected information needs to be pumped into the data center, stored, and analyzed. At present, chip design and development in China has reached a certain level, but the problem of self-control and manageability is serious. China's Internet domain name system and address, as well as barcodes on goods all use foreign technical systems and coding. The Internet of vehicles cannot use similar techniques in vehicle identification. The Ministry of Public Security has introduced a special electronic label with a recognition rate of over 99.9%, which can be installed on the windshield of a vehicle to form a unique identifier for the vehicle identity and location information. However, to track vehicle information, a certain density of data collection facilities needs to be deployed in the surveillance area.

There has been no standard for interconnection technologies, especially V2V communications. With regard to the competition between DSRC and LTE, DSRC was in a dominant position over 4G LTE from the very beginning, as the former could achieve real-time point-to-point communication within 200 m with less than 1 s's delay, while the case for the latter was 6–7 s. However, with the development of 5G LTE or LTE-V, the industry will gradually unify their viewpoint after completely solving real-time issues. LTE-V was proposed by China's telecommunications companies at the end of 2013 with certain independent IPRs. Huawei and Datang currently lead the initiative on LTE-V standardization, and are the main drafters of the 3GPP LTE-V Study Item (SI) and Work Item (WI) reports. Huawei is still in competition with Qualcomm, and has also taken the lead in cooperating with domestic and foreign OEMs to conduct real vehicle testing. It's fair to say that China dominates in the field of telecommunications, with few differences from the best practices of the world.

Unlike the US, Japan and Europe, China lacks the support of large-scale national projects in this field, fails to form a joint force among enterprises, thus leading to relatively slow progress. The US, Japan, Europe and other countries have become unified in schedule and standards for the development of V2X technology. A strategic alliance has been formed, and China might potentially lose the right to speak in the future.

The competition to dominate the computing platform of autonomous driving system remains largely among international giants, with NVIDIA holding the upper hand. In China, Horizon Robotics, a company with awe-inspiring fame among artificial intelligent startups, is also researching the Brain Processing Unit (BPU) based on FPGA framework, and the computing architecture IP codenamed "Gauss," expected launch date by the end of 2017.

Autonomous driving not only requires perception and algorithm, involves automotive dynamics, automotive engineering and many other technical disciplines, but also requires the support of vehicle control actions, such as braking, steering, lighting, and throttle opening. Therefore, domestic and foreign EV and smart car related companies likeBaidu, NextEv, Yundu and Faraday Future are likely to become new upstarts of the autonomous vehicle industry. At present, most of auto driving start-ups can be regarded as companies at the level of algorithm integration. In China, this includes such companies as Uisee, Momenta, Tusimple, Idriverplus, and Falcon.

Intelligentization is the most prominent feature of a new round of technological and industrial revolution. The future of science and technology must also be advanced in artificial intelligence. AI will provide the finishing touch to the evolution of autonomous driving technology. At this point, China currently has little gap with the world's top level in the AI field, as both its academic and functional capabilities rank among the best. In 2016, it was proposed in the US National Artificial Intelligence Research and Development Strategy Plan for China to spearheaded many fields. China is truly not falling behind in source technologies. While technology giants are busy with AI development, China's technological innovations will stand at the commanding heights of the industry. AI studies are still in the early stage, with its main application is concentrated on voice and image recognition. However, China has led the research with huge technological breakthroughs. For example, the voice recognition rate of Baidu, Iflytek and other companies has exceeded 95%. Gaussian Face/Deep ID technology developed by TANG Xiao'ou (Vice President of Shenzhen Institutes of Advanced Technology of the Chinese Academy of Sciences and professor of the Chinese University of Hong Kong) has become the world's first computer algorithm to surpass the ability of human eyes. Moreover, China's AI technology has its own unique competitive advantages, especially in Chinese speech and semantic recognition.

In the context of a booming global market, the domestic market is nothing but breath-taking. In the past two years, BAT—Baidu, Alibaba and Tencent—have made frequent pushes towards AI development. Treating AI as the top priority in its future development, Baidu allocated over RMB 10 billion as R&D fund in 2015, covering deep learning, image recognition, autonomous driving, speech recognition, robotics, and intelligent medical treatment. In 2016, it launched the "Baidu Brain" program. In 2015, Alibaba launched China's first AI platform, DTPAI. It is now planning to create an ecosystem of "Smart IoT". Tencent has a comprehensive layout from image recognition to semantic analysis, from robots to the IoT. In the future, it will integrate QQ IoT and WeChat through the "TOS+" strategy to realize the parallel development of the two in the field of intelligent hardware. Of course, China's AI pathfinders are far more than BAT. While large companies are speeding up their layouts, other companies, whether they plan to lead the trend, delve deeper in research, or find unique approaches, are confronted with different opportunities and challenges in this profitable field for years to come. Although the industry is already in an oligopoly situation in some subdivisions, there are still untouched areas. The classification, capacity and distribution rules of the entire market are waiting to be established. At present, there are many outstanding enterprises in China, both in terms of technology platforms and industrial applications. These emerging forces cannot be underestimated.

China is likely to dominate the world when AI changes the world, in the next 5–10 years or even before 2045, thanks to its sound Internet infrastructure, computing capability and massive data, and its urgent desire to improve efficiency. China's campaign for AI development occurs at the right time and place with the right people, yet does not go without challenges. China needs to start with core technologies, high-end equipment and applications and basic theories, and fully use its advantages as an Internet superpower by transforming data and user advantage resources into AI technological advantages. Finally, the promotion and application of AI technology should be deepened to strengthen and enlarge the computing industry. At the same time, AI education and popularization should be strengthened to cultivate high-quality talent teams, to prepare for the full outbreaks of the AI industry—the first of which will probably include the widespread use of autonomous vehicles.

Zhu Hongren, Chief Engineer of the Ministry of Industry and Information Technology (MIIT), said at an industry conference in September 2016 that China should boost the level of intelligence in many aspects. Smart cars require breakthroughs in R&D and the industrialization of ADAS core technologies, industry standardardization, and the promotion of ICV demonstrations. As for supporting facilities, it is necessary to promote the formulation and implementation of infrastructure and roads supporting the development of smart cars, as well as the coordinated development of ITS and ICVs.

In 2015, the McKinsey Global Institute released the China Effect on Global Innovation, a preview of an in-depth study, with a more granular, nuanced approach to determine if China is on track to convert itself from an "innovation sponge"—absorbing and adapting technology, best practices, and knowledge 1—into an innovation leader. In this analysis, McKinsey looked at all kinds of innovations that have been commercialized successfully, from pure scientific discoveries to engineering breakthroughs to new business models to efficiency improvements. China sits apart from its peers by virtue of its uniquely dynamic and massive consumer market, its unparalleled manufacturing ecosystem, and the willingness of its government to invest in unprecedentedly large engineering projects. Yet the country has yet to make an internal-combustion engine that could be exported and lags behind developed countries in sciences ranging from biotechnology to materials. By better understanding the way innovation works, Chinese business leaders, academics, and policy makers can more effectively focus their efforts to promote it. Building on the success of today's innovators, they can create policies that support innovation in each of the four archetypes. In this way, China can continue to evolve into a more mature, productive, and innovation-based economy and may even provide a model for countries around the world.

In any case, China's auto market has become a hub of innovation. The influx of a large number of Internet automakers has created a thriving scene in the development of ICVs. Traditional automakers, under such stimulation, have sharpened R&D efforts and even introduced the concept of Internet vehicles. Traditional industries and tech companies combined will fill the gap between existing technologies and innovations. At the same time, the participation of many Internet automakers makers will promote cooperation among multiple companies around the world, break the boundaries of the industry and create a complex and brand new car ecosystem in China. As for whether China's auto industry can take the lead in this transformation, it largely depends on how traditional automakers and Internet companies will compete and collaborate.

2 The Roadmap for China's Intelligent Connected Vehicles

2.1 Under the Umbrella of Made in China 2025

The Internet of Vehicles (V2X) and autonomous driving technology are like a car's ears and eyes. They complement each other, forming the car's intelligent brain where a correct judgment is made by analyzing the surrounding environment information received. The V2X emergence has also brought upgrading opportunities to the fields of automobile manufacturing, contents delivery and mobile communications. On one hand, it has elevated the auto industry from pure hardware sales to a new model of combined services and contents. On the other hand, it allows operators and service providers to locate customers and provide products and services quickly. In addition, the Chinese government requires new energy vehicles to be "capable of remote monitoring", which facilitates the V2X development across two strategic emerging industries.

To speed up the development of V2X and autonomous driving technologies in China, it is necessary to provide a sound infrastructure and national policy support. In China, policy support from the government paves the way for the widespread use of ICV technology. In 2015, China established the "Intelligent Connected Vehicle Branch of the National Technical Committee of Auto Standardization" to gather multiple parties to draw up ICV-related standards, so as to promote the sector development. In the same year, the Chinese government formulated "Made in China 2025," the first 10-year program of action to build a strong manufacturing country, which set development goals and made clear requirements for the development of smart cars and intelligentization of automobiles. By 2020, the goal is to form an independent ICV innovation system which is enterprise-centered and market-oriented, through close combination of government, manufacturers, education bodies, research institutes and markets and coordinated development across industries. By then, the share of self-developed ADAS should reach 50%, and the installation rate of connected ADAS should reach 10%. By 2025, the goal is to establish an independent industrial chain of ICV and ITS, with self-developed ADAS reaching 60% and the installation rate of connected ADAS reaching 30%. At that time, ITS solutions for vehicles will be proposed to improve traffic efficiency and reduce traffic accidents and carbon dioxide emission.

Extended Reading

China—in the Third Stage of the World's Manufacturing Industry

Miao Wei, Minister of Industry and Information Technology (MIIT), pointed out in a comprehensive interpretation of Made in China 2025 that the global manufacturing industry has basically formed a pattern of four-stage development: the US is the only country in the first stage, which has been leading the global scientific and technological innovation since the end of World War II; the high-end manufacturing countries like Germany, Britain, France, Japan and other developed EU countries are in the second stage; those middle and low-end manufacturing countries like China and other emerging developing countries are in the third stage; the fourth tier consists mainly of resource exporting countries.

Miao Wei has also said that there is still a big gap between China and advanced countries. The main deficiencies are its weak independent innovation ability, insufficient basic supporting ability, unsatisfied product quality and reliability, lag in brand building and unreasonable industrial structure. China is in the third tier, and this situation cannot be fundamentally changed for many years to come. It might take China at least another 30 years' hard work to rank among manufacturing powers.

In September 2016, the MIIT announced that in China, a foundation for the rapid development of smart cars had already been laid and a collaborative innovation mechanism formed. China Industry Innovation Alliance for the Intelligent and Connected Vehicles has been established, carrying out R&D and testing, standard formation, laws and regulations study, exchanges, and cooperation. Manufacturing and Internet enterprises have considerable innovation vitality in the development of ICV technology, especially in the field of communication technology, which has strong international competitiveness.

In October 2016, the China Society of Automotive Engineers (SAE-China) was entrusted by the National Manufacturing Strategy Advisory Committee and the MIIT to release the Technology Roadmap for Energy Saving and New Energy Vehicle, a document that took more than 500 top auto experts in the country about a year to finish. The contents of this action plan indicates that China shall "continue to support the development of electric vehicles and fuel cell vehicles; master the core technology of low-carbonized, informationized and intelligent vehicles; enhance the engineering and industrialization capabilities of core technologies of battery, driving motor, high-efficient internal combustion engine, advanced transmission, lightweight materials and intelligent control, thus forming a complete industrial and innovation system from key parts and components to the whole vehicle and bringing the energy saving and new energy vehicles of China's independent brands to an internationally advanced level."

The roadmap's overall goal is that the total carbon emission of the auto industry should reach its peak in 2028, ahead of the national commitment of "reaching the peak in 2030". Also, by 2030, NEVs will gradually become mainstream products, and the auto industry will realize the transformation to electrification primarily. ICV technology will produce a series of original scientific and technological achievements for widespread and effective application. The technological innovation system should be mature, and the continuous innovation ability and the parts industry will be internationally competitive. On this basis, the Roadmap further proposed seven sub-roadmaps for seven fields including energy-saving vehicles, BEV and PHEV, FCVs, ICVs, battery, lightweight of automobile and automobile manufacturing.

2.2 Technology Roadmap on Intelligent and Connected Vehicles

Professor LI Keqiang, then Head of Department of Automotive Engineering in Tsinghua University and Chairman of Expert Committee of CAICV (China Industry Innovation Alliance for ICV), has led the formulation of the Technology Roadmap on Intelligent and Connected Vehicles. In his remarks, the significance of the technology roadmap is to support the formulation of ICV-related industrial development strategy and technical development strategy, help reach a consensus on ICV technologies, promote development ICV action plans and establishment of national key projects, as well as promote the accumulation of resources from all walks of life, collaborative research, and R&D of relevant technologies and demonstration operation under top-level architecture.

The Technology Roadmap on Intelligent and Connected Vehicles has made a plan for the development of China's ICVs in the next decadeand set the development goals for 2020 and 2025.

In the start-up period (2016–2020): An ICV independent innovation system in which enterprises are the main players and the political, production, academic and research circles are closely combined to achieve market-oriented and cross-sectoral collaborative development is to be preliminarily developed. More than 50% of new cars are equipped with DA, PA and CA, and 10% are equipped with network connected driving assistance system. The construction of SmartWay cities is started. Vehicle traffic accidents are reduced by 30%, traffic efficiency is increased by 10%, and fuel consumption and emissions are reduced by 5%, respectively.

In the development period (2021–2025): An industry chain of independent ICV and an ITS of passenger cars and commercial vehicles basically takes shape. 80% of new cars are equipped with DA, PA, CA and HA/FA, and 25% are equipped with PA and CA. HA/FA automated driving vehicles are introduced into the market. Vehicle traffic accidents are reduced by 80%, traffic efficiency on ordinary roads is increased by 10%, and fuel consumption and emissions are reduced by 20%, respectively.

ICVs include two technology levels—intelligence and connection. Their grades can be distinguished based on these two technology levels. In terms of intelligence, based on the grading definition of the authoritative SAE and considering complexity of road traffic conditions in China, the roadmap has classified the intelligence grades by Driving Assistance (DA), Partially Autopilot (PA), Conditionally Autopilot (CA), Highly Autopilot (HA) and Fully Autopilot (FA). In terms of connection, there are three grades—auxiliary information interaction through connection, collaborative sensing through connection, and collaborative decision-making and control through connection based on content of connected communication(Fig. 1).

From a chronological perspective, the roadmap follows a progressive process in terms of intelligence and connection. The phased objectives and milestones of intelligent and connected passenger cars are as follows:

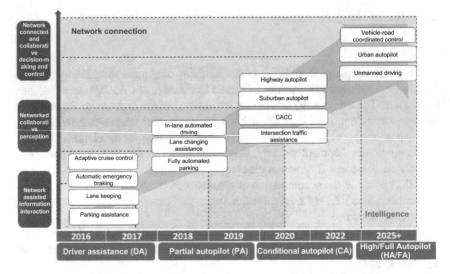

Fig. 1 Milestones of intelligent and connected passenger cars. *Source* Technology Roadmap on Intelligent and Connected Vehicles

(1) In 2016 or so, DA-level intelligence becomes a reality. Single driving assistance features are realized through autonomous environment perception, and typical systems include automatic emergency braking (AEB), lane keeping assistance (LKA), adaptive cruise control (ACC), and parking assistance (PA).

(2) In 2018 or so, PA-level intelligence becomes reality. Autonomous environment perception remains dominant, and intelligent information guidance based on network connection is provided. Typical systems include in-lane automated driving, automated parking (AP), and lane changing assistance (LCA).

(3) In 2022 or so, CA-level intelligence becomes reality. Network-connected environment sensing is materialized to adapt to the automated driving environment under complicated operating conditions. Typical systems include highway pilot, urban pilot, collaborative adaptive cruise control (CACC), and intersection traffic assistance;

(4) After 2025, HA/FA-level intelligence becomes reality. Network connected and collaborative control between vehicle and other traffic participants is materialized; automated driving is possible on highways, suburban highways and urban roads. On the basis of this, automated driving under all road conditions is realized.

The technical architecture is complex and can be regarded as "Three-Horizontal and Two-Longitudinal" technical architecture. "Three-Horizontal" refers to technologies of vehicles, information interaction and basic support which ICVs mainly involve; and "Two-Longitudinal" refers to vehicle platform and infrastructure condition which support the ICV development (Fig. 2). The Roadmap will eventually

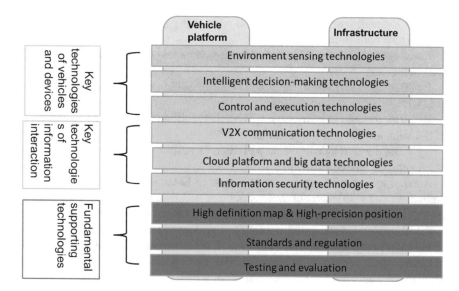

Fig. 2 "Three-Horizontal and Two-Longitudinal" technical architecture of intelligent and connected vehicles. *Source* Technology Roadmap on Intelligent and Connected Vehicles

achieve the development vision of substantial reduction in traffic accidents and casualties; significant improvement in traffic efficiency; effective reduction in transportation energy consumption and pollution; improvement in driving comfort and driver-free technologies; and enabling the elderly and the disabled with the right to drive.

To realize a coherent technical development route of intelligent and connected vehicles, the general technology roadmap respectively considers the logical relationship between mileages and stage objectives and between key components and critical generality in technical development route, subdivide according to the five categories of key component system and six critical generality technologies necessary for development of intelligent and connected vehicles, and describes technical evolution schedule of each component or each key technology.

Five categories of key component system are Environment Sensing Technology, High-precision Positioning and Map, Vehicle-mounted Intelligent Terminal HMI Product and Integrated Control and Execution System. And six critical generality technologies are Multi-source Information Fusion Technology, Vehicle Collaborative Control Technology, Communication and Information Interaction Platform Technology, Electronic and Electrical Architecture, Information Security Technology, Man–Machine Interaction and Joint Driving, Road Infrastructure and Standards and regulations.

Professor Li says that key technologies (mapping technology, communication technology, etc.) from foreign countries cannot be adopted because the overall environment is different. Therefore, domestic manufacturers need to carry out research and development independently. That is what many domestic car companies and Internet tech companies are doing now.

Based on the Roadmap, China's autonomous vehicles will develop at the following pace in the next 15 years:

By 2020, the auto industry scale will reach 30 million units, and the market share of DA/PA vehicles will reach 50%.

By 2025, the auto industry scale will reach 35 million units, and the market share of HA vehicles will reach approximately 15%.

By 2030, the auto industry scale will reach 38 million units, and the market share of FA vehicles will be close to 10%.

According to the Roadmap, we still need to break through multiple technical limitations and restrictions of various policies and regulations to have the highest level of intelligent connected vehicles on roads in China. Whether the future can be achieved step by step according to this Roadmap depends on the specific performance of automakers and Internet tech companies. It is worth noting that the Roadmap will be continuously updated with minor changes every year, and new updates in three years to accommodate the rapid development status.

Extended Reading

BAT's Layouts for V2X Applications

Currently, Baidu, Alibaba and Tencent (BAT) have set up their layouts of V2X platforms respectively. In 2014, Baidu launched Carnet, which evolved into Carlife in 2015. Baidu also launched My Car, a third-party accommodation platform providing car services. It combines V2X with Baidu's LBS service and O2O business, upgrading it to an entry to life services. In 2016, Tencent launched the open platform "Tencent Auto Intelligence (TAI)", and released "TAI ROM", "TAI APP" and "MyCar" service, which allows connection with automobiles to be made through WeChat or QQ. Tencent also established various modes of open industrial cooperation. For example, "TAI APP" can be connected through a variety of third-party connection channels, which makes it accessible to more vehicles. Tencent also allows its partners to access "TAI APP" through independently developed apps, and to port them on its V2X platform. Alibaba has also launched its own system, "Ali Yun Os for Car", which is already used on SAIC Motor's Roewe RX5. Obviously, BAT has already realized the huge market potential and space it leaves behind.

Industry experts believe that when it comes to the implementation of the national industrial strategic innovation project on ICVs, the R&D Program of SIP (Strategic Innovative Program) Automated driving System developed by Japanese Cabinet Office in 2014 can be taken as a reference. Its technical route is very clear, and it is crystal clear which fields require competition or cooperation. China can make concerted efforts to achieve breakthroughs in key ICV technologies through national government support, enterprise support as well as cooperation and sharing.

Professor Li also calls on China to make use of its institutional advantages and technical expertise to build an intelligent network system with Chinese character-istics. According to his judgment, China is likely to have its own characteristics in the following aspects to form its core competitiveness and make development breakthroughs:

Intelligent vehicle terminal using BeiDou Satellite System. This vehicle-mounted terminal should be promoted, in combination with the development of the national BeiDou Satellite System and the BeiDou ground-based augmentation systems.

LTE-V/5G and other independent communication systems with Chinese charac-teristics. China has occupied the technical and standard highland in the field of LTE. LTE-V/5G is likely to become China's independent communication system for V2X, which is conducive to the coordinated development of China's communication and automotive industries.

ICV basic data interaction platform (basic information platform). Relying on the advantages of the national system to build a basic data interaction platform is conducive to realizing big data sharing in the truest sense and can improve the efficiency of industry supervision and the level of coordinated control.

The Technology Roadmap on Intelligent and Connected Vehicles is a milestone for the development of China's autonomous vehicles, yet it is difficult to put the Roadmap into practice. As representing the guiding ideology and basic principles of Chinese mid-term and long-term economic society development, ICV is undertaking the mission of China economic strategy transformation, breakthrough and to build the future innovative society. As an important measure of implementing "Made in China 2025", ICV will provide strong strategic support for China to build a powerful manufacturing country and realize "intelligent manufacturing". On the other hand, there are problems of road safety, traffic jam, energy shortage and environmental pollution nowadays, so intelligent and connected vehicles will provide the possibility of building a healthy auto society. Besides, ICV may become the may path of making China's auto industry to realize the change from being big to being strong.

3 Chinese Enterprises and Their All-Round Explorations

China's research on autonomous driving dates back to the 1980s. At that time, China began research on intelligent mobile robots. In 1980, the state established the "Remote-Controlled Anti-Nuclear Reconnaissance Vehicle" project, in which Harbin Institute of Technology (HIT), Shenyang Institute of Automation (SIA) and National University of Defense Technology (NUDT) participated. During the "8th Five-Year Plan" period, ATB-1 (Autonomous Test Bed-1) unmanned vehicles were jointly developed by five institutes, including the Beijing Institute of Technology (BIT) and NUDT. This was the first test vehicle in China that operated autonomously, at the speed up to 21 km/h. The birth of ATB-1 marks the start of China's autonomous driving sector. Unmanned vehicle ATB-2 was successfully developed during the "9th Five-Year Plan" period. Compared with its predecessor, ATB-2's function was greatly enhanced and its driving speed in a straight road could reach a maximum of 76 km/h.

The NUDT has played a key role in the R&D of ATB series unmanned vehicles, representing the highest R&D level in autonomous vehicle in China. Since 2001, the NUDT has been working together with FAW to develop autonomous vehicles. In 2003, Hongqi CA7460 was developed—on a highway with normal traffic conditions, the lane change can be automatically performed according to the situation of the vehicle with the surrounding obstacles. After that, the NUDT developed the second-generation autonomous vehicle Hongqi HQ3 in 2006. In 2011, Hongqi HQ3 completed its first full-range highway self-driving test.

In 2008, the National Natural Science Foundation of China launched the Major State Basic Research Development Program—"Cognitive Computing of Visual and Auditory Sensations", which has been promoting the research in unmanned vehicle and its key technologies every year. It has held the "Intelligent Vehicle Future Challenge of China" event annually since 2009 to promote R&D exchanges and industrialization through integrated innovation and development of autonomous vehicles.

In 2012, "AMT Mengshi III" developed by the Academy of Military Transportation(AMT) traveled 114 km in an unmanned state with a peak speed of 105 km/h. On August 29, 2015, a Yutong bus departed from Zhengzhou to Kaifeng, and completed the autonomous driving test in a real highway environment, driving a total of 32.6 km with a top speed of 68 km/h without any human intervention throughout the journey. However, there was a driver seating in the bus in order to ensure safety. Now the autonomous driving test accepts passengers on board. In early December 2015, Baidu carried out a fully autonomous driving test in the Fifth Ring Road of Beijing, which also created a great sensation.

Since the 13th "Five-Year Plan" listed new energy and intelligent vehicles as one of the important contents, the Ministry of Industry and Information Technology (MIIT) has received dozens of non-automobile enterprise qualification applications. At present, there are few well-known Internet companies that do not want play roles in the profit-making industry of Internet automobile manufacturing. Even recently-emerged start-ups want to catch this once-in-a-lifetime opportunity.

Given of the huge opportunity in the field of intelligent travel brought by autonomous driving, domestic Internet giants and traditional automakers all want to get a head start. At present, FAW, SAIC Motor, BAIC, GAC, BYD, Changan, Baidu and NextEV have entered this field, and some of those companies' autonomous driving technology has been applied in road test. For a time, companies have showcased one after another and a picturesque scenery has been unveiled in the R&D field of ADAS and autonomous driving in China.

3.1 Baidu: I Can Do Whatever Google Does

(1) A Pioneer in Autonomous Driving in China

With its unique core competitiveness, Baidu has full firepower in autonomous driving. As a leading Internet company, Baidu has not only a solid foundation on big data, and but also a software team that is good at data analysis and algorithm design, giving it a unique development advantage in the field of artificial intelligence—a prerequisite for the autonomous driving technology to reach Level-4. In addition, Baidu also has enviable relations with the government and map resources, which makes it very likely to take the lead in launching HD map platform in China.

Baidu's autonomous driving story begins in January 2013, when Robin Li, CEO and cofounder of Baidu, formed the Institute of Deep Learning, the first AI research laboratory in China. In July 2014, Baidu Research Institute was formally established with four branches in Beijing, Shanghai, Shenzhen and Silicon Valley and three laboratories: the Silicon Valley AI Lab, the Beijing Deep Learning Lab (formerly Institute of Deep Learning) and the Beijing Big Data Lab.

Baidu Silicon Valley R&D Center has become the center of Baidu's high-tech talents, gathering outstanding Chinese scientists in Silicon Valley. Its most successful headhunt brought in Andrew Ng, known as one of the top three AI scientists in the world. (At the time of this book's publication, Andrew Ng has left Baidu, with no stated reason.) It is the most important person a Chinese Internet company could employ to date. Once the news was announced, it became a topic of concern in the international scientific and technological community.

In June 2016, Baidu ranked second in the "50 Smartest Companies in the World" by MIT Technology Review. "Baidu will lead an era of innovative software technology for better understanding the world," the prestigious US journal wrote with passion. In October 2016, the American journal Fortune published an article Why Deep Learning Is Suddenly Changing Your Life? And selected four top enterprises in deep learning—Google, Microsoft, Facebook and Baidu. Baidu is the only Chinese technology enterprise out of the four AI giants.

(2) Baidu's Partners

Baidu first established a cooperative relationship with BMW in September 2014 to jointly promote research on high-level autonomous driving technology. At that time, they signed an agreement with a three-year term. On December 10, 2015, Baidu's unmanned car, a 3 Series GT modified by Baidu in cooperation with BMW, ran its first fully autonomous trip in the mixed conditions of urban roads, expressways and highways in China.

Later, Baidu had numerous incidents with its partners. It "divorced" BMW, and then "hooked up" with Ford. The two jointly invested in Velodyne LiDAR, requesting that the company drop the LiDAR's price to $500 in five years. Shortly after that, Baidu announced a partnership with NVIDIA to advance its unmanned car project to build a driverless taxi fleet, announced by Jensen

Fig. 3 Baidu's unmanned fleet in Wuzhen

Huang, CEO of NVIDIA. Then in October 2016, Baidu reached a comprehensive strategic cooperation with Foton, a domestic heavy-duty truck maker, on ICVs. The latter had just released its "Super Truck" at 2016 IAA CV Show in Hanover, Germany.

(3) Test Driving in Wuzhen

In November 2016, the third World Internet Conference (WIC) was held in Wuzhen. During the second WIC, Baidu made a static display of its driverless cars—"Yunxiao Unmanned Vehicles"—which had become famous for the personal visit of Chinese President Xi Jinping. This time, Baidu launched 18 unmanned vehicles modified from Chinese brands in Wuzhen. It used dynamic displays and let the cars run openly for people to test drive (Fig. 3).

The autonomous driving module on the "Yunxiao Unmanned Vehicles" is a "composite sensor system" composed of a 64-line LiDAR, three 16-line LiDARs, GPS/IMU navigation, positioning system and millimeter wave radars installed on the roof, and binocular cameras on the front. With the help of this system, Baidu's unmanned vehicle's response time has reached only 200 ms, much lower than the normal person's reaction time (about 600 ms). Baidu's Chief Architect once boasted that its sensor is better than Tesla's, thus Baidu's vehicle would not cause the kind of fatal accidents that Tesla cars have made.

After strategic investment, mergers and acquisitions in both software and hardware fields, Baidu may form its own competitiveness differentiated from traditional auto suppliers like Bosch and Delphi. Baidu may set out its own technical standards and system in regional markets and even strive to form technical barriers.

(4) Smart Car Strategy

Compared with the vigorous test drives in Beijing and Wuzhen, Baidu's "Smart Car Strategy" lacks popularity. This is another plan related to the Internet of Vehicles that Baidu disclosed in March 2016. Baidu announced that it has signed a strategic cooperation agreement with Changan Automobile to jointly develop smart cars. Shortly after, Zotye became the second automaker to team up with Baidu. Here is Baidu's logic: the core functions of the first two generations of terminals—represented by PCs and mobile phones—are to search the web and communicate. As the car becomes the third generation of connected terminal, travel will becomes its core function. In this context, the demand for maps and navigation will expand dramatically. Maps will be the main entry point for Baidu's "Smart Car Strategy", and may even become the key point of Baidu's car-making business, and hold the highest priority.

According to data released by Baidu in 2016, Baidu Maphas accumulated 300 million active users over seven years. Of these, about 60 million are car owners, and the daily positioning requests have reached more than 30 billion times. Baidu hopes that feedback on locations and driving behaviors from its massive user database will provide big data support for the intellectualization of Baidu's smart cars. The "car ecology" based on Baidu Maps will include smart maps, smart driving and smart services.

(1) Smart maps. Driven by the need for smart driving, maps have begun to evolve from standard-definition maps that provide users with simple location and geographic information, to ADAS maps which clearly understand the curvature and slope of the road, the number of lanes, the 3D positions of the traffic lights and 3D modeling of buildings, eventually evolved into high-definition maps (HD maps) that are formed by combining the data from high-definition cameras and the point cloud generated by LiDAR scanning and can be accurate to 10–20 cm. These maps are not only high-definition, but also "live". At present, Baidu is constructing a "Learning Map", distributing 250 data acquisition vehicles all over the country to collect road information and send it back for mapping, to ensure the real-time accuracy of the maps.

(2) Smart driving. When the real-time HD maps constantly provide detailed road information to vehicles, the "smart driving" aspect of "car ecology" comes into being. At present, related products are divided into two layers—the upper layer includes human—machine interaction, while the lower level includes HD maps, environment perception, self-positioning and in-vehicle decision-making system. The establishment of dialogue between the upper layer and the lower layer is the overall product line architecture of Baidu Maps. Meanwhile, Baidu's R&D in the field of artificial intelligence and deep learning over the years will also be shared by many of its businesses. The current collection of samples and deep learning network models will also serve in the R&D of smart driving.

(3) Smart services. After a positive feedback loop is created between smart maps and smart driving, smart services will function as a kind of Data-as-a-Service (DaaS), and focus on the analysis and mining of big data to provide useful information to automakers, dealers and tier-1 suppliers. This will further endow auto industry' big data capabilities. In the ideal system, the above three aspects would form an effective closed loop, and a smart car ecosystem based on Baidu Maps could be established.

3.2 NextEV: "Blue Sky Coming"

NextEV is a company engaged in the R&D of high-performance smart EVs. It was founded through investments of hundreds of millions of dollars by well-known Internet companies and entrepreneurs, including internet giant Tencent, Founder of BitAuto Li Bin, Founder of AutoHome Li Xiang, Founder of JD.com Liu Qiangdong, and others. NextEV was created by numerous Internet celebrities with car-making dreams who want to reshape the auto industry. They are the new army entering the field of car manufacturing.

A couple of years ago, China's auto industry was still dominated by traditional fuel powertrain, traditional technology and traditional talents, despite the stir surrounding Tesla in the market. But "newborn calves are not afraid of tigers", as the Chinese saying goes. NextEV has never planned and developed products following the traditional ways of thinking since its founding. The philosophy of Internet economy is to learn and expand rapidly, reinventing itself in the process and constantly adjust strategies to meet the needs of the market. This strategy not only prevents delays caused by the lengthy multi-level discussion of a traditional company structure, but also greatly explore everyone's ability, drawing the maximum potential of people at the start. It is full of Internet genes and able to attract the automobile elites to join in. Its co-founders includeJack Cheng who once was the General Manager of GAC Fiat Automobile Co., Ltd., and Qin Lihong, once the Deputy General Manager of Chery Automobile Sales Co., Ltd. We can find traces of Tesla in this down-to-earth car manufacturer.

NextEV is not just a car company. Based on smart EV products, NextEV aims to redefine all the processes of serving users, and provide users with a full-scale experience beyond expectations. Its mission is to make more people like smart connected EVs through better products and experiences. Only a better experience can escalate the auto industry and inspire users. NextEV's idea is to perform R&D and services independently. It will take the new user relationship in the era of mobile Internet as a starting point, and prioritize user experience and services in its corporate goals. NextEV hopes, in China particularly, to "make up" for Tesla's shortcomings, such as its inability to solve the problem of private charging.

NextEV's R&D team is located in San Jose, Munich, London, Shanghai, Beijing, Hong Kong, Nanjing, and Hefei, integrating its global resources. Its early stage teams are not modeling design or HMI, but work on supply chain, manufacturing, engineering and R&D.

NextEV officially reached a manufacturing cooperation agreement with JAC in April 2016, with an overall scale up to 10 billion yuan. The two parties plan to upgrade JAC's existing production line, targeting an annual production capacity of 50,000 cars. NextEV will provide technical information, parts and components as well as bill of materials (BOM) to assemble the cars, while JAC will assist NextEV in the product's start of production (SOP), such as the production of sample cars, providing facilities, equipment and personnel required for the trial production.

On May 17, 2016, on the first anniversary of NextEV's establishment, Shanghai XPT Technology Limited was founded. This parts maker is fully owned by NextEV, and can be regarded as a long-term tool to further strengthen its production and manufacturing, and to master the independent R&D for corporate development. XPT will produce motors, battery systems and ECUs for NextEV, and then extend to intelligent driving, HMI and other related fields. As a production base of EV key components, XPT invested 3 billion yuan to build a high-performance motor and electronic control system production plant in Nanjing in April 2016.

On October 14, 2016, NextEV announced that it obtained road test licenses for unmanned vehicles issued by California government, joining the ranks of Google and Tesla to become one of the early-bird companies with such licenses. NextEV has been recognized for its technical capabilities, finding itself in the first tier for the R&D of autonomous driving technology in Silicon Valley. NextEV also received a $10 million tax credit from the "California Competes Tax Credit Program".

On November 22, 2016, choosing London for its first ever launch event, NextEV created a public sensation by releasing its English brand "NIO", its new LOGO, and EP9 (Fig. 4). EP9, the world's fastest EV, is attractive and unique, and is obviously designed by veteran automotive insiders. It is said that EP9 IS benchmarked by

Fig. 4 NextEV NIO EP9

star sports cars of well-known brands, such as Ferrari LaFerrari, McLaren P1 and Porsche 918 Hybrid. The car is a celebration of creativity in terms of style, power, and performance, with a total cost of approximately $1.2 million. EP9 has an astonishing peak power output of 1360 horsepower, and a 0–100 km/h acceleration time of less than 3 s. The data qualifies the car as a member of the super sports car club. On February 23, 2017, NextEV EP9 set a speed record of 257 km/h—the highest ever for an EV in the world—in the unmanned driving test at the Circuit of the Americas in Texas. Setting this speed record symbolizes NextEV's rapid advancement in the field of autonomous driving technology.

NextEV must also keep up on the road of intelligence. According to the company's plan, NextEV will deploy more autonomous driving functions to its products by 2020. "In the future, NextEV will use in-car cameras to capture driver's expression and movements. It can be used not only to hold video meetings, but also to monitor the driver's fatigue. There will be an account system for the car to be fully connected with IoT systems at home. The system will be able to turn on the apartment's air conditioner in advance, close the window automatically in case of rains, and even charge or change the batteries automatically," says Jack Cheng, Co-founder and Executive Vice President of NextEV.

On March 10, 2017, NextEV released its North America Business Strategy and its first concept car, EVE, in Austin, Texas, USA. EVE is an unmanned mobile living space with the design concept of "second living room" that allows users to enjoy pleasant and relaxed trips. The integration of vehicle and environment, people and environment can be realized through interactive technologies such as panoramic cockpit and smart holographic screen. NextEV also launched "NOMI", an AI companion system. NOMI can learn the habits and interests of users and meet their personalized needs in different scenarios. From granting space to occupants to freeing up time, from fighting against the environment to coexisting with the environment, from being a pure machine to serving as an emotional partner: this is the concept of what future cars might be, and what NextEV's EVE has showed us.

Looking back at NextEV's entrepreneurial path, Li Xiang, NextEV's co-founder, says that the path NextEV has taken to enter the auto market is similar to that of Lexus and Tesla. In other words, it starts with high-end products, and continues to explore the low-end with business model innovation. In practice, NextEV Formula E (its very first product) won the first annual championship ever in the ABB FIA Formula E Championship in 2015, making the company reputable in terms of brand and technology. The next product, super sports car EP9, strengthened its reputation in brand and technology. Finally, it will develop EVs suitable for domestic usage through its solid technical foundation and strict production management, then promote them in the market with affordable prices and future-oriented technological configurations.

Global footprint, high-end products and funding: NextEV's business model enables the public to accept its legitimacy.

3.3 Changan: Rising to Fame with One Vigorous Trip to Beijing

In recent years, the performance of Changan Automobile in the domestic market is eye-catching. Changan sold over 3 million vehicles in 2016, of which nearly 60% were self-owned brand vehicles. Among automakers in the "3 million club" in China, Changan is the only auto group that has sold more than 50% of vehicles out of its own brands. Changan has also set up R&D centers in the US, Italy, the UK and Japan, in addition to China, making it superior to other competitors in terms of global layout.

Currently, Changan has mastered more than 60 intelligent technologies in three categories, namely intelligent interconnection, intelligent interaction and intelligent driving, and has taken a leading position in the R&D of intelligent vehicles in China. In the next ten years, Changan will invest 20 billion yuan to set up an intelligent R&D team with more than 2000 staff. It will launch Highly Autopilot(HA) vehicles in 2020, and autonomous vehicles into market in 2025. Zhu Huarong, President of Changan with technical background, said in March 2017, "in view of challenges of the future, Changan will transform form a traditional auto manufacturing enterprise to a modern manufacturing-service enterprise. With its intelligent automobile products as a carrier, Changan will upgrade from a product provider to a provider of products + services + travel solutions."

Changan has a clear strategy for future vehicle development, including autonomous driving technology. In 2009, Changan begun to deploy intelligent and V2X projects. Following the global trends, it has formulated a development plan for smart car technology for 2025, called the "Changan 654 Strategy". By building six key platforms and mastering five core technologies, it will realize the industrialization of intelligent technology in four stages.

In terms of intelligent network, Changan is striving to integrate global resources, with a goal to forma global R&D system on ICV of "coordinated development in three countries and four places, each with its own emphasis". At present, Changan has set up the Intelligent Network Technology R&D Center in Chongqing Research Academy, the Intelligent Network R&D Department in Changan US R&D Center, as well as the Advanced Innovation Lab in Silicon Valley. It is also preparing to build an R&D Center in India.

On November 30, 2016, Changan officially signed a contract with PlugandPlay (PNP), a Silicon Valley technology incubator, which marked the entry of Changan into Silicon Valley. Known as a "Global Innovation Platform," PNP explores potential opportunities for business development, investment, licensing trade and acquisition through its industry-specific "accelerator" projects and innovative startups. The cooperation between Changan and PNP has enriched Changan's global layout and accelerated the development of autonomous driving technologies up to Level-3 and Level-4.

Changan will leverage the Silicon Valley eco-sphere platform to quickly establish a system for intelligent R&D. Its Advanced Innovation Lab in Silicon Valley will achieve Level-3 and Level-4 automation in three years and through two stages. In the first stage, intelligent interaction, driving and network materials were acquired and business began in 2016. In the meantime, the sub-technologies of inCall4.0 and the products using sub-technology of Level-3 automation were explored and introduced. In 2017, resources were utilized, core technology capabilities were built through investment or shareholding, and innovative ideas and solutions, such as intelligent interaction design, new electronic architecture applications, software development, AI and big data analysis applications were continuously introduced. The second stage is to establish a R&D team in 2018, to master Level-3 and Level-4 autonomous driving technology through continuous technical exploration. The new cooperation will enable Changan to take the lead in trying out the most advanced technologies in the industry. It is expected to provide an innovative resource pool for the implementation of 1–2 "initiating" and "brand-new" technologies annually from 2018 and beyond.

At present, Changan, is developing Full Autopilot (FA) cars with its domestic and foreign partners. By the end of 2016, Changan had completed the project technical approval of TJA in structured roads and integrated ACC for its self-driving cars, and 60% of concept design of Level-3 automation technologies, as well as the structural road verification. The test is expected to be accomplished in 2020 and SOP is planned in 2025.

In terms of intelligent network, Changan has included the applications of in-vehicle network, V2X and cloud network, aiming to providing convenient, interactive and pleasant driving experience. Up to now, more than one million ICV models have been put into production. Changan is also the first to successfully synchronize on-board maps and mobile phones, enabling real-time updates of traffic conditions. As for V2V and V2I communication, Changan completed eight national projects from 2012 to 2015 and joined in the intelligent alliance of Mobility Transformation Center (MTC) on behalf of China. After several years of testing, the application of nine typical V2V and V2I scenarios were completed. Changan also realizes remote online functions through its software management platform. Its latest digital service is to communicate directly with online vehicles through the system.

It's fair to say that Changan is taking the lead in the field of V2X technology in China. In May 2016, Changan invited a group of Chinese journalists to visit the headquarters of MTC, and experienced the latest technology of Changan's intelligent research, in a sensational event. Changan also joined Tsinghua University and GM in leading the formulation of the application layer standards for China's V2X technology.

In April 2016, Changan completed a long-distance test of its autonomous driving technology, which was once a headline event in the country (Fig. 5). The whole journey was nearly 2000 km, longer than tests run by any other automaker in China. The five-day test yielded a considerable amount of data. Changan's technicians revealed 38 enriched road scenarios and 13 additional domestic special models resulting in 13% growth of the car's intelligence level. The cars that carried out this intelligent driving project were equipped with forward-looking cameras, forward-looking radars, LiDAR, HD maps and sophisticated cutting-edge technology equipment, successfully completing a demonstration of reactions to complex practical

Fig. 5 Changan unmanned vehicle

driving environments. They were able to realize core functions like ACC, AEB, LKS, LCA, low speed front vehicle following, RSR, voice control, etc., within 120 km/h. However, manual operations were needed when the car drove in and out of toll stations, and drove into urban area at night. Tests show that Changan has R&D and application capabilities for many cutting-edge technologies such as automatic control, artificial intelligence and visual computing.

This is the result of Changan's open-mindedness to meet the market challenge. In addition to gaining independent IPRs through R&D in-house, it also has complementary advantages through cooperation with leaders in multiple fields. Through in-depth cooperation with leading suppliers, cross-border Internet companies, universities and research institutes as well as MTC, Changan has enhanced its intelligent technology and accelerated the industrialization process. The 2000 km road test, for example, is a result of collaboration with Bosch. It is highly likely that Bosch will also play a part as an important technical partner in Changan's production cars in the future. According to Bosch, they have provided the technologies of "TJA" and "highway assistance" for the road test vehicles.

Thanks to all the R&D accumulations made in autonomous driving, Changan has installed technological equipment including AVM (Around View Monitor), LCA and LDW on production models Reaton and CS75. Among these, LCA and AVM were first launched by Chinese OEMs. Full-speed ACC, AEB, FCW and automatic parking assist system (APA) were officially applied to Reaton, CS75, Eado, CS35 and other models in 2016.

In March 2016, Changan became the first Chinese automobile brand to cooperate with Baidu on its "Smart Car Strategy", the two parties working closely on smart interconnection, maps and services. In the future, Changan plans to apply processes covering vehicle intelligent interconnection, remote vehicle control, and intelligent voice interaction to fully cover the rest of the driving process, providing users with better human–machine interconnection solutions and driving experience.

In March 2017, Changan announced the establishment of a joint laboratory with Iflytek for research on intelligent onboard voice technology, big data analysis, image recognition and onboard application platforms. At the same time, based on the superior resources of both companies, strategic cooperation would be carried out in three major aspects: intelligent voice, AI big data analytics and V2X operation platform.

3.4 GAC Group: In the Vanguard of the Trend but with a Pragmatic Attitude

GAC Group closely follows future development trends towards electrification, intelligence, and lighter weight, laying out plans for V2X, ADAS and smart cars. Breakthroughs have been made for certain key task projects and in-depth research work, enabling technology spillovers and industrialization.

In terms of V2X technology, GAC has completed the independent development of Telematics remote control system in 2013. This was first installed on production models of the Trumpchi GA3/GA3S, making GAC the first OEM in China to complete the independent development and commercialization of Telematics. At present, Telematics and T-Box system have been promoted and applied in all GAC Trumpchi vehicles. From this base, GAC is also developing advanced technologies, such as new-generation V2V and V2I communications.

In terms of vehicle safety and ADAS, GAC has independently developed intelligent systems such as Around View Monitor (AVM), Lane Departure Warning (LDW), Blind Spot Detection (BSD), and Forward Collision Warning (FCW), all of which are installed in its production cars. Its next step is to develop and apply advanced technologies such as Night Vision (NV), Automatic Emergency Braking (AEB), Adaptive Cruise Control (ACC) and Head-Up Display (HUD).

As one of the first domestic auto enterprises to enter the field of research on driverless vehicles, GAC launched China's first autonomous vehicle ever developed wholly by an OEM with its new-energy vehicle platform GA5 PHEV. GAC possess all the IPRs for this car, which can serve as the best test platform for autonomous driving.

So far, GAC has the capability to independently develop automotive intelligent technology, enabling it to carry out research on key technologies such as intelligent human–machine interaction, high-precision navigation, V2V/V2I/V2P wireless communications, driving environment recognition based on multi-sensor data fusion, whole-trip/partial-trip and driving strategy automatic planning, and automatic execution control. Environment recognition technologies are now installed on the GAC's smart cars, allowing for fully autonomous driving between any two

Fig. 6 WitStar, GAC's autonomous concept car

preset points in local areas. These technologies include identification of lane lines, road boundaries, traffic signs and traffic lights, and intelligent decision-making and line-controlled execution based on multi-sensor data fusion, functions, like automatic parking, active collision avoidance, lane keeping, cruise following, active and passive lane changing, and pedestrian protection. As automatic driving technology continues to develop, and the automatic parking technology gradually improves, it will eventually be applied to production models.

WitStar, an autonomous concept car developed by GAC R&D Center, made its world debut at Guangzhou International Automobile Exhibition in 2013, and showcased overseas at the North American International Auto Show in 2015 (Fig. 6). It attracted the attention and coverage of the mainstream US media outlets (such as USA Today). In January 2016, GAC's intelligent vehicle was recognized by MTC, which admitted GAC as its member. In April 2016, Xinhua News Agency made detailed on-site report on GAC's first-generation intelligent vehicles driving autonomously in different conditions, such as highways and urban roads.

GAC also verifies its product performance in competition. In November 2016, the eighth "China Intelligent Vehicle Future Challenge" was held in Changshu, Jiangsu Province. Two unmanned vehicles jointly developed by GAC and Xi'an Jiaotong University ranked third and fourth respectively out of 24 teams, with the total score of second among all the groups.

In order to maintain its competitive advantage in the autonomous driving field, GAC will increase future investment in R&D and industrialization. At present, it has formulated a development plan for autonomous driving for the next ten years. This plan is divided into the start-up stage (2017–2020) and the development stage (2021–2025). The start-up stage will primarily establish a standard system of autonomous vehicles, an independent R&D system, a production supporting system, and a data security technical standard, master the key technologies, including sensors and controllers, of autonomous driving systems for passenger cars and commercial vehicles. This will ensure that the supply capacity meets the needs of the vehicle production scale, that the product quality will remain high, and that costs will maintain market competitiveness. The development period will see the establishment of a complete standard system for autonomous vehicles, an independent R&D system, a production supporting system and industrial clusters. As key technologies in sensors and controllers reach an advanced level, and key technologies in the actuators are mastered, the product quality and price will become strongly competitive in China. The entire life cycle of automobiles will be digitalized, networked, and intelligentalized. The transformation and upgrading of the auto industry will essentially be completed. Autonomous vehicle information security certification will be compulsory, and GAC will be domestically competitive in the field of autonomous driving.

Based on the ICV grade requirements, GAC's phased development goal of autonomous driving is as follows. Around 2018, the level PA will be reached. It will focus mainly on intelligent environment perception, and in-lane autonomous driving, Automatic Parking (AP) and Lane Changing Assist (LCA), based on connected information systems. Around 2022, the level CA will be reached. The car will have the ability to perceive the networked environment, and will be able to adapt to complex working conditions autonomously. Typical scenes will include Highway Pilot, Urban Pilot, Cooperative Adaptive Cruise Control (CACC) and Intersection Movement Assist (IMA). After 2025, the level HA/FA will be realized. With the connected and collaborative control ability between the vehicle and other traffic participants, the autonomous driving on highways, suburban roads and urban roads can be a reality. Afterwards, full autonomous driving in all road conditions will become a reality as well.

GAC will also integrate resources to establish information and energy cloud platforms, completing the smart NEV demonstrative operation and the industrialization of key components. Through mobile communication technologies such as smart phones, 4G network and Telematics, IOT integration of V2V, V2I and V2P will take shape in the era of "Internet+ " and big data.

3.5   SAIC Motor: Targeting "Advanced Technologies" with Phased Approach

In its development strategy, SAIC Motor has focused on "advanced technologies" to transform itself from "technical follower" to "technical leader". In SAIC's view, researching advanced technologies requires it the abandonment of the "closed-door" approach, irrespective of external circumstances. However, an internal innovation environment is yet to be created. Through long-term efforts, SAIC has explored an advanced technology R&D mode differentiated by its ability to adapt to its own development. Advanced technological breakthroughs, a "differentiated" advanced R&D mode, and the combination of "technology" and "capital" through the establishment of venture capital companies and innovation centers in Silicon Valley allow the company to explore needs for future development in fields like new energy, new materials, and intelligent interconnection.

Intelligent driving is one of SAIC's research directions. In terms of strategic development goals, intelligent driving technology encapsulates intelligentization development in technologies and systems for the next 5–10 years, through its movement towards electrification, lighter weight and intelligence. This will allow it to complete its autonomous driving function on structured roads in five years, then in all road conditions in ten years.

In 2015, SAIC displayed its self-developed intelligent vehicle iGS at Auto Shanghai. iGS can observe the surrounding environment through its cameras and radars, transmit data to the control software for analysis, and then give instruction autonomously. iGS has thus reached Level-3 autonomy. As verified by road tests, iGS can accomplish functions such as remote-control parking, automatic cruise, automatic following, lane keeping, lane changing and autonomous overtaking in working conditions at a speed of 60–120 km/h. Currently, SAIC is working on the in-house development of its second-generation MG iGS.

Alibaba's V2X plan has been deployed for a long time. It launched YunOS for Car, an operation system specially designed for automotive scenes, in early 2016. On this basis, SAIC and Alibaba jointly launched Internet vehicles. Based on the development of SAIC's complete vehicles and components, Alibaba provides the car with a "brain", as well as extensive resources and capabilities. In October 2016, the monthly sales volume of Roewe RX5, China's first Internet car jointly built by SAIC Motor and Alibaba, exceeded 20,000 only four months after its market premier. Although priced over 140,000 yuan, the Roewe RX5 premium—with Internet functions—accounted for 70% of the total sales, a fact unforeseen by many. This would not have been possible without the ecosystem of YunOS and the title of "first Internet car."

3.6 BAIC Group: Crowdfunding in Technology and Collaboration in Innovation

In 2016, BAIC Group maintained its title as the fifth largest automaker in terms of sales volume in China, achieving sales of 2.847 million units with a yearly growth of 15%, and operating revenue of 406.1 billion yuan with a yearly growth of 17.9%. It also achieved historical breakthroughs in all key operations. Among the top five first-tier auto groups in China, BAIC Group was the only one to increase its market share. BAIC's house brand also experienced massive growth in 2016. BAIC Senova gained cumulative annual sales of 224,500 vehicles, for an increase of over 80% from the last year. To facilitate the development of self-owned brand vehicles, BAIC bought the rights to Saab's 9–3 and 9–5 models and powertrain technologies at the end of 2009. BAIC's meteoric rise is the result of years of careful study.

In the context of a technology boom, BAIC has proposed two development directions to promote intelligent vehicles—intelligent driving and connected services. BAIC divides its intelligentization into three stages: elementary, intermediate and advanced. The elementary stage of intelligentization has more or less been achieved, as automatic parking and following of other vehicles on the road with clear road markings is now feasible. Other breakthroughs have been made in the intermediate stage, where cars are semi-autonomous, with more sophisticated ADAS functions. At the advanced stage, the car will be virtually driverless—a long-term goal under hot pursuit.

BAIC established the BAIC New Technology Research Institute in 2015to support the R&D of smart cars. One of its priorities is to assist in the development of NEVs to become intelligent, energy saving and environmentally friendly, through a combination of intelligentization and electrification. In the future, BAIC's intelligent development will rely on its own R&D team to establish a new crowd intelligence innovation platform by adopting a variety of modes, such as crowd innovation, crowd sourcing, and crowd funding. "BAIC will realize technology crowdfunding on this platform, break the boundaries for R&D personnel, and promote the R&D of all intelligent, personalized functions." says Xu Heyi, Chairman of BAIC.

BAIC has recruited talents in the field of intelligent vehicle technology, simultaneously strengthening ties with Internet and IT companies. These reliable strategic partners will aid in open and cooperative innovation, and accelerate breakthroughs in key technologies. In early 2016, BAIC officially began its 2.0 era. In terms of R&D, the "2.0 era" aims to make products of the next generation by creating products with higher technology advancements, that are closer to the market and needs of users. By 2020, BAIC's own-brand cars will reach the 3.0 stage of intelligent driving, and achieve the Conditional Autopilot level. In addition, Nova-Link, Nova-Pilot, Nova-Space and other smart travel systems will also be launched for Senova cars, according to BAIC's plan.

In July 2016, BAIC New Technology Research Institute and Panjin Municipal Government signed a strategic cooperation agreement for unmanned automobiles. In October, BAIC-Panjin unmanned vehicle made its global debut, and the project started. More than 100 unmanned cars of three types would run at full capacity on the 22 km road of the Red Beach Scenic Zone in Panjin.

The 2016 Baidu World Conference, unveiled the BAIC-Baidu unmanned vehicle, the latest joint R&D achievement between BAIC New Energy and Baidu. With Level-4 autonomy, it could perform fully "unmanned driving" within the restricted areas.

BAIC is working hard, and heading into the era of autonomous driving.

3.7 FAW Group: Move Ahead with You on a "Smartway"

FAW Group is one of the first domestic automakers engaged in the R&D of autonomous driving technology. In 2007, FAW cooperated with the National University of Defense Technology (NUDT) to manufacture a prototype based on Hongqi HQ3 that could drive autonomously on the highway. On July 14, 2011, unmanned vehicle Hongqi HQ3 completed 286 km of a unmanned highway driving from Changsha to Wuhan, spending 3 h and 22 min at the average speed of 87 km/h. Only 0.78% of distance of travel was handled through human intervention, setting a new record for China's self-developed unmanned vehicles in complex traffic conditions. In 2013, FAW launched Hongqi H7, a car with PA functions. The Hongqi H7 is equipped with ESP (9th generation from Bosch), ACC, LDW, FCW and other technological equipment, providing drivers with a relaxed and enjoyable driving experience.

In April 2015, FAW officially released its "Smartway" technology strategy, making it one of the first domestic OEMs to issue a complete ICV strategy. Facing the "Internet+ " trend in the industry, FAW is committed to make "Smartway" a leading ICV technology in China.

In order to achieve such a goal, FAW established five technical support platforms, namely, an integrated technical platform based on "Internet+ ", combining design, manufacturing and service; a powertrain and chassis mechatronics technical platform for energy saving and new energy vehicles; an embedded vehicle and assembly control software technical platform; an intelligent mobility technical platform; and a D-Partner+ information service technical platform.

In April 2015, FAW held a seminar to let participants experience "Smartway" technology by interacting with the real car in Tongji University, giving people their first chance to experience ICV technology firsthand. This experience mainly included four intelligent functions including "mobile car hailing, automatic parking, following at congestion and autonomous driving".

FAW has formulated a detailed technical development plan to ensure that "Smartway" evolves from the current "Smartway 1.0" to "Smartway 4.0" over the next 10 years. At present, "Smartway 1.0" has been applied to Hongqi cars, with ADAS functions such as ACC, AEB, LDW and FCW. "Smartway 2.0" will be available in 2018, and Hongqi intelligent connected passenger cars and Jiefang intelligent connected commercial vehicles will be unveiled, featuring intelligent functions such as temporary single-task intelligent piloting and D-partner 2.0 information service, and completing the layout of the ICV ecosystem. In 2020, "Smartway 3.0" will be available. Highway self-driving, in-depth perception and urban self-driving technologies will also be released, including functions like multi-task long-term piloting and smart city solutions. By 2025, "Smartway 4.0" will be available, and the commercial operation of the smart service platforms will be realized. The penetration rate of vehicles with HA technology will also reach over 50%.

3.8 Dongfeng Motor: A Unique Approach to Cross-Border Cooperation

Following trends in automobile intellectualization, Dongfeng Motor has planned a new layout and approach suitable for its own goals.

In October 2014, Dongfeng and Huawei signed an agreement to start cross-border cooperation in the hope of accelerating the research process of vehicle-to-human, vehicle-to-vehicle, vehicle-to-cloud connectivity, making smart cars and V2X a reality, and enabling smart cars to ride emotionally and autonomously. Dongfeng focuses on product integration and application, while Huawei focuses on development and production. In the future, Huawei's new product ME909T, featuring high compatibility, integration, quality and stability, will be installed on Dongfeng's communication-enabled autonomous vehicles. The strategic cooperation between Dongfeng and Huawei will be carried out in three steps. The first step is that Huawei develops WindLink products for Dongfeng, which have since been applied in Dongfeng passenger cars, such as Fengshen AX7 andAX5. The second step is to jointly plan future V2X products to make vehicles intelligent and connected. The third step is to integrate resources to create a truly unmanned vehicle.

In-vehicle smart interconnection system WindLink communicates with vehicles through electronic equipment to enable wireless communications between people, vehicles and the world. It can provide 29 services in 9 systems including security protection, remote control, speech recognition, and artificial navigation. WindLink's speech recognition system enables voice control over navigation, air conditioner and various applications, and utilizes nuance speech recognition technology. It works through local and cloud technology, and is complemented by industry-leading dual-MIC hardware noise reduction to greatly improve the accuracy of speech recognition. In terms of security, the WindLink smart protection system is online 7–24, with functions such as one-touch SOS, automatic SOS when the airbag bursts, and vehicle tracking (to assist the police). In terms of remote control, WindLink can also control the vehicle through a mobile app, such as turning on the AC (for heating, cooling, defrosting and defogging) before entering the car, lifting or lowering the windows, and locking and unlocking the doors, improving driving comfort and safety.

In early 2017, Dongfeng, Huawei and China Academy of Transportation Science (CATS) held a tripartite seminar in Beijing on the world's first and new-generation connected autonomous driving technology based off LTE-V/5G, with demonstration. The three parties discussed future development direction, and reached a preliminary agreement to cooperate on "Technical Verification and Trial Operation of Intelligent Connected EVs". The seminar demonstrated applications such as car hailing through a mobile phone, dynamic speed limiting, cooperative lane changing, going through tunnels, intelligent intersection scheduling, dynamic formation, dynamic path planning and automatic parking. This activity shows that tripartite R&D on ICVs has yielded initial results. Through this endeavor, the three parties have developed key technologies towards the latest and most cutting-edge ICVs in China, enabling small-scale demo operations. Further cooperation will be carried out, with

Dongfeng focusing on vehicle control research, Huawei conducting communication construction and CATS building the test environment. These ICVs are expected to be demonstrated in the next three to five years.

At present, domestic and foreign auto giants, IT and telecommunication tycoons have joined hands to explore the development paths of future automobiles. As pillar enterprises in the national industry, Dongfeng and Huawei's strategic partnership will work towards further future development.

3.9 BYD: An EV Icon and Its Way to Intelligence

Since its humble beginning as a producer of mobile phone batteries, BYD "accidentally" broke into the automotive industry, achieving remarkable results since then. BYD has seen 20 extraordinary years, with a mixed bag of failures and successes. In this period, it suffered from excessively expansion, but began to take off following the electrification of the auto industry. From being ridiculed to becoming an industry leader, BYD proved the hard power of being a "Chinese Brand". The new energy empire that BYD is building is also an example of the soft power of "Made in China".

Now, BYD is transforming itself from a carmaker into a provider of low-carbon travel services. In early 2013, BYD and Beijing Institute of Technology (BIT) signed a cooperation agreement. More than three months later, they announced the birth of a co-developed experimental wire-controlled autonomous car, which implemented the external CAN bus control of the chassis and used LiDAR. The car won first place in the national unmanned vehicle competition that year.

In February 2014, BYD signed a cooperation agreement with the Institute for Infocomm Research (I2R) under Agency for Science, Technology and Research (A*STAR) of Singapore. I2R is the largest information and communication technology research organization in Singapore, with strong capabilities in intelligence and communications systems. The two parties planned to consolidate their respective resources in the fields of electric vehicles and autonomous driving, and jointly establish laboratories to build smart EVs, with plans to use electronic wire-controlled driving, and control of other systems. In addition, BYD provided more than 100 electric taxis and partnered with I2R to promote ITS in Singapore. I2R would carry out intelligent monitoring and management of driverless EV fleet according to real-time vehicle and road conditions, to create future-oriented intelligent transportation and develop the management system of autonomous cars.

In January 2016, Wang Chuanfu, Chairman and President of BYD, publicly stated that it would team up with Baidu to develop unmanned driving technology and use data from Baidu's HD maps to build the future V2X model. This legendary figure in auto industry also pointed out that BYD is now more focused on reducing car accidents through intelligent control and increasing the existing traffic safety rate by tenfold, which is at the top of its development agenda.

In April 2016, "Yuan" followed "Song" as BYD's second model equipped with Baidu's CarLife solution. CarLife is a service platform launched by Baidu V2X to connect cars and the mobile phones. By connecting your mobile phone to the car, the two screens synchronize to enable voice interaction, basic navigation, music entertainment and other functions. At the World Internet Conference held in Wuzhen in November 2016, an autonomous vehicle, co-developed based on BYD's "Qin" by Baidu and BYD, was unveiled togreat fanfare.

BYD reveals that it has developed a strategic plan based on V2X. The plan is divided into three phases. The first phase of "cloud services" has been implemented, and second phase "cloud computing" is under development. The third phrase will be "intelligent and connected." In the "cloud computing" phrase, BYD identifies three key aspects, "terminal, network, and cloud". "Terminal" refers to the automobile terminal. "Network" means the interconnected network, and "cloud" is BYD's private cloud. With regards to the automobile terminal, it ensures that all sensors are redefined for the cloud terminal, requiring that all sensors be open and tested for V2X. Then, sensor signals are owned by consumers and developers, which can be used on mobile phones to create truly open cars. In terms of the construction of V2V network, first-generation product Carpad is already in production. The second-generation product Cloud Carpad has completed fusion with mobile phones, while the third-generation product combines various information received in the sensors on the vehicle with V2V platform. The fourth generation consists of a self-learning system, in which the vehicle acts as the "brain". As for a private cloud, BYD has built a cloud platform that aims to deliver the functions users want. During the driving process, the in-vehicle sensors will collect the user's preference information as much as possible, and the big data generated and analyzed will customize the car accordingly. Through the creation of "terminal, network, and cloud", BYD will build a platform of connectivity to allow everything to be connected.

3.10   Foton: Building China's Super Truck

Compared with passenger cars, commercial vehicles and especially trucks, could use the convenience brought by V2X and autonomous driving. This is directly reflected in transportation costs, and is related to operational efficiency. From this point of view, CVs rely more on V2X and smart driving than PVs. As the most diversified and largest CV maker in China, Foton has long known and planned for this. According to its strategy "Foton Auto Industry 4.0", Foton plans to build a customer-centric system based on data from V2X, big data and cloud platforms, management its enterprises intelligently through big data, and enable large-scale customization services through smart products, factories and manufacturing. Foton is confident in future development. As it believes that since CVs are different from PVs in nature, it also believes that China's CVs are likely to challenge or even replace competitors from Japan and South Korea in the future. If China can cultivate world-class CV makers, Fotonwill be one of them.

During the "12th Five-Year Plan" period, Foton invested 10 billion yuan on verification and testing, and R&D personnel to establish a two-level R&D system and hardware facilities. These investments have greatly improved its technical level, in terms of product reliability, intelligence, safety, and electronics. This has laid a foundation for the "13th Five-Year Plan", facilitating its goal of becoming a world-class enterprise in that period. In the meantime, Foton's global layout has gradually taken shape. Autonomous driving technology is the highlight of Foton's internationalization and global layout. The "super truck" built by Foton with Daimler's DNA has achieved all-round breakthroughs in R&D, quality and manufacturing.

2016 was a vital year for Foton "Super Truck". In January, the first concept truck was launched held in Guangzhou. In April, Foton's "Super Truck" was announced at Auto China auto show. In June, Foton teamed up with Daimler, Cummins, ZF, Continental, and Baiduto set up the Super Truck Global Innovation Alliance in Europe. In September, Foton "Super Truck" made world debut in Hanover, Germany (Fig. 7), withstanding rigorous tests by German testing agencies like ATP, DEKRA and FEHRENKOTTER under the supervision of European professional commercial vehicle journal Tracker. It won the attention of truck users around the world. In early November, Foton invested in the establishment of the first Super Truck Global Innovation Center with estimated cumulative investment over 10 billion yuan.

On October 10, 2016, Foton signed a strategic cooperation agreement with Baidu in Beijing, targeting a comprehensive cooperation on V2X, big data, intelligent vehicles and autonomous driving to create the future intelligent connected commercial vehicles. This cooperation autonomous driving aims to achieve Level-3 automation, based on Baidu's HD maps. The two sides will jointly launch a family of autonomous "Super Trucks" including heavy, medium and light ones.

With Foton as its partner, Baidu will enhance its data collection capabilities and improve big data on CVs. The two parties will jointly develop CV-bonded MyCar,

Fig. 7 World debut of Foton "Super Truck" with V2X features

Co-Driver and Car-Guard. On this basis, they will cooperate to build a CV V2X operation service platform, sharing operating income. Through technical cooperation with Baidu in V2X, Foton Motor will also establish a data cloud platform to analyze the user's usage and optimize the monitoring program. This will not only save fuel, but also reduce any of the CV's hidden dangers. In addition, V2X technology will also impact Foton's financial services. The big data platform will detect the user' behaviors and remotely control the vehicle through the V2X to avoid possible financial risks.

Foton self-driving vehicles are equipped with the "iFoton", a V2X cloud platform, to achieve real-time communication with outside. Radars and cameras are used to identify the surrounding environment, and their fusion with HD maps allow for vehicle positioning. Self-developed control strategies complete the driving process through integration of core technologies of low carbonization, informationization and intelligentization. Foton "Super Truck", labeled as "environment friendly, efficient, energy-saving, safe and intelligent", aims to reduce vehicle fuel consumption by 30% (or BEV), reduce vehicle carbon emission by 30% (or zero emission), and improve freight efficiency by 70%. According to reports, Foton self-driving trucks are expected to reach mass production in 2025.

Extended Reading

"Super Truck" Project

In January 2010, the US Department of Energy announced research on advanced technologies to improve fuel economy and transportation efficiency of heavy-duty trucks in the US, and set up a "Super Truck" project to achieve energy-saving and emission reduction for Class 8 heavy-duty trucks. The project was implemented over five years, with a total investment of $270 million. Through cooperation with excellent heavy-duty trucks and powertrain manufacturers in North America, the government and enterprises jointly funded innovative research in the fields of vehicle and engine, and finally achieved the goal of reducing the total fuel consumption of a single vehicle by 50% in 2015. Having realized the importance of vehicle diesel for energy security, energy conservation and emission reduction, the US and Europe set up national "Super Truck" strategic projects. In Europe, the "Super Truck" project, represented by Germany's Daimler Benz, aims to "Reshape Transportation in the Future". Inspired by this, Foton has integrated global resources to build connected "Super Trucks" that combine autonomous driving, new energy and V2X.

Chapter 9
A Brighter Future Beckons

The auto industry is entering an unprecedented period of change, as we strive to build the bones of the new from the ashes of the old. Even our understanding of what connotes an "automobile" has radically transformed. No longer necessarily powered by a gas engine, an automobile may run on electricity or even solar energy. Furthermore, it may become "the fourth screen" around us. Change is inevitable for automotive products and companies, and even the industry or sector as a whole. In 2016, Carlos Ghosn, then CEO of Renault-Nissan, optimistically predicted "major technological disruptions in the next five years, which will change the product mix. You'll see more electric cars, more self-driving and more interconnectivity." Mark Fields, former CEO of Ford, has echoed Ghosn's stance on the field, declaring that "automated driving will define the next ten years. We predict that it will bring great impact to the society, just like the production line that Ford invented more than 100 years ago." Field's association between autonomous vehicle technology and assembly line perhaps highlights the significance that Ford—and the rest of the auto industry—has invested into this technological advancement.

In 2015, Goldman Sachs—the world's leading investment bank—published a report on the auto industry of 2025, predicting the global auto industry to soon undergo a profound transformation within the next decade, with technology as its catalyst. Automobiles, manufacturers and consumers will all bear witness to this dramatic change.

The 2015 report identifies four key technologies that will shape this change, predicting that cars of the next decade will be more environmental-friendly, convenient, safe and affordable. Even amid consumer and regulatory pressures on automakers to reduce carbon dioxide emissions, the transportation industry accounts for 22% of global greenhouse gas emissions. At present, there are 37 mega-cities in the world. Nine of them are in China, with an average population of 10 million or more. From 2010 to 2025, the world's urban population will increase by 50%, and cities will continue to grow as traffic conditions deteriorate. Considering that the average car remains idle for more than 90% of its life time in tandem with the rising cost of car ownership, the efficient matching of cars with users who need them at

© China Machine Press, Beijing and Springer Nature Singapore Pte Ltd. 2021 289
Z. Chai et al., *Autonomous Driving Changes the Future*,
https://doi.org/10.1007/978-981-15-6728-5_9

that moment becomes a prudent business opportunity. Reducing traffic accidents has always been a priority for the auto industry; this only becomes even more pertinent as the average age of the global population increases. A projected 10% of the world's population will be over 65 in 2025, increasing safety needs. In most economies, car ownership grows considerably when per capita income rises from $10,000 to $20,000—a level which many developing countries will reach for the first time by 2025, generating huge demand for smaller and cheaper cars. Goldman Sachs reports global automobile sales potentially reaching 120 million units in 2025. With 35 million projected purchases, China would continue to lead the global market; while India's projected 7.4 million units would gain it the title of the world's third largest auto market.

Goldman expects global emissions regulation and continued pushes toward CO_2 reduction to play an essential role in shaping the auto industry between now and 2025—particularly through thermal efficiency, lithium ion batteries and fuel cells.

Goldman Sachs identifies seven major changes that will dominate the next decade of the automotive sector. These include hybrid and electric vehicles (EV), advanced lightweight materials, autonomous driving, supply chain changes, new competitors, connected vehicles and a shift to emerging markets. Five of these seven traits (apart from a shift to lighter vehicles and focus on emerging markets,) can be subsumed under the larger category of autonomous vehicles. Each of these five traits has been discussed extensively in the past chapters of this book.

On September 19, 2016, then-US President Barack Obama published an article entitled "Self-driving, yes, but also safe.", This article announced the White House's new policy for self-driving cars on the road, and called on the government and private companies to address the application of this new technology. In the article, Obama argues that the concept of driverless cars from science fiction movies is gradually becoming a reality, and the accessibility of safer driving tools contributing to less congestion and pollution. He states "this technology is not just about the latest devices or Apps—it is about making people's lives better."

On the premise that autonomous driving will continue to advance, the role of traditional automakers must also be redefined, shifting from manufacturer to travel service provider. As a sector, the auto industry chain will expand and widen in scope, greatly extending its downstream sector. The whole industry will be reshaped and integrated across borders, strengthening the position of its dominant players. The participation, interaction and entanglement of these forces will profoundly affect our future travel, life, society, energy, environment, and science and technology.

The auto industry is capital-intensive, especially with regards to autonomous driving. Whether an emerging player or veteran corporation, the company that masters practical driverless technology will undoubtedly reshape the future of the industry, and become its face—just as Microsoft has become ubiquitous with Windows and its PC monopoly, or Apple with its iPhone. If humankind's basic need for travel were to be monopolized by a single company, it would likely become the first trillion-dollar corporation in human history.

The continuous development of autonomous driving technology will also greatly impact employment, business models, and even government finances. For example, there would be a downtick in enrollment to driving schools, and shrinking markets for partner training. It would make null designated drivers, taxi drivers, and auto maintenance and insurance. It would even enable a decline in hospital orthopedics or reduction in fines for traffic management departments. MIT scholars Andrew McAfee and Erik Brynjolfsson explain this in the book "Race Against the Machine": The advances in computing technology included in Google's smart cars represent the next wave of new technologies that will reduce job opportunities for the whole of society. This would render the value of many skills once considered very secure—such as driving—worthless.

PricewaterhouseCoopers (PwC) predicts that autonomous driving will leave only 1% of the cars on the road, reducing the number of cars on the road from the current 245 million to 2.4 million. According to their predictions, most traditional automakers will shutter by 2030. Peripheral industries such as the auto insurance market (current market value (CMV) an estimated $1980 billion,) the auto financial market (CMV $98 billion,) the parking market (CMV $100 billion,) and the after-sales market (CMV $300 billion) would also be eliminated.

Currently, houses and cars are the two most valuable means of living for the average person to own. With the advent of the autonomous driving era, both will undergo subtle changes. Real estate values change based on factors such as proximity to workplaces and schools. If people can work effectively anywhere or are equally productive even on the way to and from work, they may prefer to move out of cities and live in the countryside. At the same time, urban transportation infrastructure must transform to meet the demand of autonomous driving, impacting building design concepts (likely to include fewer parking spaces) and real estate valuation. Car sharing will become a deep-rooted institution, available to anyone with a smartphone in their hand, making personal cars more or less obsolete.

No one can accurately imagine what the future of autonomous vehicles looks like, but it surely behooves prediction. The era of autonomous driving has already arrived in inexorable ways. Though it still facings challenges in fields such as technology, infrastructure, regulations, business models, and ethics, autonomous vehicles will surely overcome these challenges, riding the wave of progress to a bright future. As Li Bai, a great poet in China's Tang Dynasty, has espoused, "Hard is the Way of the World," eventually "the time will come to ride the wind and cleave the waves, hoisting my cloud-sail high, I venture the deep blue sea."

1   "Black Technologies": Safe, Cool and Environmental-Friendly

Autonomous driving will not only bring us a safer traffic environment, but also save energy and reduce pollution. It will also provide a cool technological experience.

Technologies have developed far beyond our imagination, and the automobile has become a place where engineers can let their creativity flourish. Some ideas are closer to reality while others are further away, but who can say if whether or not they will one day enrich our experience behind the wheel? We will take a look at some of these vehicular "black technologies" below.

1.1   Biometric Technology

According to the report "Biometrics in the Global Automotive Industry, 2016–2025 "on health, wellness, and wellbeing (HWW), advanced driver assistance systems (ADAS), and security for biometrics in the auto industry, nearly one-third of automobiles will be equipped with biosensors to control the vehicles through biometrics by 2025. With the development of smart car technology, biometrics is both more common and more commonly used in automobiles. In addition to speech and fingerprint recognition, advanced biometric technologies such as gesture recognition and facial recognition are gradually becoming prevalent. These technologies substantially improve driving experience, occupant health and driving safety.

Continental AG has launched vehicle-borne biometric recognition technology, which enables drivers to personalize seat position, rearview mirror angle, music, temperature, navigation and other on-board functions. Through facial recognition, the car can automatically restore the driver's setting preferences. The engine can only be started after passing fingerprint verification, greatly improving car safety. Bosch's biometric technology is similar. When the driver enters the car, a camera quickly recognizes their face and customizes their experience. The system is controlled by a tactile feedback display and an innovative gesture control system, both of which provide the driver with real-time perceptible feedback. If the system finds that the driver is asleep or distracted during driving, the car issues a warning to help prevent a accident.

1.2   Virtual Touchscreen

At the 2017 Consumer Electronics Show (CES) in Las Vegas, BMW exhibited a virtual touchscreen system called the BMW Holo Active Touch (Fig. 1) that brought the scenes from the sci-fi movie into the real world. The virtual touch screen suspended in the air becomes an interactive interface between the driver and his car. The driver operates through gesture control and tactile feedback, while in-vehicle cameras recognize the fingertip movements. When the fingertip touches the virtual floating screen, a pulse signal is sent to activate the corresponding function. It combines the advantages of BMW's head-up display system, gesture control system and touchscreen operating system. The full-color interface that emerges in the air is formed by the principle of reflection. It is the first human-computer interaction system ever to be controlled without physical contact, and unique in that it preserves the visual and usage experience of traditional touchscreens.

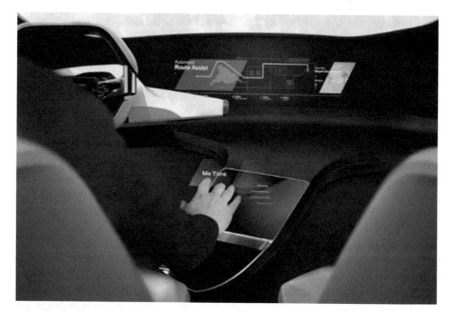

Fig. 1 BMW Holo Active Touch virtual touchscreen. *Source* BMW

Fig. 2 Mercedes-Benz's IAA concept car. *Source* Mercedes-Benz.

1.3 Flexible Material

Vision Next 100, a concept car meant to represent BMW's next 100 years of technologic advancements, adopts a unique technology—a kind of flexible material called Alive Geometry, which is mainly used for the car's dashboard and front fender. The car's tires are completely hidden inside the car housing. This flexible material looks like the scaly hide of a living creature. When the wheel turns, the "scales" formed by the material stretch and flex to accommodate the tire positioning. BMW's official explanation points out that this design helps to reduce the car's air resistance during driving. Of course, those who fear reptiles might want to stay away from it.

1.4   Deformable Cars

Mercedes-Benz's IAA concept car (Fig. 2) debuted at the 2015 Frankfurt Motor-Show showcasing numerous black technologies, the most powerful of which is deformation. IAA is the acronym for "Intelligent Aerodynamic Automobile", thus the car was named "Intelligent Aerodynamic Concept Car". The IAA transforms automatically at 80 km/h or at the press of a button, while moving, from its sedan-like design mode to an ultra-sleek, gas-powered fish of a car. When it enters an aerodynamic mode, eight segments at the back of the car extend out, adding over 15 inches to the overall vehicle length, extending its teardrop-shaped tail. At the front of the car, flaps in the bumper extend out nearly an inch, while smaller flaps extend on the back bumper. These keep air away from the wheel arches, where turbulence can become an issue. Of course, Mercedes has also fitted the car with "Active Rims" that "alter their cupping from 55 mm to zero."There's even an active aero aid inside the front bumper, where a louver extends backward by 2.3 inches to smooth out the air traveling underneath the car. Those features cut through air resistance, dramatically reducing the car's drag coefficient to 0.19, the lowest ever in the world.

1.5   Dual Thumb Controls

Mercedes-Benz's Dual Thumb Controls (Fig. 3) replace traditional control buttons with sensitible touch control technology on the steering wheel. It only needs the fingers to come in contact with the induction touch panel on the steering wheel

Fig. 3 The Dual Thumb Controls. *Source* Mercedes-Benz

to "fully control" the dashboard and multimedia system on the display screen, so that the driver can focus only on the road and steering wheel when driving. The left-hand button adjusts the function options on the dashboard, while the right-hand button manages the central display. The system has perfect control sensitivity and accuracy, so that commands can be input by slightly sliding the thumb horizontally or vertically. The return / home button next to the touch panel can also be used to return to the main menu. Drivers can even adjust the touch induction speed into three levels—"fast, medium and slow"—according to their preferences.

1.6 Transparent Vehicle

Usually, when driving on a flat road or uphill, the engine hood will obstruct the driver's view, a potentially hazardous position. The Transparent Bonnet system (Fig. 4) developed by Jaguar Land Rover uses the camera to project the image blocked by the engine hood onto the windscreen, creating an "illusion" of transparency in the bonnet. The Urban Windscreen system is another black technology developed by the same company, and is a combination of camera, ultrasound, radar and laser ranging technology. It eliminates blind spots by building a "transparent" C-pillar, turning the car into a see-through "ghost", and detecting objects within 5 m around the car at 360 degrees. This includes the space above and below, allowing it to detect suspended branches or obstacles of certain height, identify terrain such as asphalt, grass, gravel, sand or snow, and transmit the information to the "terrain feedback adaptive system"

Fig. 4 Transparent Bonnet System. *Source* Jaguar Land Rover

Fig. 5   Color-changing windows. *Source* Continental AG

to automatically change the driving mode. If the terrain is deemed too difficult to pass, an alarm will be sound.

1.7   Color-Changing Windows

Honda and Continental AG are developing a color-changing window (Fig. 5) that can be adjusted with the touch of a finger. The user can start the color change by touching the window, and adjust the coloration by sliding their finger over its surface. The color results from the transmittance of a layer of light across the glass. This function can not only replace the sun visor of the windshield, but can also be used for side windows to eliminate the need for bulky blinds and their motors. The only drawback is that it may result in fingerprints and buildup on the glass, much like a smartphone screens. Better keep a bottle of Windex in the trunk.

1.8   Streaming Video in Rearview Mirror

Traditional glass rearview mirrors have two problems. One problem is that the field of view is affected by the shape of the rear window of the vehicle. The other is that any passenger sitting in the center rear row will block said window and pose a safety hazard. GM first launched the "Streaming Video in Rearview Mirror" on the production car in 2016, replacing the traditional glass rearview mirror with the "camera + LCD display" (while retaining the traditional mirror function). The video image replaces the mirror reflection, expanding the field of view by 300% (80~140°) and eliminating rear blind spots. It not only has higher definition and a wider field of vision, but is also unaffected by the size of the rear window, and the occupants. However, as the camera is located externally, it may be obstructed by dirt or a target of vandalism. Therefore, the Streaming Video in Rearview Mirror is somewhat risky in some countries and regions with poor security or environment.

Additionally, BMW launched a Mirror less Concept Car based on it Model i8 line at the CES 2016. It's evident from name alone that this car is devoid of mirrors.

2 Transport Ecology in the Future: Harmonious, Efficient and Shared

If the industrial revolution increased productivity, the development of autonomous driving technology shows promise towards large-scale social and urban transformation. Autonomous vehicles can free individuals from the physical and mental task of driving, allowing the car to become an extension of one's home or office. Additionally, this would increase accessibility to the elderly and disabled.

If autonomous vehicles were to become reliable enough, the government may one day ban humans from driving. In the future, passengers sitting in driverless cars might even view human drivers as "old-fashioned," dangerous and even barbaric.

Were a driver no longer behind the steering wheel, she would likely occupy herself with work, a movie, games or even a nap. With her attention no longer concentrated on road conditions, other vehicles and traffic, the car becomes a peaceful space, isolated from the anxiety on the road. By combining autonomous vehicles with intelligent transportation and cloud computing, we can build an intelligent urban vehicle command and dispatch service center, share traffic resources, achieve optimal travel planning, and turn cars into smart moving nodes that provide services for the public. This in turn could greatly reduce the number of cars in the city, ameliorating traffic conditions for a more peaceful city.

By that time, people would likely be more willing to support the government in readapting urban streets. This would result in the gradual transformation of conventional lanes and street parking into bus lanes, bicycle lanes, small parks and wider sidewalks. Road design would become more pedestrian-friendly, such as longer crossing time and sidewalks without curbs. The popularity of autonomous vehicles would also mean that government contributions to transportation infrastructure such as ultra-wide lanes, guardrails, speed bumps, wide shoulders and even stop signs could be greatly reduced. In short, autonomous vehicles give planners and engineers the opportunity to create a more comfortable and pleasant urban environment. A city will require autonomous vehicles to use centralized parking, refueling and charging facilities, thus allowing planners to reduce parking spaces in most new buildings. This will significantly reduce the construction cost and rental fees of the buildings, making the city more affordable for both individuals and businesses. At the same time, it will also increase the city's available space.

Furthermore, autonomous driving will have another effect—the demise of traffic lights. Traffic lights were originally designed for horse-drawn carriages—an era far-removed from that of autonomous driving. The function of traffic lights is to convey a visual message to drivers entering the intersection: stop at the red light and go at the green light. A transportation system realized by autonomous driving technology and supported by the Internet of Vehicles would allow each vehicle to exchange wireless information almost at the speed of light, easily communicating its position and statistics in real time. Traffic lights would thus become obsolete.

Researchers at the MIT Sense able City Lab have studied the urban effects of autonomous driving. One of the great effects of driverless cars on urban life is to blur the boundaries between private and public transport. For example, every morning, "your" car will carry you to your company, and then send other family members or other unrelated strangers to their destinations. Cars will no longer remain idle in the parking lots for more than 20 h, but move around to transport passengers by relying on autonomous driving technology. By doing away with the concept of a "private car" in the future and instead using seat- and car- sharing, each passenger can arrive at their destination on time, while reducing the number of cars on the road by 20%. For large cities like New York and Beijing, the effects of reducing congestion are significant. Secondly, if the number of cars on urban roads were reduced by 80%, the impact on future urban development would be massive, especially when considering factors such as the environment, transportation, traffic efficiency and parking. After reducing the number of vehicles on the road, land originally used as parking spaces could be released and reused. Of course, the cutting of vehicle population would also reduce the cost of road construction and maintenance (and related energy consumption). An engineering study shows that autonomous driving technology can quadruple the capacity of existing highways.

We may not fully accept the argument of these MIT researchers—such as their hypothesis on the decline of private cars—as high-income, suburban and rural residents continue to have strong practical demands for cars. However, in densely-populated urban areas, where most people do not own a car, it is very likely that the driverless cars will function much like Uber, as a solution to the residents' travel demand.

A study conducted by Columbia University shows that were Uber to replace all the taxis in New York with 9000 unmanned cars, the passenger's waiting time would be reduced to only 36 s, at a price of about $0.5 per mile. With Uber's current valuation, it is entirely capable of replacing 171,000 taxis across the United States. If each car costs $25,000, it will cost about $4.3 billion to replace all the taxis.

Perhaps in the future, autonomous vehicles will be seen as the next generation of smart hardware—a kind of mobile device similar to today's smartphones, if a bit larger. As a mobile node or end device for car sharing and intelligent services, every autonomous vehicles in the same area can be controlled by the same super smart terminal. Theoretically, the departure place and time and destination of any car can be pre-determined. This would make it entirely possible to intelligently and dynamically assign a unique path to the car within a given time period, meaning its probability of running into traffic jams and accidents would be close to zero. In this way, the efficiency of the entire transportation system is all but guaranteed. All cars can travel at the same speed on roads with the same specifications. "Car trains"— dense formations of high-speed driverless cars—would make efficient use of road space by operating at the same fast rhythm as an automatic production line in a factory. In the event of an accident, all the vehicles on the road can brake simultaneously to avoid follow-up accidents.

Based on the concept of urban super smart terminal, some people have even further envisioned that the changes to the production and maintenance of automobiles could be disruptive. Extant autonomous vehicles can detect the surrounding environment by integrating data from cameras, radars, LiDARs and so on, operating as an independent computer. If every autonomous driving unit in the same city were to be controlled by one urban super smart terminal (much like the many processes running in parallel in an operating system) then most of the sensors on each vehicle can be removed. Terminals would communicate with each other, much like mobile networks to ensure everything runs smoothly. In this way, the chassis of the car is stripped from its control system and moving components, as well as the car body. An ordinary 2–4 seat car uses a chassis and a car body; a 6–9 seat car uses two tandem chassis and a larger car body; a medium-sized bus uses 4 chassis; a freight truck uses 8 chassis; even a marine vessel or a freight train use chassis. Only one kind of chassis is to cover all kinds of ground transportation. By that same standard and specifications, the production and maintenance becomes extremely cheap and efficient. It is possible that, like the PC motherboard and CPU, there is no distinction between good or bad car, but only between usable and unusable car. The latter would be replaced immediately, because producing a new one could be cheaper than repairing even minor damage. However, this plan is easier said than done. For example, the smartphone and EV charging interfaces are not standardized. Though seemingly a minor plug dispute, it reflects a huge interest struggle.

The benefits brought by autonomous driving technology are also reflected in the reduced demand for parking spaces and subsequent traffic efficiency. In many metropolitan centers, up to 30% of daytime traffic can be attributed to drivers looking for parking spaces. If driverless technology is implemented, the demand for downtown parking spaces can be reduced and traffic flow can be maximized for efficiency.

What economic returns could autonomous vehicles generate? If every person does something economically valuable in lieu of driving, (assuming that each person gains an extra hour a day at the conservative estimate of $5 per hour,) then considering 100 million office workers, working 200 days a year, the overall economic benefits are at least $100 billion. In addition, if traffic is no longer congested and autonomous vehicles can run faster and connect with each other, the world's road capacity will greatly enlarge.

In the USA, the high fuel efficiency and extensive use of alternative fuels within autonomous vehicles will likely result in the reduction of highway discretionary funds, making many branches of the US interstate highway system redundant.

This will also give planners the basis to dismantle and reduce extraneous freeways and suburban main roads, allowing for the extra budget to be allocated to other infrastructure, such as affordable, efficient mass transit public transport.

The era of autonomous driving seems to be serene, but the picturesque ideal holds minor flaws. One potential problem might be if a commuter who owns a private driverless car travels from the suburbs to the city centre every morning, they may ask their car to drive back home or circle the streets, in order to avoid paying for parking—and this will no doubt have a negative impact on congestion and emissions. The best way to solve this problem is to charge a relatively high road usage fee, similar to the new policy that has been considered in Oregon State and Washington State to replace the decreasing gasoline tax. Cities just like London and Singapore also need to impose congestion charging policies in the areas of highest density to prevent empty cars from wandering around the streets and wasting fuel.

Autonomous vehicles also affect the public transport system. Many track systems have now achieved driverlessness, but larger impacts will strike on conventional or BRT routes in large cities. This shift will happen quickly once public transport managers realize that replacing drivers with an autonomous driving system can save considerable costs. Wages and employee benefits account for nearly half of a bus company's operating budget. If this part of the budget is saved, it will greatly increase service frequency, passenger facilities and system capacity, and make public transport more accessible and affordable, especially for those who can't or don't want to drive. It is true that the planning and maintenance departments still need human employees. However, in the next decade or two, most buses will become completely driverless.

3   Imagine A Day in the Autonomous Driving Society

Thomas L. Friedman, a columnist for The New York Times who has gained global fame for his book The World is Flat, published a new book on global energy and environmental issues—"Hot, Flat and Crowded" in 2008, which depicts us a future world after the green technology revolution. Writing as journalist but sounding much like a rabbi offering counsel, Friedman asked, "When do we feel best about ourselves as Americans? It's when we are doing things for others with others. Leading the green technology revolution would enable us to do just that." Friedman's advice to leave no carbon footprint gains integrity by his own personal commitments, which he described in some detail including building an energy-efficient home and driving a hybrid vehicle.

In that world, electricity is the universal standard energy source that replaces thermochemical energy, and each user has the dual functions of user and supplier, meaning everyone is involved in the energy revolution. Friedman wrote in his book, "When you arrive at the office, your car is parked in a parking lot where your car battery can be recharged or you can sell electricity to the grid… at 2:32 pm, when the temperature reaches 87 Fahrenheit, your car battery still contains most of the electricity charged last night. Through calculation your car judges that the electricity price on the smart grid has reached its peak and it is a good time to sell electricity. Your smart car first calculates your general electricity needed to get off work on Wednesday—including sending kids to the football field and stopping at the supermarket to buy things—plus additional 10% of the electricity in case you change the original route and then it sells the surplus electricity for $0.40/kWh. The power company purchases 5 kWh of electricity from your car through the parking lot interface. This not only allows the power company to meet the peak load demand, but also maintains a flat load curve for the entire system. You earned $2, the solar panels and the parking lot that connects your car to the grid have also earned a small portion. Your car has earned $24 for you by storing and selling electricity this month. At the same time, this month's car battery charging cost is only $47, because you choose to charge at night when the electricity price is low. This means that you probably spend about $1.50 per gallon".

Friedman considered the car's electrification and intelligence trends in his vision of the future when he wrote this book nine years ago. However, he likely did not realize that the development of self-driving cars and car-sharing would be so near, nor so popular. In the future, the car may no longer be private property. Its earnings, (far more than the "chump change" described by Friedman) could even come closer to balancing its operating costs while alleviating pressure on the electrical grid plans. Optimistically, these driverless cars will become common by 2025, far sooner than Friedman (and anyone else) could have imagined in their wildest dreams.

The Boston Consulting Group's research predicted that intelligent vehicles would hit the roads on a large scale in 2017; the created market value is predicted to reach $42 billion by 2025. At the time of the report, the cost of vehicle-equipped driverless technology ranged from $2,000 to $10,000. As utilization rate increases over the next 10 years, it will fall by 4% to 10. By 2035, it is likely that 18 million vehicles in the world will be equipped with some autonomous driving functions, and another 12 million units to become fully self-driving. By then, China will be the largest market for these vehicles.

There are four necessities of modern daily life: "clothes, food, shelter and travel." Though the former are already colloquially known as the three necessities of life, for the modern person, one could argue, "travel" is equally important. (With the exception of housebound individuals)Due to the rapid advancement of autonomous driving technology and the popularity of car sharing, the field of "travel" is expected to undergo significant changes, bringing with it a revolution in the way that each of us goes about our daily lives.

What does the autonomous driving society mean in a practical sense for a citizen of the future? Imagine, if you will, a day in the life in the year 2027.

Your day starts at midnight. It's Friday. You've just returned from a trip abroad on a long-distance flight. The plane arrives late due to bad weather. By the time you get your luggage, it's already 1:00 AM and your connecting flight has left without you. The next flight is at 8:00 AM. You can't just stay in a hotel for the night—you promised to take your daughter to the beach tomorrow—no, later today. Instead, you call a driverless taxi through the app on your phone. Ten minutes later, the autonomous car pulls up at the curbside, greeting you with its blinking lights.

You quickly set down your luggage and get in. This is a car of the future, without driver, without steering wheel, and even without brakes and accelerator pedal. The car's interior resembles a small living room, with seats that toggle between facing the road in front, or each other. The car windshield has been replaced by a huge OLED screen. Its augmented reality function includes all the features of a personal computer or a smartphone.

After you sink down into the back seat, an elegant figure appears on the screen, greeting you with a smile and the details of your trip. Meanwhile, the car rolls into the dark night, at the start of its hundred-mile journey.

"They" are not a real person, but a synthesis in VR—her appearance determined by the company, and automatically calibrated to the passenger's preferences. Backed by strong artificial intelligence, "they" behave like a human. "They" can hold a conversation—with good humor, or in all earnesty—all in accordance with the preferences of the passenger, whose background has long been excavated by big data, cloud computing and artificial intelligence algorithms.

Your data in the "cloud" has undoubtedly been scraped and used to create an online profile, then grouped with millions of others and put through a thorough AI algorithm for analysis and categorization. Though horrifying, it will be the reality of the future. Based on an agreement between the taxi company and the data analysis company, your personal information has already traveled through the OTA via the "elegant figure" in VR. In fact, it's almost fair to say that "they" knows you better than yourself. When you loaded the app up to call the car, the company already knew to arrange your favorite white SUV for you, with interiors washed with a soothing light blue color. To accommodate your long-haul midnight drive, the seats of this car also transform into a bed.

The VR figure also chats with you about the country you've just visited. "They" ask if you have visited the country's famous forest park, where visitors can travel by boat up the river to the famous waterfall that "flows straight down three thousand feet". "They" also express regret for the country's recently-impeached female president. This female president was unpopular amongst her people, who held her personally responsible for the economic depression—all indications that her impeachment was on her own account. However, the complex AI is aware of your political leanings and approval of this "Iron Lady" president: "they" echo your sentiments and commiserate.

The "elegant figure" poses a query. "Mr. Zhao, my records show that you've arranged for a white minivan for your next trip. It will wait for you at 8:30 AM, outside your residence at 316 Danling Street. Your final destination will be the beach. Seeing as you are still on the road so late today, I'd like to confirm whether or not you'd like to delay your departure time?"

"No, thanks. I can take a break in the car."

The "elegant figure" says, "Alright. We will transform the seat configuration into a bed. If that is all, then I will not bother you further. Good night." After that, "they" disappear from the screen.

At this time, the self-driving car reduces its speed. When the safety within the cabin is ensured, the seats automatically recline to become a comfortable bed. You are prompted to take out a pillow and a blanket from a previously-unseen compartment. The car then slowly accelerates to its normal speed. After nine exhausting hours on the plane, you now fall into a deep asleep. As you sleep, the car maps a route through cities and towns, communicating with other cars and road facilities through V2X. It discovers an accident ahead, where heavy rain caused a landslide an hour ago, shutting down two lanes on the three-lane highway. Though still passable, the smart car decisively takes a detour instead. Though an extra ten kilometers in distance, the time saved more than makes up for it. In the blink of an eye, five hours have passed. The OLED gradually brightens while soft music plays, waking you up. You're finally home. A wave of your smartwatch completes the payment for the ride. You retrieve your luggage and walk up the driveway. Your wife is already up to greet you at the door—thanks to an automatic alert from the car to her smartphone. After all, it is a night trip, and the self-driving car understands well the worries of the family.

Its job completed, the self-driving electric car seeks out a charging station to park at. Thanks to the substantial increase in the energy density of batteries, an electric vehicle with an endurance mileage of more than 500 kilometers is no longer rare. In any case, the car charges itself, sending its <vacant> status update to the company's dispatching network. From there, the company makes plans for its next trip based on proximity, for after the battery is fully charged.

It's a little past dawn, and the eagerly-awaited weekend family beach trip is here. Your 9-year-old daughter is overjoyed to see you. Your parents, who live in your home, lovingly watch their son and granddaughter play. After a while, your wife calls you in for breakfast. The meal was actually prepared entirely by a humanoid robot waiter, using its pre-programmed knowledge of nutrition, and following the personal preferences set by your wife. Now freed from household duties, she has more time to take care of her daughter and her studies. The kid's exam results have improved rapidly. To her slight embarrassment, she is praised by both teachers and classmates.

At exactly 8:30 AM, a white minivan pulls itself up to your house and waits quietly by the front yard. The temperature inside has been adjusted to a perfect 22 °C. You lead the whole family downstairs and walk towards the car. Just like last night's car rental, you enter the password sent to your phone on the pad on the door. The vehicle is now unlocked and at your service.

Three rows of spacious seating fill the 4.7 meter-long minivan. At the click of a button, the seats start to flip, fold, and shift—finally revealing a small meeting room, with a coffee table in the center. Upon closer inspection, you find that the coffee table actually consists of four screens, covered by a delicate yet wear-resistant glass. The "table" can be used in this form; alternatively, the monitors can each be adjusted, allowing passengers to surf the internet, read reports or watch movies on the go. Seating around the table holds four adults and one child. After the configuration is completed, the whole family enters "the car meeting room", and the car drives off to the beach.

The whole family enjoys the gathering, even in the car. Your daughter fires off question after question about your overseas trip, while you patiently describe everything you saw and did and ate and heard. As you talk, you also connect to the "cloud", selecting some of the photos taken you took to show to your family on the desktop screen. Everyone excitedly remarks on the beautiful natural scenery and the foreign cultures. The little girl can't help but show off a couple of the prettier nature photos taken on "Daddy's business trip" with her friends on social media. In just a short while, she receives a lot of thumbs-up on her Wechat moments.

The government's newly revised traffic regulations have already banned manned vehicles from getting on the road. However, a relatively broad grace period is still set to protect private property. Through careful policy guidance, manned vehicles are expected to be gradually eliminated. Most of the vehicles currently on the road are unmanned. What few manned vehicles are on the road are identified by the bolded signs installed into the body of the vehicle itself. This would allow them to be recognized by driverless cars via special electronic identification codes. Manned vehicles are only permitted to drive on specific road lanes, to maximize the separation of the two types of vehicles and improve traffic safety. The advanced AI allows autonomous vehicles to understand and predict human driving behaviors: a cyclist's hand-signals when he changes lanes, or a policeman's direction of traffic, etc. However, human unpredictability mean it would be advisable for autonomous vehicles to remain further away from human-driven cars.

Even in such mixed traffic conditions, vehicles could rely on V2V technology to communicate with each other at any time to maintain a relatively safe distance, greatly reducing the probability of accidents. The roadside communication facilities visible at any time along highways also rapidly transmit traffic information to each vehicle, allowing it to judge the route of travel and avoid areas of high traffic or emergency. Safety and efficiency are the most prominent features of intelligent transportation system.

Everything has been smooth sailing for the minivan—until suddenly, a human-driven car in the adjacent lane swerves into the lane, cutting your car off. The driver of the other car was about to miss his exit, and desperately tried to shuffle over in a panic. Though the two cars were only meters apart, your minivan swiftly maneuvers to avoid the would-be accident. Inside the car, the family experienced the slightest of swaying, and continued with their chatter, blissfully unaware. The minivan not only decisively and skillfully responded to avoid impending collision with the human-driven car, but also sent a warning to the surrounding vehicles through V2X communications: "that car is behaving erratically, please stay away from it."

One and a half hours remain until the minivan reaches its destination. The family has almost finished chatting, too. You suggest using the time wisely, to maximize your time at the beach. While your daughter does her homework, you work on your report on your recent business trip. You both start working on the monitors in front of you. As you write your report, you draw on graphs and pictures from the "cloud" to support your argument. Your daughter logs onto her primary school network and her class's group page and begins to work on her homework. Your wife keeps a watchful eye on her progress, while doing your grocery shopping for the next week from an online supermarket. The goods will be delivered only a couple hours after the order is placed; every aspect, from order entry, to goods collection, loading, and unloading all are automated through the Internet of Things. Decisions like which orders are loaded on which delivery vehicle; the route design, etc. are routed through network and processed by an optimization algorithm. The last "one kilometer" is delivered by a drone mounted on the truck. That way, even if nobody is at home, the groceries can be received by the robot helper. The robot will even report back to you after confirming the delivery!

Your own parents find something else to do as well. They watch an old classic from 2016, called "Hacksaw Ridge." Even watching the film again after ten years, the old couple is still deeply moved.

By 11:00 AM, your family has arrived at the beach. Everyone's gotten something done. Your business trip report, your daughter's homework and your wife's shopping list have all been sent or submitted. Now the whole weekend is yours, and you spend a leisurely day on the beach.

The sun sets sooner than you even realize, dying the sky in its colors. A sheen of gold casts itself over the beach and its visitors. Looking at the sparkling sea with your happy family by your side, you can't help but let out a contented sigh. This wonderful life comes not only from the nourishing warmth of family, but also with the help of ever-changing scientific and technological progress.

Isn't such an autonomously driving society something to yearn for? For this goal, we should arm our minds with knowledge, and prepare ourselves to embrace this exciting day not too long in the future!

Printed in the United States
by Baker & Taylor Publisher Services